METALLURGY AND METALLURGICAL ENGINEERING SERIES

ROBERT F. MEHL, PH.D., D.SC., *Consulting Editor*

The Principles of

METALLOGRAPHIC
LABORATORY PRACTICE

Metallurgy and Metallurgical Engineering Series

ROBERT F. MEHL, *Consulting Editor*

MICHAEL B. BEVER, *Associate Consulting Editor*

The Principles of

METALLOGRAPHIC LABORATORY PRACTICE

BY

GEORGE L. KEHL

Associate Professor of Metallurgy, Columbia University

THIRD EDITION

New York   Toronto   London

McGRAW-HILL BOOK COMPANY, Inc.

1949

THE PRINCIPLES OF
METALLOGRAPHIC LABORATORY PRACTICE

VII

To Kay and Karen Jo

PREFACE TO THE THIRD EDITION

In the preparation of this third edition, the author was resolved that the revision should in no way depart from the original concept of providing a basic treatment of the principles associated with general metallographic laboratory practice. As in former editions, no attempt was made to include or even to suggest laboratory experiments that might illustrate the principles discussed therein. It was believed that if such experimental demonstrations are warranted, they should then be devised by the individual teacher to meet his own particular aims.

Since the publication of the second edition of this textbook in 1943, many developments have been made in metallographic techniques, and many new designs of equipment pertaining to metallographic laboratories have been introduced. In part, these improvements and new developments are directly related to the Second World War and the attendant ramifications of such a state of affairs on scientific progress. Others represent the successful conclusions of scientific investigations developed under a more normal and less accelerated pace.

This third edition has been completely rewritten to present in logical fashion the most up-to-date laboratory techniques and related metallurgical laboratory equipment, as well as to include the older accepted and well-established techniques and procedures. As in former editions, emphasis has been placed on the principles associated with techniques and procedures rather than on operational directions. In many cases, however, owing principally to the nature of the subject under discussion, it was believed desirable not only to discuss principles, but to describe in some detail certain fundamental operations. Failure in this respect would have defeated the intended purpose of this text.

In addition to the inclusion of new subject material, the text has been expanded to include a more thorough discussion of certain subjects only briefly treated in former editions. Because of this, more illustrations of laboratory equipment and line drawings and photomicrographs illustrating certain principles of techniques and operations have necessarily been included. The appendix, containing tabular data relating to electrolytes for electrolytic polishing, etching reagents, photographic developing formulas, etc., has been completely revised to include the latest data appropriate to each subject. The data have been arranged and presented in a manner that should prove useful to the reader.

One of the pleasures in writing a textbook is finally to acknowledge the assistance given the author by industrial organizations and technical laboratories, technical societies, and individuals, who have gained so little through their willing cooperation and valuable suggestions. It is a pleasure to acknowledge, and to credit specifically in the text, the many illustrations and technical data furnished by manufacturers of metallographic and metallurgical equipment, and the permission given the author by the American Society for Metals, the American Society for Testing Materials, the Society of Automotive Engineers, *Materials and Methods*, and others, to reproduce technical data first appearing in their respective publications.

GEORGE L. KEHL

NEW YORK, N.Y.
April, 1949

PREFACE TO THE FIRST EDITION

In this volume the author attempts to fill what is believed to be a very definite need for a textbook of fundamental principles involved in metallographic laboratory practice—a book that will bridge the gap between theoretical physical metallurgy and its practical application in the laboratory.

As the title implies, this book is primarily a treatment of laboratory principles. In many instances, however, the statement of principles has been supplemented by explicit recommendations and precautions so that many important laboratory manipulations will be properly executed. Wherever it seemed desirable, the author has illustrated, side by side, the results of proper and improper techniques and the effects of correct and incorrect procedures.

The author has refrained from including definitely outlined experiments, or even suggestions of experiments, as it is believed that such exercises can best be devised by the individual teacher after consideration of his own particular aims and laboratory facilities. Certain subjects have been discussed which are not strictly germane to metallography, but their inclusion in the text seemed essential to a thorough understanding of laboratory principles. It was believed that, if these subjects were omitted, the intended purpose of the book would not be fulfilled. In an attempt to make the volume nearly self-contained, an appendix consisting of thirty-eight tables has been included. These tables deal with etching reagents, hardness numbers, thermoelectric temperature equivalents, and many other important data.

Acknowledgment is made for the assistance rendered by many commercial and industrial organizations which have contributed photographs and technical data included in the text and credited therein; and for permission granted by the American Society for Metals, the American Society for Testing Materials, and the U. S. Bureau of Standards to reproduce illustrations and data first presented in their technical publications.

It is a real pleasure to acknowledge the valuable assistance given the author by Dr. V. O. Homerberg and Dr. M. Cohen of the Massachusetts Institute of Technology in critically reviewing the manuscript and in contributing many helpful suggestions. For encouragement in the preparation of this volume the author is grateful to Professor Allison Butts and Dr. G. E. Doan, Lehigh University.

To Professor Bradley Stoughton, Dean of the College of Engineering, Lehigh University, the author expresses his sincere appreciation for unfailing advice and counsel.

GEORGE L. KEHL

BETHLEHEM, PA.
January, 1939

CONTENTS

Tukon Hardness Test—Penetrators and Loads—Method of Measuring Impression—Hardness Scale—Condition of Test Surface—The Shore Sclero- scope Hardness Test—Vertical-scale Type, Model C-2—Principles—Condition of Test Surface—Minimum Thickness of Test Section—Standardization— Dial-recording Type, Model D—The Monotron Hardness Test—Penetrators and Hardness Scales—Condition of Test Surface—Minimum Thickness of Test Section—The Herbert Pendulum Hardness Test—Time Test—Scale Test— Time Work-hardening Test—Scale Work-hardening Test—Damping Test— The Microcharacter Hardness Test—Cutting Tool and Loads—Method of Measuring Cut Width—Microhardness Scale—Condition of Test Surface— The Eberbach Hardness Test—Penetrator and Loads—Method of Measuring Impression—Hardness Scale—Condition of Test Surface—Hardness Conver- sion—References.

CHAPTER 7

Methods for Determining the Austenitic Grain Size in Steel—Ascertaining the Austenitic Grain Size—Appropriate Cooling Methods—Hypoeutectoid Carbon Steels—Eutectoid Carbon Steels—Hypereutectoid Carbon Steels—Fully Hard- ened Steels—Heat Etching—Fracture Method—Carburizing Method—Oxida- tion Method—Methods of Grain-size Designation—The ASTM Comparative Method—Grain-size Measuring Eyepiece—Graff-Snyder Intercept Method— Methods for Determining Actual Grain Size in Nonferrous Metals and Alloys— The ASTM Comparative Method—Jefferies' Planimetric Method—Heyn's Intercept Method—The McQuaid-Ehn Test—Method of Carburizing—Inter- pretation of Results—The Jominy Hardenability Test—Standard Test Speci- mens—Hardness Measurements—Interpretation of Test Results—AISI "H" Steels—The Shepherd Penetration-fracture Test—Preparation of the Test Specimens—Hardening Treatment—Determination of Grain Size—Determi- nation of Penetration—P-F Designation—The Spark Test—Principles—Abra- sive Wheels—Pressure between the Wheel and the Test Section—Examination of the Spark Stream—Magnetic Test Methods—Magnetic Analysis—Con- tinuity Tests—Fluorescent Test Methods—Supersonic Test Methods—Prin- ciples of Operation—Depth of Penetration—Selected References on the Prin- ciples of X rays and Crystal Structure—References.

CHAPTER 8

Thermoelectric Pyrometers — Thermocouples — Principles — Peltier Effect — Thomson Effect—Thermoelectric Potentials—Cold-junction Correction— Requirements and Preparation of Thermocouples—Thermocouple Protection Tubes—Calibration of Thermocouples—Primary Calibration—Secondary Cali- bration—Millivolt-temperature Equivalents—Emf Measuring Instruments— Potentiometers — Millivoltmeters — Temperature Controllers — Mechanical- operating Types—Photoelectric-operating Types—Temperature Recorders and Controllers—Suggestions on the Proper Use of Thermoelectric Pyrometers— Thermocouples—Indicating Instruments—Extension Lead Wires—Resistance Thermometers—Principles—Callendar Equation for Temperatures above 0°C —Equations for Temperatures below 0°C—Resistance-measuring Methods— Wheatstone-bridge Principle—Three-wire Compensation Method—Four-wire

CHAPTER 1

PREPARATION OF SPECIMENS FOR MICROSCOPIC EXAMINATION

The science of metallography is essentially the study of the structural charactertistics or constitution of a metal or an alloy in relation to its physical and mechanical properties. One important phase of this study is known as macroscopic examination and involves the visual observation of the rather gross structural details of a metal, either by the unaided eye or with the aid of a low-power microscope or binocular. Because the attending magnifications are of low order, usually under $10\times$, macroscopic observations are somewhat limited as to the kind of metallurgical data revealed. This circumstance notwithstanding, macroscopic examinations, when appropriately carried out, are of considerable importance in many instances—in fact, some metallic characteristics are best determined by such studies—and details of them will be discussed more fully in a later chapter.

Without doubt, the most important part of metallography deals with the microscopic examination of a prepared metal specimen, employing magnifications with the optical microscope of from $100\times$ to as high as $2000\times$. Such microscopic studies are of much broader scope than are macroscopic examinations, and under appropriate conditions of observation there will be revealed to the trained metallographer an abundance of constitutional information concerning the metal or alloy under investigation. There will be defined clearly such structural characteristics as grain size; the size, shape, and distribution of secondary phases and nonmetallic inclusions; and segregation and other heterogeneous conditions—all of which profoundly influence the mechanical properties and behavior characteristics of the metal. When these and other constitutional features are determined by microscopic examination and the extent to which they exist in the microstructure is known, it is then possible to predict with considerable accuracy the expected behavior of the metal when used for a specific purpose. Of equal importance is the fact that within limits there is reflected in the microstructure an almost complete history of the mechanical and thermal treatment that a metal has received. It has been only through diligent study of metals microscopically that many perplexing problems of physical metallurgy have been solved, and it may be safely predicted that the con-

1

tributions that will be made to the field of physical metallurgy in the future will depend in part, or solely, upon structural evidence revealed by the microscope.

Our experience indicates that little can be learned regarding the structural characteristics of a metal by microscopic examination unless the surface that is to be so examined is first prepared according to more or less rigid and precise procedures. With the use of the modern metallurgical microscope and precision optical parts where the obtainable resolution may be as great as a fraction of the wave length of light used to illuminate the specimen, it is evident that perfect specimen preparation is of the greatest importance. Improper preparation is likely to remove all-important inclusions, erode grain boundaries, or temper hardened steel specimens, ultimately producing a structure, superficially at least, which upon microscopic examination will appear entirely different from that which is truly representative and characteristic of the metal. Obviously, an examination of such a prepared specimen will lead only to erroneous interpretations and unreliable conclusions.

In general, the procedure of specimen preparation consists of first obtaining a flat, semipolished surface by means of grinding the specimen on a series of emery papers of decreasing grit size, or by grinding on suitable abrasive laps, followed by fine and final polishing on one or more cloth-covered lap wheels. These operations ultimately produce a flat, scratch-free, mirrorlike surface—the required condition of the specimen surface before it can be etched and the metallographic structure appropriately revealed.

Specimen preparation is in principle a relatively simple procedure, but in actually carrying out the manipulations many difficulties may be encountered. At the outset, it must be emphasized that specimen preparation is still a technical art, and that to carry out the manipulations properly requires skill and finger dexterity on the part of the technician. Such skill can be attained and mastered only after continued practice.

One of the most essential factors leading to a highly developed technique is the care with which the specimen is handled throughout the subsequent stages of preparation. Of equal importance is the necessity for clean working conditions during the entire preparation procedure, as experience has shown that one particle of grit or foreign material on the polishing cloth may ruin an otherwise perfectly polished specimen. Hence at frequent intervals during the procedure the specimen undergoing preparation, as well as the hands of the technician, should be thoroughly washed with soap and water. This cleansing procedure is necessary to prevent the surface from becoming scratched by adhering debris from the emery papers, and it is particularly an essential operation when a change is made from one grinding paper to another one of finer grit size.

Sampling. The choice of a representative metallographic sample which is to be examined under the microscope is an important step in the preliminary consideration of the examination and, unfortunately, it is not always given the careful consideration that is necessary. Specimens selected should be characteristic of the material and serve the object of the

Fig. 1. Type of metallurgical cutoff machine wherein the specimen is completely submerged in a liquid coolant during sectioning. Top cover and wheel guard not shown. (*Courtesy of Precision Scientific Co.*)

examination. For example, if a structural member has failed in service and the cause of failure is to be determined by metallographic examination, the specimen should be selected and removed from that particular region of the fracture where it will reveal the maximum amount of information. For comparative purposes a specimen of this nature should be supplemented by one taken from a sound and normal section of the whole. In the examination of rolled shapes it is recommended that both a longitudinal and a transverse section be examined. This is desirable, as inclusions and other important characteristics may not be revealed satisfactorily by an examination in only one direction. Cast or annealed metals should be examined over their entire cross sections so that differences of structure from the outer edges to the interior may be observed.   Particular attention should be paid to examination of the edges of rolled and forged sections in order to detect the presence of decarburization and other adverse surface conditions.

If the section that is selected to serve as the specimen is relatively soft it may be removed from the piece by either manual or power hacksawing.   In

FIG. 2.   Type of metallurgical cutoff machine wherein the specimen is cooled during section-ing by a recirculatory stream of liquid coolant.   (*Courtesy of Buehler, Ltd.*)

the case of brittle metals and alloys, such as cast irons and tin-rich bronzes, the piece may be appropriately fractured by a hammer blow and a suitable fragment selected as the specimen.

Specimens of hard materials which cannot be easily sawed, such as fully hardened steel or age-hardened nonferrous alloys, must be secured from the whole by means of cutting with an abrasive wheel, known as a "cutoff" wheel.   This wheel is essentially a thin composition disk ($\frac{1}{16}$ in. in thick-ness is recommended) which is impregnated with a suitable cutting abrasive, such as emery, carborundum, or diamond dust. The disk is rotated at comparatively high rotational speeds, and provision is generally provided in the apparatus to cool the specimen adequately during cutting. Cooling is secured either by cutting the specimen while completely submerged in water or other liquid coolants, or by directing a steady stream of cooling fluid to the area of cutting. Metallurgical cutting-off equipment is shown in Figs. 1 and 2.

Unless care is taken in the selection of a cutoff disk,[1] and ample cooling of the specimen is provided during cutting, the original structure of the specimen may be radically altered, at least superficially at the cut surface, due to evolution of frictional heat. This circumstance is strikingly illustrated in the case of fully hardened steel, shown in Fig. 3. It is, indeed, excellent technique never to permit the specimen to feel warm to the hand, either during sampling by whatever method is used or during the subsequent grinding operations to be described. This of course eliminates the possibility of securing a specimen by oxyacetylene cutting or by similar methods, unless the heated parts are removed or discounted.

Whenever possible, the specimen should be of a size that is convenient and comfortable to handle. Specimens that have large surface areas will usually require prolonged polishing to produce a satisfactory surface, whereas specimens that are small, unless mounted, will tend to rock on the emery papers during grinding, with resultant rounded edges and curved surfaces. Specimens too small to be handled conveniently may be mounted according to one of the methods described on page 49. A specimen of a size and shape easy to manipulate is one from ¾ to 1 in. across the polished surface, either round or square, and approximately ½ in. high. This relation between the breadth and the height of the specimen will assist in maintaining a flat surface with the least amount of fatigue to the technician during subsequent grinding and polishing.

HAND-GRINDING TECHNIQUE

Rough Grinding. The surface of the specimen that is to be examined is first made plane by means of a file or, preferably, on a specially designed motor-driven emery belt, shown in Fig. 4. If the surface area of the specimen is large, a slight rotation of the specimen while grinding on the belt will assist in producing a plane surface in a surprisingly short time. In order not to alter superficially the structure of the metal undergoing grinding, particularly if it is fully hardened, the specimen should be kept cool by frequently dipping it into water during the grinding operation.

---

[1] Wheel Nos. 4008 and 120C1½, manufactured by The Allison Company, Bridgeport, Conn., are ideal cutoff wheels, particularly suited for cutting fully hardened material without alteration of the structure through excessive heating.

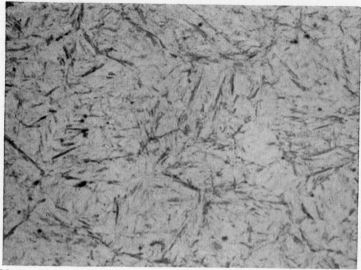

a. Martensitic structure preserved by cutting off under a liquid coolant. 1000 ×.

b. Martensitic structure altered by cutting off in the absence of a liquid coolant. 1000 ×.

FIG. 3.   Superficial alteration of the structure of fully hardened steel arising from inadequate cooling during sectioning.

Excessive pressure during rough grinding, as well as during subsequent hand grinding on the emery papers, will not only form deep-seated scratches that will be difficult to remove but will promote the formation of considerable amounts of disturbed metal on the surface of the specimen (see page 60). This disturbed or distorted metal may in many cases extend inward from the surface to a very considerable depth and, as will be shown later, unless this distorted metal is completely removed, its presence may greatly alter the appearance of the metallographic structure.

Fig. 4. Belt-type specimen grinder used principally for rough-grinding operations. *(Courtesy of Buehler, Ltd.)*

Although distortion of the surface metal cannot be prevented entirely from occurring during grinding and polishing, it nevertheless can be minimized by the use of proper techniques. This is accomplished primarily by using light contact pressure at all times during grinding and polishing and by eventually removing what disturbed metal is formed by alternately polishing and etching several times, as described on page 63.

It is advisable during the rough-grinding operation to bevel the sharp edges and corners of the specimen to prevent tearing the emery papers and polishing cloths in later operations. In cases where it is necessary to examine the polished surface out to the very edges, the specimen must be appropriately mounted before grinding, as described on page 56.

When the specimen appears to be flat over the entire surface and grinding has removed all surface imperfections, this operation is considered to be completed. The specimen, as well as the technician's hands, should now be washed thoroughly with soap and running water to prevent carrying over onto the first grinding paper particles of coarse grit from the roughing belt.

Intermediate Grinding. Emery papers used in the preparation of metallographic specimens must be of high quality, particularly with respect to

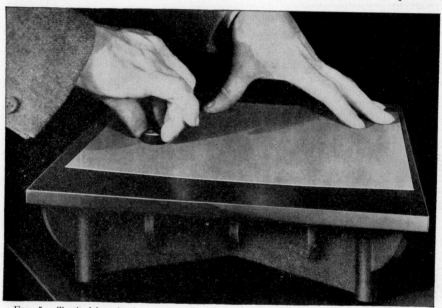

Fig. 5. Typical bench plate for hand grinding metallographic specimens illustrating the preferred method of holding emery paper and specimen during grinding. (*Courtesy of Francis Lucas.*)

size uniformity of the emery particles. Many excellent papers of this kind are available, such as French-Hubert, Behr-Manning, Aloxite, and Electrocut papers, all of which have been found satisfactory for metallographic purposes.

It is always desirable to use a new emery paper of appropriate grade for each specimen undergoing preparation. A paper that has been used, for example, in the preparation of a steel specimen may have embedded on its surface sufficiently hard debris to cause coarse and deep scratches on the surface of a softer specimen, such as brass, during subsequent grinding. Furthermore, worn emery papers should be avoided, unless grinding is done on these for a special purpose, because the dull grit on such papers will readily promote distortion of the surface metal.

The first grinding paper used after the roughing operation is usually a No. 1 or a No. 0 French-Hubert paper or its equivalent. The emery paper is placed on a bench plate, as shown in Fig. 5, or on any clean, hard, level surface, such as a sheet of plate glass. The specimen and paper are held as shown, and under moderately applied pressure the specimen is gently drawn back and forth across the entire length of the paper. While being ground, the specimen is held so that the new, finer scratches being introduced on the surface are approximately at right angles to the old scratches resulting from the previous flattening operation. Adhering to such a procedure makes it quite easy to recognize when grinding is complete on this particular paper, as all the old, coarse scratches will be ground out completely and replaced by a series of new, finer scratches. This condition of the surface may be quickly ascertained by the unaided eye or by the use of a $6\times$ or $8\times$ magnifying glass. It will be noted in Fig. 6 that the preceding coarser scratches in each case have been completely removed in grinding the specimen on successively finer grades of emery paper.

In grinding many heat-treated alloys, and in particular many of the soft metals, as will be noted later, is it necessary that the grinding papers be first thoroughly wetted with a suitable lubricant. Many solutions have been proposed for this purpose, some of which are oils, gasoline, paraffin in kerosene, liquid soaps, glycerin, and glycerin and water mixtures. These liquids not only serve as lubricants and minimize smearing of the soft metals but they also act as coolants for heat-treated alloys whose structure may be superficially altered if ground dry.

Fine Grinding. The subsequent grinding operations and the accompanying manipulations are the same as previously described. Generally two additional emery papers of increasing fineness are used—No. 00 and No. 000 French-Hubert papers or their equivalents. In special cases, however, as in the preparation of magnesium alloys and lead (see pages 35 and 37), grinding is continued through the No. 0000 paper.

During grinding on each successive paper, the specimen is held as described for intermediate grinding, *i.e.*, with the newly formed scratches at right angles to those which were introduced on the preceding paper. A change from one paper to another or from the last paper to the lapping wheel requires, in keeping with good technique, a thorough washing of the specimen in running water. If upon visual examination the surface of the specimen appears to contain more or less uniform scratches from the fine-grinding operation, and if the coarser scratches from the intermediate grinding operation have been completely removed, the specimen is then ready for the fine and final polishing operations.

MACHINE-GRINDING TECHNIQUE

Paper Grinders. The hand method of specimen grinding as described is a more or less tedious and laborious procedure and, when a considerable number of specimens are to be prepared, it is found to be time-consuming. More efficient grinding may be carried out by mechanically grinding the specimen on a rotating disk, similar to that shown in Fig. 8. Circular-cut

a. Scratches introduced by belt grinder. Grit size—120 equivalent mesh. 75 ×.   *b.* Scratches introduced by No. 0 Behr-Manning emery paper. 75 ×.

c. Scratches introduced by No. 00 Behr-Manning emery paper. 75 ×.   *d.* Scratches introduced by No. 000 Behr-Manning emery paper. 75 ×.

FIG. 6.   Illustrating the characteristic appearance of scratches introduced on the specimen surface by grinding on successively finer grades of emery paper.

emery papers of the type and grades used in hand grinding are mounted on the disk and held in position by a ring fitted around the periphery. The lap is rotated at approximately 600 rpm for the first paper (No. 1 or No. 0), and preferably at a decreased speed for papers of finer grit size.

Considerable care must be taken during mechanical grinding to prevent the specimen from becoming overheated because of excessive contact pressure between the specimen and the rotating emery paper. Not only may the structure of the specimen be altered superficially (if it is in a hardened condition), but, as mentioned heretofore, heavy pressure will result in the formation of considerable amounts of disturbed metal which will be difficult to eliminate. In general, the precautions set forth for the hand-grinding method apply equally to specimen preparation by mechanical grinding methods.

Paraffin Grinding Laps. Grinding of metallographic specimens may be effectively carried out on a series of mechanically driven paraffin laps, the units being similar in design to those used in paper grinding. This method of grinding, as compared to either hand or mechanical paper grinding, produces a minimum amount of disturbed metal, satisfactorily preserves nonmetallic inclusions (probably the main advantage of this method), and maintains a reasonable flat surface out to the edges of the specimen.

A paraffin lap is essentially a polishing disk covered with a layer of high-melting-point paraffin or a disk covered with either billiard or canvas cloth impregnated with paraffin. In the first case, the layer of paraffin must be machined flat, and preferably grooved as in the case of lead laps (see Fig. 7) in order to retain an abundance of the abrasive suspension on the surface of the lap during grinding.

The grinding laps are usually charged before and during grinding with an aqueous soap solution of emery abrasives, ranging in fineness from 200 to 600 equivalent mesh. The technique employed for grinding a specimen is similar to that used for lead-lap grinding and is described on page 12.

Lead Grinding Laps. A method of grinding specimens for metallographic examination, proposed by Ellinger and Acken,[1] consists in grinding the specimen on a series of lead laps, each lap respectively impregnated with a suitable emery abrasive of decreasing particle size. The advantages of this method over that of the usual hand or mechanical grinding methods employing emery papers are many, the most important being (1) the ability to retain during grinding a nearly perfectly plane surface to the very edge of the specimen, (2) retention of nonmetallic inclusions, (3) formation of a minimum amount of disturbed metal, and (4) the rapidity with which a specimen may be prepared. Because of the low degree of surface distortion produced by this method, lead laps are ideally suited for the preparation of metals and alloys that exhibit rapid work-hardening characteristics, such as austenitic grades of stainless steel.

A lead lap consists essentially of a polishing disk faced with a layer of pure lead or of an alloy consisting of 50 per cent lead and 50 per cent tin. The lead-tin facing is preferred because it is harder than one of pure lead,

[1] Ellinger, G. A., and J. S. Acken: A Method of Preparation of Metallographic Specimens, *Trans. ASM*, Vol. 27, 1938.

and as a consequence it will withstand wear by abrasion much better. The face of the lap is purposely grooved in the form of a spiral in a direction counter to the rotation of the lap, as shown in Fig. 7, in order to retain on the surface during grinding an abundance of the water-emery abrasive suspension.

The functioning principle of the lead lap is essentially one involving the temporary impregnation of a relatively soft metallic matrix with an appropriately graded metallographic abrasive. By virtue of the manner by which the abrasive is charged onto the lap during grinding, as described below, many of the abrasive particles become partially embedded on the

FIG. 7.   Lead-tin alloy grinding lap.   (*Courtesy of Buehler, Ltd.*)

surface of the lap, and are thus more or less rigidly, but temporarily, held in fixed positions. The exposed portions of the embedded particles protruding from the surface of the lap effectively serve as the cutting medium when they are passed beneath the specimen surface as the lap is rotated. Many of these particles become dislodged during grinding, and these along with other abrasive particles that never were thoroughly embedded are eventually washed from the surface of the lap with other debris and deposited in the waste bowl. If found necessary, additional abrasive may be added to the lap during grinding to compensate for that which was lost.

The actual manipulations in grinding a specimen are relatively simple. The first lead lap that is used after the rough-grinding operation is usually charged with a relatively coarse abrasive, such as emery No. 180 or No. 302.[1] The lap is rotated at approximately 300 to 600 rpm and the specimen, under moderately applied pressure, is held in contact with the revolving lap so that the newly formed scratches are at right angles to those introduced on the surface previously. During grinding, the specimen is moved continuously back and forth from the center to the edge of the lap in order to prevent uneven wear of the lap surface. Usually it is necessary to charge the lap only once or twice while grinding one specimen, and this may be done conveniently by means of a shaker bottle containing about 25 g of the appropriate emery powder in 500 ml of water. If the lap becomes dry during grinding and sufficient abrasive is already present, as determined by feel and associated degree of drag between the specimen and the rotating lap, clear water should be added in small amounts.

Grinding is continued on the first lead lap until all the previous coarse scratches are replaced by a set of new finer scratches characteristic of the abrasive used. This usually requires only a few minutes for a specimen having a surface area of about $\frac{1}{2}$ sq in.

Intermediate and final grinding are carried out on lead laps charged, respectively, with emery powders No. $303\frac{1}{2}$ and No. 305. Final grinding with No. 305 emery produces scratches of such fineness that they may be removed without difficulty in the final polishing operation to be described, without first having to resort to preliminary polishing. In order to prevent contamination of the laps when a change is made from one lap to another one of finer emery impregnation, the specimen as well as the hands of the technician must be thoroughly washed in running water.

Regardless of the care that is taken while grinding specimens, the lead lap will eventually become out of plane. To prevent this and attending troubles, the lap must be resurfaced at frequent intervals with a cast-iron plate. This plate is essentially one of the same size and shape as the lead lap, the face of the plate being accurately machined and ground plane. The cast-iron plate is held in contact with the revolving lap with an abundance of the proper emery-water mixture between the two. Remaining debris on the lead lap after this operation is completed may be removed with a stiff-bristle brush and clear water.

POLISHING TECHNIQUE

Fine polishing of a metallographic specimen is for the purpose of removing from the surface of the specimen the fine scratches introduced during the last grinding operation, and of ultimately producing a highly polished scratch-free surface.

[1] The grades of emery powder referred to are those manufactured by the American Optical Company.

The success in final polishing, as well as the time consumed in this operation, depends largely upon the care that was exercised during the preceding grinding operations. It must be emphasized that both time and effort will be wasted if an attempt is made to final-polish a specimen containing deep and coarse scratches that were not removed completely in the last grinding operation. In such cases the specimen should be reground, with more attention paid to the uniformity of scratch fineness before final polishing is again undertaken.

FIG. 8. Bench-type metallographic polishing unit, equipped with 8-in.-diameter lap and two-position motor control—150 and 250 rpm. (*Courtesy of Buehler, Ltd.*)

Whenever possible, arrangements should be made to conduct the polishing operation in relatively dust-free surroundings. This is to prevent contamination of the polishing laps, and in most cases, unless lead grinding laps or other wet methods of grinding are used, this arrangement requires that polishing be done in a room separate from that used for grinding. Experience indicates that, unless this precaution is taken, debris and loose emery powder from the grinding papers may be circulated by air drafts and deposited on the polishing units.

Metallographic Polishing Units. The preliminary and final polishing operations on ground metallographic specimens are performed on one or more polishing wheels or polishing laps. Such laps are essentially brass or bronze disks, 8 to 10 in. in diameter, covered with an appropriate grade of polishing cloth (see page 21). The wheels are generally rotated in a horizontal plane and are usually individually motor driven, as shown in Fig. 8, to facilitate control and adjustment of rotational speed when such changes are necessary. The design of modern polishing units permit their use on a bench top (as in Fig. 8), or they may be flush-mounted into a table top, as illustrated in Fig. 9, to secure a convenient and permanent polishing arrangement.

Fig. 9. Metallographic polishing table containing two polishing units. (*Courtesy of Buehler, Ltd.*)

One of the latest developments in metallographic grinding and polishing equipment is the automatic polishing unit which uniquely prepares one or more specimens simultaneously. In one design of automatic polisher, illustrated in Fig. 10, two to six specimens are held in contact with the revolving lap wheel by means of a circularly shaped specimen holder (*A*). The polishing lap rotates at a speed of 96 rpm, and the specimen holder in an opposite direction at a speed of 90 rpm. Because of this difference in rotational speed and direction, and because the centers of the lap and specimen holder are offset from one another as noted in Fig. 10, no specimen on any one lap repeats contact with the lap at the same spot. As the specimen holder makes one revolution, each specimen partially runs off the lap and thereby removes to the bowl used abrasive and debris. Different contact pressures may be applied to the specimen holder, and a suspension of appropriate grinding or polishing abrasive is automatically applied to the rotating lap through a reservoir located at the center of the specimen holder.

Fig. 10.   Illustrating the essential features of one type of automatic grinding and polishing unit for metallographic specimens.   Note specimen holder *A* and lead-tin alloy grinding lap. (*Courtesy of Precision Scientific Co.*)

Without question, automatic grinding and polishing units can save considerable time in routine preparation of metallographic specimens and, in particular, effort on the part of the technician.   It is claimed by many metallographers, however, that in the use of automatic units as contrasted to hand-controlled techniques, it is more difficult to observe the progress of specimen preparation and particularly to control the degree of final polishing to the nicety required in high-quality preparational work.

Metallographic Polishing Abrasives. Critical considerations of the kind of polished surface required on a metallographic specimen indicate that abrasives appropriate for metallographic grinding and polishing must produce ultimately a polished surface by uniformly removing or cutting away surface metal with little or no attending metal flow. This mode of securing a polished surface is fundamentally different from that of burnishing, where a high degree of polished luster is obtained principally by flow and distortion of surface metal.

Physically, an abrasive considered ideal for metallographic purposes is one that possesses reasonably high hardness; the external shape of the individual particles is such that numerous sharp cutting edges and corners are exposed; the particles, if fractured during use, expose new and additional cutting edges and corners; and the nature of the abrasive will permit accurate sizing and separation by levigation into fractions of uniform particle sizes. Grinding and polishing media having particles that are more spheroidal in shape or that quickly lose their cutting edges and corners during use and thus become and remain dull are not considered suitable for metallographic work. The particles of such abrasives, for the most part, merely roll between the specimen surface and the medium supporting the abrasive, causing an undue amount of surface distortion and metal flow and effective dislodgement of nonmetallic inclusions.

Diamond Dust. Perhaps the closest approach to an ideal metallographic polishing abrasive is uniformly graded, unadulterated diamond chips. This medium has been used extensively in the past for specimen preparation of very hard alloys, such as sintered tungsten and boron carbides (see page 43), and recently its use has been extended with considerable success to final polishing of more common metals and alloys.[1]

Commercial grades of diamond dust suitable for metallographic preparational work can be secured in particle sizes ranging from about 120 to 0–0.5 microns. Various lubricants are recommended for use in diamond-dust polishing, some of which are olive oil, peanut oil, oil of lavender, mineral oils of different viscosities, and carbon tetrachloride. The choice of any one vehicle depends largely upon the kind of metal or alloy being prepared and the nature of the lap surface on which polishing is being carried out.

Diamond dust mixed with an appropriate vehicle to the consistency of a thin paste (known as Diamet Hyprez) is especially suitable for metallographic polishing. It may be secured in three grades with regard to range of particle sizes (4 to 8 microns, 0 to 2 microns, and 0 to 0.5 microns), each

[1] Woodside, G. C., and H. H. Blackett: Polishing Metallographic Specimens with Diamond Dust, *Metal Progress*, Vol. 52, 1947.

grade being conveniently supplied in an individual applicator gun resembling a large hypodermic syringe.

A series of well-graded diamond compounds, suited for both metallographic grinding and polishing, is available from the Magnus Chemical Company, Inc. Similar to Diamet Hyprez, the Magnus diamond particles are suspended in a viscous petrolatum and, to facilitate dispensing when used, each grade is contained within a collapsible tube. The diamond compounds are available in eight grades with regard to average particle size, ranging from 120 to 1 microns.

Alundum. For intermediate or preliminary polishing of metallographic specimens, when this operation in the procedure is followed, an abrasive of alundum (fused aluminum oxide) is generally used, or at times carborundum (silicon carbide) and boron carbide, all of about 500 to 600 equivalent mesh. Any of these abrasives are generally applied to the lapping cloth by means of a shaker bottle containing an appropriate amount of abrasive suspended in water.

Magnesium Oxide. Magnesium oxide is generally used as the abrasive in final polishing of soft metals, such as aluminum, magnesium, and others, or as an alternate abrasive to alumina in polishing cast iron and other relatively hard metals. Grades of magnesium oxide prepared specially for metallographic work are available under the proprietary name of Shamva or as special magnesium oxide powders available from most chemical suppliers.

One of the most satisfactory grades available for metallographic polishing is Merck's *heavy* magnesium oxide. The heavy grade, as contrasted to the ordinary light and fluffy oxide, contains a preponderance of uniformly sized, octahedrally shaped particles of considerable hardness which exhibit well-defined cutting edges and corners. Heavy magnesium oxide, like those special grades of magnesia described above, contain no water-soluble alkalies as do some of the more ordinary grades. This circumstance is particularly important with regard to metallographic polishing of aluminum and certain aluminum alloys, since the presence of free alkalies would severely stain and chemically attack the surface undergoing preparation.

Magnesium oxide of all grades, when left exposed to moist air or when in contact with water, slowly reacts to form magnesium hydroxide. Sufficient carbon dioxide is present in the air and in ordinary tap water to continue the reaction further and to form basic magnesium carbonate. The presence of carbonate in otherwise unaffected magnesium oxide is very undesirable because of the hardness, coarseness, and nonuniform size of the magnesium carbonate particles formed—a circumstance that practically ruins the remaining magnesium oxide for further use as a final polishing abrasive. It is imperative, therefore, that only fresh, unexposed magnesium oxide be used as a polishing abrasive and that it be used only in conjunction with distilled water—never with ordinary tap water.

Contaminated magnesium oxide can be reclaimed through decomposing any magnesium carbonate present by calcining the dry powder at about 1000°C (1832°F), and finally grinding and appropriately sizing the calcined product. Resorting to this reclamation procedure, however, is usually uneconomical because of the relatively low cost of magnesium oxide.

The proper technique in using magnesium oxide in final polishing consists of applying a small quantity of fresh, dry powder to the lapping cloth and adding *distilled* water in sufficient amounts to form a thin paste. The paste is then worked into the cloth fibers with the finger tips, and after so charging the cloth and during subsequent polishing, additional distilled water is added when necessary to keep the lap moist.

Polishing cloths on which magnesium oxide has been used require special cleansing and storage procedures, and are described on page 22.

Alumina. Alumina (aluminum oxide) is probably the most satisfactory and most commonly used final polishing abrasive employed for general metallographic work. It is procurable in the form of dry powder as "levigated alumina," or under various proprietary names, such as Linde polishing powders, grades A and B, Norton Alumina, etc.; and in the form of pastes or water suspensions under such trade names as Gamal, AB polishing compound, Rite-Tonerde, Fisher Polishing Alumina, and others.

Alumina is known to exist in three basic crystallographic forms. Of the three modifications, designated as alpha, beta, and gamma, the alpha and gamma forms are most widely used as polishing media.

Linde A alumina is representative of the alpha form, whose crystallographic arrangement is hexagonal. This particular abrasive has an average particle size of about 0.3 micron and a hardness, in massive form, of 9 on the Mohs scale of hardness. The particles are usually so uniformly small that levigation of the dry powder is unnecessary before it is used.

The gamma modification of alumina, of which Gamal and Linde B powder are representative, is cubic in crystal structure. Linde B powder has a uniform particle size of less than 0.1 micron and, accordingly, is well suited for final polishing operations. In physical shape, Gamal alumina consists of thin, flat platelets of exceptionally high hardness. The stresses imposed upon these platelets, owing to contact pressure during polishing between the specimen surface and the rotating lap, are sufficiently great to fracture the platelets into many small cubical-shaped particles or fragments thereof. These particles, because of their fractured shape, exhibit well-defined cutting edges and corners and thus are ideally suited as a metallographic abrasive.

A synthetic-sapphire polishing compound, known as Precisionite, has been introduced recently for final metallographic polishing. The average size of the alumina particles is about 0.3 micron, and the abrasive performs well when used on Forstmann's or similar types of polishing cloths. It is recommended for polishing most metals and alloys, excepting aluminum, which becomes badly stained and tarnished.

Some grades of dry alumina powder supplied for metallographic polishing in the levigated condition require relevigation to assure uniform and small particle sizes. Levigation merely consists of suspending a small quantity of alumina in clear water in an appropriately tall glass vessel and, after the agitated particles have been allowed to settle from about 1 to 10 min, the supernatant alumina-water mixture is siphoned off, yielding a fine suspension of abrasive. The remaining sediment may be relevigated in the same manner to yield an additional alumina suspension slightly coarser than the first fraction removed, or it may be discarded.

An ideal alumina for metallographic polishing, comparable in quality to the well-known but unavailable Austrian Tonerde, may be prepared as follows:[1] Hydrated alumina, No. C-730,[2] is first converted to alumina by heating a loosely packed layer of the powder (about 2 in. deep) for about 2 hr at a temperature of 925 to 1100°C (1700 to 2012°F). When the powder has cooled to ordinary temperature, the alumina is levigated in proportions of 100 to 150 g of alumina powder to 1000 ml of water, preferably using distilled water. The time of levigation necessary to produce a very fine suspension is about 10 min; this time may be increased to 15 or 20 min if an exceedingly fine particle size is desired, suitable for polishing soft metals such as aluminum, lead, tin, etc. When levigation is complete, the supernatant suspension is siphoned off and the remaining coarse sediment is discarded.

Because certain water-soluble alkalies associated with the original hydrated alumina are not decomposed during initial heating, the suspending liquid of the supernatant suspension is decidedly basic. If stored for any length of time in this state of high alkalinity, the suspension will settle eventually and form a hard cake that will be difficult, if not impossible, to break up by agitation. To agglomerate the alumina into a curd, and thus prevent caking and attending difficulties, the pH of the suspending liquid is lowered by carefully adding to it, drop by drop, dilute hydrochloric acid. Using phenolphthalein as an indicator, sufficient dilute acid is added to the suspension to change the characteristic red color of the suspending liquid to a faint pink or water-white color. When the end point is reached, the suspending liquid is still slightly alkaline since the pH-indicating range of

[1] United States Steel Corporation, U.S. Patent No. 2,377,057.
[2] Manufactured by the Aluminum Ore Company, East St. Louis, Ill.

phenolphthalein is between 8.3 and 10.0. This circumstance, however, is desirable in that slightly basic suspensions more readily wet a specimen surface during polishing than those having a pH of 7 or slightly lower. Inasmuch as the suspending liquid is not buffered, the pH of the liquid may, upon standing, slowly increase over a period of time. This change, if occurring, may readily be detected by observing the color of the suspending liquid. If it is decidedly red in color, more hydrochloric acid may be added in appropriate amounts.

To reduce the bulk of the alumina suspension for ease in storage, the suspension is allowed to settle for a few days and the clear, supernatant liquid is siphoned off and discarded. The remaining thick alumina may then be bottled as a stock mixture, and diluted with 10 to 20 parts of water as needed for use.

Miscellaneous Abrasives. In addition to the abrasives already discussed, chromic oxide and iron oxide (jeweler's rouge) have been used with success in metallographic polishing. Iron oxide, however, has a tendency to flow surface metal, and although a highly polished surface may be secured through its use, the surface is not representative of a good metallographic polish.

Metallographic Polishing Cloths. The choice of an appropriate cloth for metallographic polishing depends largely upon the specific nature of the specimen to be polished and the purpose of the subsequent metallographic examination. There is a great variety of polishing cloths of excellent quality from which a selection may be made, some of which are obtainable under the proprietary names of Selvyt, Kitten's Ear, Vel-Chamee, Gamal, Miracloth, Microcloth, and others. In general, the surface texture of polishing cloths varies from one having no definite nap or pile, such as pure silk or airplane-wing-covering material (used in special cases), to a more general-purpose polishing cloth having a relatively deep pile, such as velvet or velveteen.[1] Intermediate between napless and deep-piled materials are such cloths as billiard-table covering, wool broadcloths of various degrees of fineness, and different weights of canvas duck.

In addition to woven fabrics, synthetic metallographic polishing cloths are available for general polishing purposes, of which Gamal and Microcloth are examples. These synthetic materials consist of short synthetic fibers uniformly distributed over a cotton backing cloth. The fibers are attached to the backing material with a flexible, waterproof adhesive, and during manufacture of the cloth, the fibers are made to stand upright by

---

[1] One of the finest all-purpose polishing cloths having a nappy surface is cloth quality No. 7760 (finished, but undyed), manufactured by Julius Forstmann and Company, Inc., Passaic, N.J. This cloth was formerly designated as quality No. 5974.

means of an electrostatic charge. Synthetic polishing cloths produce an excellent metallographic polish and exhibit exceptionally high resistance to wear, tear, and other forms of maltreatment.

Polishing cloths of better quality generally require no preliminary treatment prior to being used. However, in the case of inexpensive materials, it is often necessary to boil such cloths in water in order to soften any hard fibers present and to wash the cloths thoroughly with tincture of green soap to remove foreign material.

A cloth to be used for the first time (after being treated, if necessary, as described) is thoroughly soaked in clear water and, while still wet, it is stretched tightly over the lapping wheel and securely clamped into position. The cloth is then charged with an appropriate and selected polishing abrasive. Charging is best carried out by mixing the dry polishing abrasive with water to the consistency of a thin paste and working this mixture into the cloth fibers with the finger tips.

When a polishing cloth is not to be used for an extended period of time, it should be removed from the lapping wheel and thoroughly soaped and washed in running water. The cloth may then be dried or, preferably, stored for future use in a vessel filled with sufficient water to completely cover the cloth. The cleansing procedure will remove practically all metal debris adhering to the cloth, and wet storage will prevent the cloth from becoming caked with polishing abrasive not removed during washing.

Polishing cloths on which magnesium oxide has been used should be thoroughly washed in running water *immediately* after use and should be stored in a vessel containing acidified distilled water (2 per cent hydrochloric acid solution). Magnesium oxide or basic magnesium carbonate particles entrapped in the cloth fibers will be converted to water-soluble magnesium chloride by reaction with hydrochloric acid, and thus render the polishing cloth free of contamination.

Preliminary Lapping Procedure. The purpose of preliminary lapping is to remove the fine scratches introduced on the specimen surface during the last grinding operation. The polishing wheel for this operation is covered generally with broadcloth, billiard cloth, or a lightweight canvas duck, and is rotated at about 400 to 500 rpm. The abrasive used is either No. 600 alundum, No. 600 carborundum, or their equivalents.

Experience has shown that during preliminary lapping with the suggested abrasives, there is a tendency for the surface of the specimen to become pitted. Although preliminary polishing will undoubtedly save considerable time in producing ultimately a highly polished surface, the operation should be omitted from the preparation procedure when pitting and dislodgement of soft secondary phases (*e.g.*, lead particles in leaded brass) would be objectionable. In such cases, it is advisable to final-polish the specimen as described below directly after the last grinding operation, or resort to special polishing techniques as described on pages 45 to 47.

The technique of preliminary lapping consists of firmly holding the ground specimen against the rotating lap, and during the operation, moving the specimen continuously back and forth from the center to the periphery of the wheel. Whenever necessary during polishing the alundum or carborundum abrasive may be applied to the lap by means of a shaker bottle containing approximately 15 g of abrasive per 100 ml of water. If, however, sufficient abrasive is present on the lap and the cloth is becoming dry, clear water should be applied in appropriate amounts.

For successful preliminary polishing, it is essential to observe carefully the degree of wetness of the polishing cloth. If the cloth is too wet, the polishing action of the abrasive-cloth combination will be severely retarded, and if too dry, the polished surface may become tarnished. The optimum wetness of the cloth should be such that when the specimen is removed from the lap, the adhering film of moisture on the specimen surface will evaporate in about 1 to 5 sec.

When preliminary lapping is completed—usually requiring 2 to 5 min —the specimen is thoroughly washed in running water and swabbed with a tuft of wet cotton to remove adhering abrasive and debris; rinsed in ethyl alcohol or Petrohol;[1] and finally dried in a blast of warm air. A convenient dryer for this purpose is an ordinary portable hair dryer. The polished surface is then critically examined for remaining scratches. Scratches present on the polished surface, coarser than those characteristic of the abrasive used, may have been introduced by debris on the polishing cloth or may be carry-over scratches from one of the initial grinding operations. In the latter case, continued polishing to remove such coarse scratches will be very time consuming, and under some circumstances impossible to do. The only recourse in such cases is to regrind the specimen with more attention paid to proper techniques of grinding.

Specimens properly prepared through the grinding and the preliminary polishing stages will exhibit scratches of a fineness characteristic of No. 600 alundum or carborundum, and the polished surface will appear uniform and dull in luster.

Final Lapping Procedure. The final polishing operation is for the purpose of removing the scratches from the specimen surface introduced during preliminary lapping and producing ultimately a uniformly polished, scratch-free surface.

Usually in this operation, unless the preservation of nonmetallic inclu-

---

[1] Isopropyl alcohol, supplied by the Standard Alcohol Co., Linden, N.J., is an ideal washing medium and comparatively inexpensive. It cannot be used, however, for mixing etching reagents.

sions is paramount, the lapping wheel is covered with a well-piled polishing cloth, such as Forstmann's cloth, Gamal, Miracloth, or velveteen. Depending upon the metal or alloy to be polished, one of the conventional abrasives mentioned heretofore—levigated alumina, magnesium oxide, Precisionite, or chromic oxide—is used as the polishing medium. For most metallographic specimens levigated alumina performs superbly and is generally recognized as a more or less universal polishing abrasive for final lapping.

During the polishing operation, moderate pressure is applied to the specimen and, as polishing proceeds, the specimen is moved continuously from the center to the periphery of the polishing wheel. The suspended abrasive may be applied to the polishing cloth whenever necessary by means of a shaker bottle; but if sufficient abrasive is already present, the lapping cloth may be kept moist by the addition of clear water. Occasionally, and particularly towards the end of the lapping operation, the specimen is rotated counterwise to the rotation of the lap. This manipulation will effectively and continuously change the direction of polishing and thereby prevent the formation of "comet tails." Such formations are the inevitable consequence of directional polishing with its attendant tendency to drag out inclusions and locally abrade away metal immediately adjacent to, and on the follow side of, pits and voids left by inclusions. The appearance of comet tails is illustrated in Fig. 11a.

To minimize as far as possible the formation of additional disturbed metal, final polishing is discontinued as soon as all scratches are completely removed. Usually the removal of preliminary polishing scratches can be accomplished in 3 to 5 min. When final polishing is completed, the specimen is thoroughly washed in running water and swabbed with water-wet cotton to remove the last traces of clinging abrasive, and finally it is rinsed in Petrohol and dried in a warm stream of air.

If final polishing has been properly carried out, the surface of the specimen will appear scratch free when examined at 100× and there will be no evidence of comet-tail formation. In the event that fine scratches are still present on the surface, final polishing may be continued; it is more likely, however, that better results will be obtained by repeating the preliminary polishing operation before again attempting final polishing.

In the handling of polished specimens, certain precautions must be taken. The prepared surface must not come into contact with foreign objects that may scratch or mar it, nor must the polished surface be touched with the fingers, which would cause staining and eventual uneven attack by the etching reagent during subsequent etching.

The specimen may be etched directly after polishing, which is the more common procedure, or it may be stored for future use and examination in

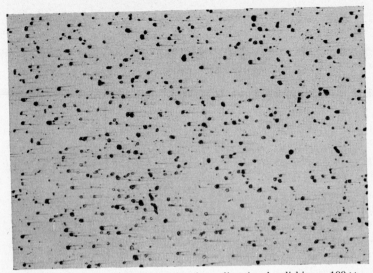

a. Formation of comet tails arising from directional polishing.   100 ×.

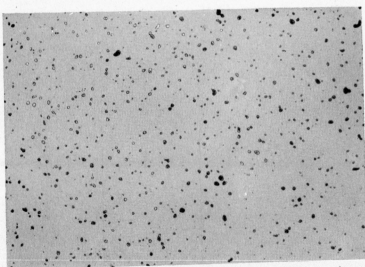

b. Elimination of comet tails by turning the specimen during polishing or by rotating it counterwise to the rotation of the polishing lap.   100 ×.

Fɪɢ. 11.   Effect of directional polishing on the appearance of an unetched metallographic specimen containing many inclusions.

the unetched condition. In any event, the surface of the specimen must be protected from oxidation and other adverse atmospheric effects by methods of storage as described on page 47.

ELECTROLYTIC POLISHING

The preparation of metallographic specimens by the mechanical methods described is a more or less tedious and painstaking task, particularly in the preparation of relatively soft metals and their alloys. Successful metallographic examination of a metal requires that the final polish on the prepared surface be of high order and that the method of preparation yield a surface that is intrinsically characteristic of the metal itself. These circumstances are not always easy to attain because of the inevitable formation of disturbed metal during polishing.

An electrolytic method of polishing introduced by Jacquet and Rocquet[1] alleviates many of the difficulties encountered in mechanical polishing. Because electrolytic polishing precludes the formation of additional disturbed metal on the ground surface of a specimen, it is a method ideally suited for the final preparation of many soft metals, single-phase alloys, and alloys that readily work-harden such as austenitic stainless steels. At present, suitable electrolytes have been developed for polishing most of the common metals and alloys.

The quality of the polished surface produced electrolytically is equal to that resulting from competent mechanical polishing, and, because the specimen in most cases does not need to be repolished and etched alternately to remove disturbed metal, the electrolytic polishing process is time saving and economical to carry out. This is particularly true in the routine preparation of a great many specimens of the same kind or in the polishing of excessively large specimens with associated great surface areas. Furthermore, any number of surfaces of a specimen may be polished simultaneously during the operation, provided such surfaces have been given the preliminary preparation as will be described.

Perhaps the main disadvantage associated with electrolytic polishing is the complete or partial removal of nonmetallic inclusions by chemical reaction between such inclusions and the electrolyte. This behavior is particularly evident with the types of inclusions found in commercially pure aluminum, aluminum alloys, and in steel. Furthermore, certain intermetallic and metalloid compounds, *e.g.*, carbides in steel, may be left standing in high relief after polishing—a circumstance arising from different rates of dissolution associated with the different structural constituents. Phases in relief are particularly objectionable when subsequent microscopic examination is to be conducted at relatively high magnifications.

[1] Jacquet, P., and P. Rocquet: *Comptes rendus*, Vol. 20, 1935.

Other disadvantages associated with this method of polishing are the staining of specimens mounted in synthetic plastics due to chemical attack of such plastics by some electrolytes, and the formation of an undulated polished surface rather than one that is plane. Under proper electrolytic polishing conditions, however, the undulation of the polished surface is of rather small magnitude, and its presence is not distracting nor objectionable during microscopic examination at either high or low magnifications.

The details concerning the principles of electrolytic polishing are not fully understood at the present time. The general mechanism, however, appears to be associated with anodic dissolution. The protruding ridges and peaks of the ground specimen surface are removed by preferential solution of metal, and the valleys existing between the ridges are either completely protected from anodic solution by the reaction products formed or, more likely, by the rate of dissolution being much less in these regions than it is at the ridges and high points.

In a proper arrangement of an electrolytic cell wherein the ground metallographic specimen serves as the anode and some suitable material serves as the cathode, direct current from an external source is carried through the electrolyte from anode to cathode by positively charged ions of specimen metal. The solution of specimen metal during the process gives rise to the formation of a boundary layer or film of electrolyte immediately adjacent to the specimen surface that is of different composition and of higher electrical resistance than the remainder of the electrolyte. General considerations of this boundary film indicate that at the protruding ridges of the ground surface the layer is thinner, the concentration gradient is steeper, and the electrical resistance is lower, than that part of the boundary film associated with the valleys. As a consequence, under a given potential the current density at the peaks and ridges will be greater than that at the valleys; and if the applied potential is sufficiently high, metal from the ridges will go into solution at a faster rate than metal from more depressed regions. This differential in the rates of dissolution will cause a gradual smoothing out, or polishing, of the ground surface.

Perhaps the most important factor contributing towards successful electrolytic polishing is the relationship established between current density and voltage for a given electrolyte and electrolytic cell arrangement. For many of the electrolytes used, successful polishing, without complications, is secured when conditions are such that a change in applied voltage has no appreciable effect on the current density. The range of such voltages is defined schematically in Fig. 12 by the portion of the current density-voltage curve, *C-D*. It is of interest to note that the plateau *C-D*, associated with electrolytes recommended for use with such metals as tungsten, magnesium, and zinc, is relatively long and may exist over a distance

equivalent to several volts.   In other cases, however, involving the electrolytes generally used in polishing copper and cobalt, the plateau extends
over a range of only a few tenths of a volt—a circumstance requiring critical
adjustment of the terminal voltage so that it falls within the limits of the
plateau.   In still other cases, particularly those of the perchloric-acid-type
electrolytes and such metals as iron, aluminum, and tin, the plateau is
either so restricted that it cannot be observed by normal methods of
securing current-voltage data, or it is entirely absent.   Restriction of the
plateau notwithstanding, successful electrolytic polishing can be secured
over a range of current densities and voltages (usually high) between limits
defined by the cessation of etching and the beginning of anode gassing and
uneven polishing.

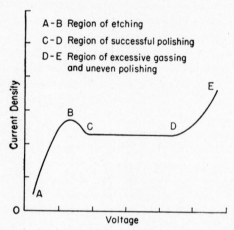

Fig. 12.   Current density–voltage relationship characteristic of electrolytes that function at
nearly constant current density.

In the region of the curve *A-B*, Fig. 12, the current density increases
proportionally with increase in applied voltage.   This proportional increase
is related to the formation of an unstable boundary film on the surface of the
specimen, whose rate of diffusion into the surrounding electrolyte is high in
comparison to its rate of formation.   In this region of the curve, surface
metal is removed unevenly with little regard to selective dissolution at the
ridges.   The specimen will appear dull in luster and etched, owing principally to chemical reaction between the specimen and the electrolyte.

Although the boundary layer is unstable for conditions of current density
and voltage defined by *A-B*, it nevertheless tends toward greater stability
as the voltage is increased over this range.   As maximum stability is
approached and the boundary layer is nearly completely formed (voltage
B), both the electrical resistance and the back emf of the film increase with

a small but significant attendant lowering of the current density as indicated by change in slope of the curve.

With increase in voltage, a current density is soon reached (point C) where an equilibrium state is established between the rate of boundary-film formation and its rate of diffusion into the remaining electrolyte. The film becomes virtually saturated with metal ions, owing to the relatively high rate of dissolution of specimen metal, and with increase in voltage to some limiting value the boundary layer merely increases uniformly in thickness. This increase in film thickness is associated with a proportional increase in electrical resistance of the layer, accounting for the current density remaining practically constant with increase in voltage up to the limiting voltage value, indicated by point D. It is in this region of the curve, C-D, as mentioned heretofore, that successful electrolytic polishing takes place with most electrolytes.

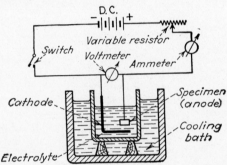

Fig. 13. Electrical circuit and arrangement of apparatus suitable for most electrolytic polishing operations.

When the voltage exceeds the limiting voltage for successful polishing, sufficient gassing occurs at the specimen surface (usually the liberation of oxygen) to break down the continuously covering boundary layer as fast as it is formed. Because of this, and as shown in the portion of the curve D-E, the current density again increases proportionally with increase in applied voltage. Owing to the severe anodic gassing and the accumulation of large gas pockets on the surface of the specimen, polishing in this region of the curve will be uneven and the polished surface severely undulated.

The procedure of electrolytic polishing is relatively simple and requires very little skill on the part of the technician. The surface of the specimen to be polished (one or more surfaces) is made plane by grinding, as described before, on a series of emery papers of decreasing grit size or on a series of three lead laps. For most metals and alloys, grinding through the No. 000 emery paper or the No. 305 lead lap is adequate preparation. When all debris has been washed from the surface of the specimen, a final rinse in ether or other appropriate solvents is recommended to ensure complete removal of oily substances that might otherwise interfere with uniform polishing.

The specimen to be polished is made the anode in an electrolytic cell, whose external circuiting system and cell arrangement may be rather simple, as illustrated in Fig. 13; or in one more complicated, as shown in Fig. 14, which is recommended specifically for use with orthophosphoric acid electrolytes. The material to be used as a cathode will depend upon the metal to be polished and the specific electrolyte to be used (see Table 1, page 409). Convenient and versatile electrolytic polishing and etching units, designed

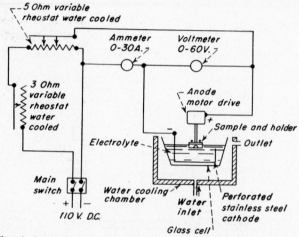

Fig. 14. Electrical circuit and arrangement of apparatus that is particularly suited for electrolytic polishing with orthophosphoric acid electrolytes. Note rotating anode. (*From ASTM Standards, Part I-B, 1946.*)

specially for metallographic specimens, are available, one model of which is shown in Fig. 15.

In electrolytic polishing the choice of an electrolyte will depend upon the composition and the structural characteristics (the number and kinds of phases present) of the specimen to be polished. Recommended electrolytes, in conformation to the principles of electrolytic polishing already discussed, are capable of forming a high-resistance boundary layer at reasonable voltages.

Electrolytes[1] suitable for electrolytic polishing of many common metals and alloys are given in Table 1, along with data pertaining to time of pol-

[1] **Warning!** Electrolytes containing perchloric acid and acetic anhydride must be used with extreme caution. Unless this electrolyte is adequately cooled during mixing and during use, a violent explosion may occur. Specimens mounted in synthetic plastic materials, such as bakelite, Lucite, or Tenite, or in bismuth or bismuth alloys, *should not* be used with this electrolyte.

ishing, temperature, voltage, and current density. The data of voltage and current density are to be considered only a first approximation since their precise values will depend upon concentration of the electrolyte, exact temperature of the reagent, and the distance between cathode and specimen. All electrolytic polishing operations require some personal observation on the part of the technician to determine the most suitable operating conditions for a particular cell arrangement and set of polishing conditions.

Although electrolytic polishing cannot inherently produce disturbed metal on the polished surface of a specimen, it is often observed that a

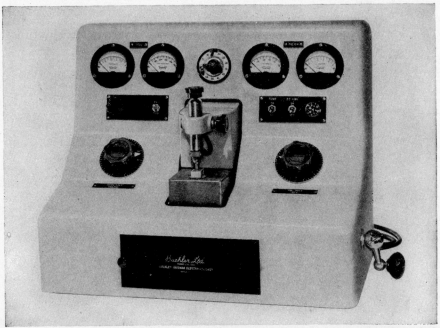

Fig. 15. Electrolytic polishing and etching unit designed specially for metallographic specimens. (*Courtesy of Buehler, Ltd.*)

specimen so polished does show evidence of distorted metal after etching. This condition may be attributed to distortion caused during the grinding operation. Although the formation of this disturbed metal cannot be entirely prevented, it can nevertheless be effectively removed by alternately polishing electrolytically and etching.

An example of electrolytically polished stainless steel is shown in Fig. 16.

SPECIAL MECHANICAL-PREPARATORY PROCEDURES

The metallographic preparation of soft metals, such as aluminum, copper, lead, etc.; the preparation of very hard materials, such as cemented

a. Specimen mechanically polished.   Note evidence of disturbed metal.   100 ×.

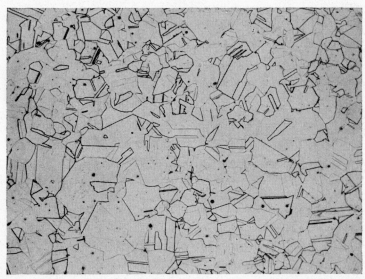

b. Specimen electrolytically polished.   Note complete absence of disturbed metal.   100 ×.

FIG. 16.   Influence of mechanical and electrolytic polishing on the appearance of 18–8 stainless steel after etching.   (*Courtesy of U.S. Steel Corp. Research Laboratory.*)

tungsten and boron carbides; and the retention of inclusions in steel and graphite flakes in cast iron require more or less specialized techniques and manipulations. The comparatively soft metals and their alloys tend to smear during grinding and polishing, and, unless great care is exercised during preparation, disturbed metal will be formed to such a considerable depth that it will be practically impossible to remove it by the usual methods. Consequently, the intrinsic characteristics of the microstructure will be completely obscured.

Very hard materials, solely because of their intrinsic hardness, are difficult if not impossible to prepare by the usual metallographic techniques and with the commonly employed metallographic abrasives. In such cases, satisfactory preparation can be secured only by the use of diamond dust and relatively hard-surfaced laps, as described on page 43.

The preparation of many of the soft metals and alloys may be effectively carried out, with a minimum amount of surface distortion, by the use of a microtome. A microtome is essentially a precision instrument that will produce a plane surface on a soft-metal specimen by removing extremely thin layers of the surface metal (usually 2 to 5 microns in thickness). This is accomplished by a movable cutting blade.

Because a microtome is not a common piece of laboratory equipment, the technique involved in its use will not be discussed.[1] Rather, the more general procedure of preparation of the soft metals by grinding and lap polishing methods will be considered.

Aluminum and Its Alloys.[2] The preparation of aluminum and aluminum alloys for metallographic examination is somewhat difficult because of the relative ease with which surface flow and distortion occur during grinding and polishing. In part, surface distortion and attending troubles may be mitigated by securing a plane surface and removing saw marks from the specimen with a medium-cut mill file, instead of grinding the specimen on the conventional emery belt.

When a plane surface has been obtained and the edges of the specimen have been beveled, the specimen is then ground on the usual three grades of emery paper—No. 0, 00, and 000—care being taken to turn the specimen 90 deg when a change is made from one paper to another of finer grit size. To maintain a bright specimen surface and to prevent debris and emery particles from becoming embedded in the surface during grinding, each paper used is coated with a solution of paraffin dissolved in kerosene (about 25 g paraffin per 500 ml kerosene). It is essential under these conditions

---

[1] For a discussion of this subject see F. F. Lucas: Application of Microtome Methods for the Preparation of Soft Metals for Microscopic Examination, *Proc. AIME*, Institute of Metals Division, 1927.

[2] In part from ASTM Standards, Part I-B, 1946.

that when a change is made from one grinding paper to another, the specimen be thoroughly washed in a suitable solvent (kerosene) to remove wax and grinding debris adhering to the specimen surface.

The first wet polishing operation is for the purpose of removing the fine emery scratches introduced on the surface of the specimen by the No. 000 emery paper. This preliminary polishing operation is usually carried out on a broadcloth-covered lap, rotating at approximately 300 rpm. The usual abrasive used is a water suspension of No. 600 alundum or a similar material.

Final polishing is for the purpose of removing the alundum scratches and producing a scratch-free surface. This operation is carried out on a polishing disk generally covered with either Miracloth, Selvyt, or Kitten's Ear broadcloth, the lap rotating at about 150 to 200 rpm. Either levigated alumina or magnesium oxide may be used as the polishing abrasive, although in many respects the latter has been found to be more satisfactory. Because of this, the polishing technique employing magnesium oxide will be described.

The polishing cloth is thoroughly soaked with distilled water (*not* tap water), and Merck's heavy magnesium oxide powder is worked into the cloth fibers with the finger tips, care being taken to remove from the cloth any large and hard particles that may be present in the abrasive. (For appropriate methods of using magnesium oxide and handling of polishing cloths upon which this abrasive has been used, see pages 19 and 22.)

During the first stage of final polishing, particular attention should be directed to keeping the lapping cloth *moist* with distilled water—never allowing it to become soaking wet or very dry. A cloth that is too dry will tend to drag out the softer constituents in an alloy of two or more phases, whereas a cloth that is too wet will tend to corrode certain of the structural constituents of the specimen and relief-polish the harder ones. Of equal importance is the contact pressure between the specimen and the revolving lap wheel. The optimum condition in this respect is best determined by trial, as the pressure, as well as the time of final polishing, depends upon the chemical composition and heat-treatment of the specimen.

When practically all the scratches on the surface of the specimen have been removed, the polishing cloth is flushed with clear distilled water in such amounts that at the end of the polishing operation the cloth is completely free of abrasive. It is advisable in this stage of the operation to rotate the specimen counter to the rotation of the polishing wheel in order to eliminate directional polishing marks and comet tails.

Copper and Its Alloys. In the preparation of copper and copper-alloy specimens for microscopic examination, a plane surface is secured by filing or by grinding on an emery wheel copiously supplied with water. The usual hand-grinding operations on emery papers of decreasing grit size, as is common practice in the preparation of other metals, may be omitted from the procedure. The specimen may be polished in three stages, directly after filing or initial flat grinding. The polishing laps rotate at 250 to 1800 rpm, depending upon the alloy undergoing preparation, the kind of polishing cloth being used, and the preference of the technician.

The first stage of polishing is carried out on a lap covered with an 8- to 12-oz canvas duck cloth, using an abrasive of either FF Turkish emery, No. 500 carborundum, or grades of alundum of No. 500 or finer. The second polishing operation consists of polishing the specimen on a wool-broadcloth-covered lap, using an abrasive of powdered tripoli; or an alternate procedure consisting of polishing on two felt-covered laps, the first charged with "RF" emery, and the second lap with "SF6X" flour emery. The third and final polishing stage is done on a lap covered with either a fine grade wool, Gamal, or Kitten's Ear broadcloth, employing as the final polishing medium a water suspension of alumina or finely powdered magnesium oxide. Alternate to this procedure, final polishing may be carried out on a chamois covered lap, using an abrasive of jeweler's rouge.

All polishing wheels are kept wet during use by a water drip or shaker bottle and the specimens, between steps, are kept wet and thoroughly rinsed free of abrasives. After its removal from the final wheel, a specimen may be immediately etched or rinsed in alcohol and quickly dried, prior to etching. In much of the routine preparation of specimens, the final etching is depended upon to remove many shallow scratches. When rouge is being used on the final wheel it is possible to remove the specimen at a critical moment when the polished face becomes clean and dry, and to etch directly.

Pure copper is more difficult to polish than are its alloys, since a nearly perfect surface is required in order to detect the presence of cuprous oxide in the unetched specimen.

Lead and Its Alloys. The preparation of lead and lead-alloy specimens is difficult because of the inherent softness associated with these materials. This circumstance gives rise to considerable surface flow and distortion of the surface metal during metallographic preparation and hence tends to mask completely the true structure upon subsequent etching. Although it is generally agreed that a surface free of cold work is almost impossible to obtain by polishing, this condition is now neither necessary nor particularly advantageous for common leads and alloys with but small amounts of other elements in solution, since recent developments in etching make the uniform removal of cold-worked material possible. (See reagents Nos. 1 and 3, Table 6, page 423.) Alloys containing hard constituents, however, present more difficulty because of uneven attack by etching solutions.

A number of methods for preparing specimens are in regular use, the choice for a given specimen generally depending upon the alloy, the size and shape of the specimen, and, perhaps, the available equipment or the operator's preference. It is always necessary, however, to remove sufficient metal to eliminate the effects of sawing, shearing, filing, or whatever means was used in securing the specimen. For the softer alloys, cutting or grinding, as described later, will be satisfactory but alloys containing hard constituents should be prepared with a microtome or its equivalent.

Bassett and Snyder[1] use a mill file to remove all traces of other working (saw marks, etc.) and to secure a flat surface. The surface is then ground smooth by hand on lubricated emery papers backed with plate glass. Although papers of various degrees of fineness may be used, Nos. 0 and 000 papers are usually sufficient. A new paper about 30 times the area of the surface of solid specimens and eight times the diameter of tube specimens should be used each time. Bumps or foreign particles on the paper are removed by scraping with a straight edge, while an abundant supply of gasoline or a solution of paraffin in kerosene is used as a lubricant. Grinding, with the specimen pressed firmly against the paper, is continued until the emery is worn smooth and a bright surface, free of prominent scratches, is obtained (5 to 7 min on each paper). The specimen is etched after it has first been wiped free of gasoline or—if another lubricant was used—after it has been cleaned in gasoline and then wiped. Ordinarily, such surfaces will show sufficient detail for examination at the magnifications generally used for lead. If, after etching, it appears that further preparation is desirable, grinding about 5 min on No. 0000 emery paper may be helpful.

Final fine polishing is not common practice with lead specimens but Vilella and Beregekoff[2] recommend a suitable procedure. The specimen is polished wet on clean broadcloth liberally smeared with tincture of green soap, a water suspension of the finest levigated alumina being used as the abrasive. When a black smudge appears, polishing should be very carefully continued until it disappears and the surface is again bright. Scratches remaining after this operation are sometimes visible after etching but can be removed by alternately etching and polishing by hand on silk velvet well soaked with soap and alumina.

Tarnishing and air oxidation of lead specimens, in either the etched or unetched condition, may be prevented for a number of days by immersing the specimens in acetone and evaporating the acetone by suction.

[1] Bassett, W. H., Jr., and C. J. Snyder: Method of Preparation of Lead and Lead Alloy Cable Sheath for Microscopic Examination, *Proc. ASTM*, Vol. 32, Part II, 1932.

[2] Vilella, J. R., and D. Beregekoff: Polishing and Etching Lead, Tin and Some of Their Alloys for Microscopic Examinations, *Ind. Eng. Chem.*, Vol. 19, 1927.

Magnesium and Its Alloys. In general the preparation of magnesium and magnesium-alloy specimens is similar to the preparation of aluminum specimens.

As in the preparation of other soft specimens, a plane surface can be secured with a microtome; otherwise, grinding successively on Aloxite cloths Nos. 50, 150, and 320 is recommended. The grinding is then finished by successive use of Nos. 0, 000, and 0000 metallographic emery paper. For convenience the abrasive cloths or papers may be mounted on disks rotating at 800 to 1200 rpm.

Polishing is carried out in three stages, the first two commonly referred to as "rough polishing" and the third as "fine polishing." The rough polishing is usually done on velvet or Vel-Chamee cloth mounted on laps rotating at 500 to 600 rpm. In the first stage the abrasive is usually XF alundum and in the second it is No. 600 alundum; in both, only enough *distilled* water is used to prevent seizure of the specimen. Fine polishing is accomplished on a velvet-covered lap rotating at 100 to 400 rpm and moistened with a suspension of specially levigated alumina. Seizure of the specimen is prevented by using a small amount of filtered liquid soap on the wheel.

The special levigated alumina referred to above is prepared by placing 150 g of the best commercial grade of levigated alumina and 2 liters of 0.001N sodium hydroxide into a 4-liter bottle and agitating the contents by a stream of compressed air for about 30 min. Two more liters of 0.001N sodium hydroxide are then added and the mixture thoroughly shaken. After standing for $1\frac{1}{2}$ hr, the upper $\frac{1}{2}$ in. of supernatant liquid is carefully siphoned off, yielding a suspension suitable for final polishing. Relevigation of the remaining sediment may be carried out in the same manner, yielding a suspension slightly coarser than that first secured.

An alternate method[1] for preparing magnesium alloys for metallographic examination consists of securing a flat surface on the specimen of interest by means of a file, and subsequently hand grinding the specimen in the usual way on Nos. 1, 0, 00, and 000 metallographic emery papers. Directly after the last grinding operation, the specimen is final-polished on a rotating lap covered with Selvyt cloth. A high-quality grade of levigated alumina is recommended as the polishing abrasive, and a lap speed of about 200 rpm is desirable.

A new polishing cloth is appropriately charged with abrasive by thoroughly working a small quantity of alumina into the cloth fibers. During charging of the cloth and during subsequent polishing a minimum of alumina is used and only sufficient water is added to keep the lapping cloth

[1] *ASTM Bull.*, October, 1944.

moist.   Maintaining a proper ratio of alumina and water is an essential part of the polishing procedure and it is a circumstance that is best determined by noting with sensitive finger tips the degree of drag between the specimen surface and the rotating lap.   If conditions of polishing are appropriate, the lapping cloth will slowly but evenly become darkened while the specimen is being polished.   It is recommended that moderate pressure be applied to the specimen, and during polishing the specimen be rotated in a direction counter to the rotation of the lap.

Tin and Its Alloys.   These metals are in a class with lead and its alloys and may advantageously be prepared in a similar manner.   Tin and its softer alloys may be prepared by microtoming, but the harder alloys may require one or two polishing steps before etching.

Tafts[1] suggests a method for the preparation of tin and its alloys which consists of grinding on the usual emery papers, with final grinding on No. 000 or No. 0000 papers previously rubbed with a piece of hard metal to remove any upstanding emery particles.   Scratches from the last paper are removed by preliminary machine polishing on Selvyt cloth, using heavy magnesium oxide as the abrasive.   The fine scratches resulting from this operation are removed by light hand polishing on Selvyt cloth stretched tightly over a hard, plane, piece of wood.   The cloth is first prepared by soaking it with benzene and rubbing into the cloth fibers a specially prepared alumina.   The benzene acts as a cleaner and prevents a smudge from forming on the specimen surface.   During polishing on this pad, distilled water is used as the lubricant.

The alumina used for final polishing may be prepared as follows: Ammonium alum of analytical reagent purity is gently heated to remove the water of recrystallization, after which it is heated in a fused silica or porcelain container for approximately 8 hr at 1000°C (1832°F).   After cooling to ordinary temperature the alumina may be used directly as described above.

An alternate method of preparing tin and its alloys for metallographic examination consists of first securing a plane surface by means of lathe turning, employing a tool having a sharp, smooth cutting edge and relatively large clearance.   The specimen is then ground on Nos. 0, 00, and 000 emery papers, each paper being lubricated with a paraffin-kerosene solution (100 g paraffin dissolved in about 200 ml kerosene).   Because of the relatively large amount of paraffin used, the solution is nearly solid at ordinary temperature.   It is necessary, therefore, that the solution be used warm when application is made to the emery papers.   To remove the paraffin layer adhering to the surface of the specimen when a change is made from one grinding paper to another, the specimen is washed in warm kerosene

[1] Tafts, H. J.: Mounting and Polishing of Tin Alloys, *J. Roy. Microscop. Soc.*, Vol. 56, 1936.

and the kerosene subsequently removed by cleansing with soap and warm water. After grinding is complete on each successive paper, the specimen is etched in 10 to 20 per cent hydrochloric acid, or other reagents as required, to remove whatever disturbed metal may have formed on the surface during grinding.

The ground specimen is usually polished in two stages. Preliminary lapping is carried out on a broadcloth-covered lap, using No. 600 alundum suspended in a heavy soap solution, prepared as described below. After preliminary polishing, the specimen is etched, and finally polished on a second wheel covered with high-quality silk velvet. The abrasive used is the same alundum suspension as that employed in the preliminary lapping operation. For proper delineation of the structure, it is essential that the specimen be alternately polished and etched several times.

The special soap-alundum suspension may be prepared by thoroughly agitating 5 g of No. 600 alundum in about 300 ml of water. After settling for about 15 min the supernatant suspension is removed by siphoning. To this supernatant suspension is added 40 g of high-purity soap and the mixture heated to facilitate solution of the soap. When the soap is completely dissolved, the mixture at ordinary temperature is nearly solid and hence when used the mixture must be warmed to restore fluidity.

Zinc and Its Alloys. The metallographic preparation of zinc is complicated by the fact that mechanical twins are readily formed during mechanical grinding and polishing, particularly in coarse-grained zinc; and the recrystallization temperature of commercial zinc may range from 100°C (212°F) to as low as ordinary temperature for very pure zinc. Owing to these circumstances, it is therefore imperative that sufficient surface metal be removed during preparation to eliminate completely induced distortion and that the preparational manipulations be carefully executed so as to avoid heating the specimen and causing attendant recrystallization. This situation in some cases precludes mounting the specimen of interest in materials requiring the application of heat to effect a mount.

The specimen is initially flattened by grinding on a belt sanding machine, with the use successively of belts of about the following grades: Nos. 30, 50, 120, and 320. It is advisable to dip the specimens in water occasionally to prevent heating. At least $\frac{1}{16}$ in. of the surface of the section should be removed in these four operations so that the distortion produced at the sheared or sawed edges of the specimen may be removed. If a sanding machine is not available, a mill file may be substituted.

Dry grinding is carried out on four grades of emery paper, Nos. 1, 0, 00, 000, resting on plate glass. Each succeeding operation should be carried out at an angle to the preceding operation. Merely removing the scratches from the preceding operation is not sufficient; experiments have demonstrated that disturbed metal (as shown by the presence of twin bands after etching) may be found to depths of at least 20 times the depth of the deepest scratch, in some cases.   Therefore, when all preceding scratches have been removed, grinding should be continued for some time.   Since there is no method of judging the depth of distortion for a given specimen, the amount of grinding must be gauged largely by experience.   It may be noted, however, that coarse-grained pure zinc is most easily distorted, while fine-grained material—particularly alloys—will give little trouble from this source except, possibly, in the final stages of polishing.

Wet polishing is carried out on four cloth-covered laps rotating from 200 to 400 rpm.   The first two are covered with white duck, the last two with broadcloth or billiard cloth.   Rodda[1] has described a suitable method for preparing abrasives for use on the four wheels, but those suitable for aluminum and magnesium can be used.   In wet polishing care is required also in removing the disturbed metal produced by the preceding operation.   The amount of metal removed during wet polishing is slight and it is more feasible to remove the bulk of the distorted metal by etching with Palmerton reagent.[2]   The following schedule is recommended:

Etch 3½ min
Polish on No. 1 wheel
Etch 1½ min
Polish on No. 2 wheel
Etch 30 sec
Polish on No. 3 wheel
Etch 10 sec
Polish on No. 4 wheel
Etch 3 sec

The etching schedule may be varied somewhat, depending upon the specimen and the purpose of the examination.   For examination at low magnification, polishing may generally be stopped at the third wheel. When the examination is to be at relatively high magnification, the etching time after the fourth wheel should generally be reduced to 1 sec, particularly if there are constituents present which are unattacked by the etching reagent.   Otherwise, the difference in elevation between etched and unetched constituents will make it impossible to bring both into focus at one time.   In the preparation of specimens containing constituents of widely different hardnesses and etching characteristics, it is advisable to eliminate the customary procedure of alternate polishing and etching.   Generally, repolishing after etching will not flatten the etched surface satisfactorily nor eliminate completely the etched structure.

[1] Rodda, J. L.: Preparation of Graded Abrasives for Metallographic Polishing, *Trans. Inst. Metals Division, AIME,* Vol. 99, 1932.
[2] See Table 5, p. 422, reagent No. 1.

Galvanized Sheet Steel. By the use of ordinary metallographic techniques it is very difficult to appropriately prepare cross sections of galvanized sheet steel, particularly when the zinc and zinc alloy layers are to be the object of the examination. The principal difficulties associated with the regular techniques are (1) rounding off of the zinc coating at the exterior edge during polishing, owing to differences in hardness between the zinc layer and the base steel; (2) staining and discoloration of the galvanized coating when polished with the usual water-suspended metallographic abrasives, *e.g.*, levigated alumina; and (3) nonuniform etching of the zinc coating when etched with reagents recommended for etching massive zinc.

To alleviate these difficulties, Rowland and Romig[1] have suggested a preparational procedure for galvanized steel and a method of preparing an alumina suspension that, because of a controlled pH, does not darken and stain the zinc coating during polishing.

To assist in supporting the edges of the zinc coating, appropriately sized galvanized-sheet specimens are coated with bakelite lacquer (No. BL-3128) and the coating thoroughly hardened by baking. The sheet specimens are then either mounted in a steel clamp or in a conventional bakelite mount to facilitate in the subsequent grinding and polishing operations.

Initial flattening of the specimen mount may be secured by grinding on a flattening belt or, preferably, by means of a machine-shop surface grinder. The flattened surface is then hand ground on emery papers, Nos. 0, 00, and 000, with final grinding carried out on a rotating lap covered with No. 000 emery paper impregnated with graphite. The lap is rotated at between 1000 and 1200 rpm, and during grinding the specimen is moved continuously from the center to the periphery of the lapping wheel.

When all hand-grinding scratches have been removed on the lap, the specimen is ready for final polishing. This operation may be carried out in either one or two stages, depending upon the quality of polished surface desired. Preliminary lapping is done on a rotating lap wheel covered with a short-piled lapping cloth, such as Miracloth or Forstmann's broadcloth. The lap is rotated at 300 to 500 rpm, and a specially prepared alumina suspension, as described below, is used as the abrasive. It is essential during polishing that the cloth be kept only slightly moistened, to prevent staining of the zinc coating when it is removed from the polishing cloth. The degree of wetness of the cloth should be such that when the specimen is removed from the lap, the adhering film of moisture will evaporate immediately with gentle blowing.

During preliminary polishing the specimen is etched at frequent inter-

[1] Rowland, D. H., and O. E. Romig: The Metallography of Galvanized Sheet Steel Using a Specially Prepared Polishing Medium with Controlled pH, *Trans, ASM*, Vol. 31, 1943.

vals—usually a total of 3 or 4 times—in a specially prepared nitric acid–amyl alcohol etching reagent (see reagent No. 3B, Table 5, page 422). Subsequent polishing, after each etching treatment, is done for the shortest length of time required to completely remove the effects of the previous etch.   Prolonged polishing is undesirable because of beveling or rounding-off of the edges of the zinc coating.

If an exceptionally high-quality metallographic surface is desired, the specimen may be final polished on a rotating lap wheel covered with a high-grade silk velvet.   Prior to being used, the cloth should be thoroughly boiled in distilled water to soften any hard fibres present and to remove the dye from the cloth, which might otherwise severely discolor the zinc coating during polishing.

The polishing cloth is appropriately charged with the special alumina mentioned heretofore, and during polishing the specimen is moved continuously from the center to the periphery of the lap under fairly heavily applied pressure.   Usually no more than about 1 min is required in final polishing on laps rotating at 300 to 500 rpm.

Immediately after preliminary and final polishing, the polished specimen must be dried to prevent discoloration of the galvanized coating.   Adhering debris and alumina may be removed by swabbing the polished surface with a tuft of cotton saturated with amyl alcohol, followed by a thorough rinsing in absolute ethyl alcohol, and finally drying the specimen in a blast of warm air.   Under no circumstances should the polished specimen be washed or rinsed in tap or distilled water, or 95 per cent ethyl alcohol.

The specially prepared alumina developed by Rowland and Romig for polishing galvanized sheets may be prepared as follows: Aluminum sulphate of technical quality $(Al_2[SO_4]_3 \cdot 18H_2O)$ is first converted to alumina by roasting at 1000 to 1100°C (1832 to 2012°F) for about 2 to 5 hr.   Some provision must be made to remove the corrosive sulphur trioxide fumes evolved during the roasting process.

When the roasted product has cooled to ordinary temperature it is mixed with distilled water, on a weight basis of one part of alumina to 40 parts of distilled water, and the suspension passed through a 200-mesh sieve to break up the curd.   The sieved material is then leached by boiling the water-alumina mixture, and after cooling and allowing the alumina to settle, the supernatant liquid is decanted and discarded.   Fresh distilled water is again added and the leaching process repeated.   In the manner described, the alumina is leached for as many times as is necessary to remove all but a trace of aluminum sulphate present.   The presence of sulphate may be detected by first adding a few drops of concentrated hydrochloric acid to about 50 ml of the leached solution.   To this is added 10 ml of a 10 per cent aqueous barium chloride solution, and the solution boiled and allowed to stand for about 5 min. The appearance of a white precipitate (barium sulphate) indicates that aluminum sulphate is still present in the alumina and that further leaching is necessary.

After the last leaching operation, the pH of the suspending liquid is adjusted to a value of between 7 and 7.6. This adjustment is secured by adding a few drops of triethanol amine (0.1 g dissolved in 200 ml of distilled water), and using either a pH meter to observe the change in the pH, or an indicator of phenolsulphonaphthalein (phenol red). If phenol red is used, 5 drops of the indicating solution (0.001 g of the dye dissolved in 25 ml of a 50 per cent aqueous solution of absolute ethyl alcohol) in 10 ml of the alumina suspension will be faintly pink in color when the proper pH value is obtained by additions of triethanol amine.

Cemented Tungsten Carbide and Other Hard Alloys. In preparing very hard alloys for metallographic examination, such as cemented tungsten or boron carbide specimens, the usual methods of preparation are unsuitable owing to the inability of the common metallographic abrasives to cut such materials.

Hoyt[1] has devised a preparational procedure for cemented tungsten carbides that, with some modifications, is the method generally used today and which has been extended with success to the preparation of other hard and friable materials. Briefly, the method consists of initially flattening an appropriate specimen on the flat side of a silicon carbide grinding wheel and subsequently grinding the specimen on a copper-faced lap charged with kerosene and boron carbide of 140 to 200 mesh. During the course of grinding, the boron carbide particles are fractured into progressively smaller sizes which, near the end of the grinding operation, produce scratches on the surface of the specimen that are fine and uniform. These fine scratches are then removed by polishing the ground specimen on a wooden lap rotating at 150 rpm, using an abrasive of diamond dust (1.5 to 2.0 microns average particle size) and a lubricant of almond oil.

A method of preparing cemented carbide specimens has been developed by Kehl[2] that is ideally suited for preparing not only normal specimens, but also specimens having a high degree of porosity and those that are imperfectly bonded and contain carbide particles that may be easily dislodged. The preparation procedure differs from that of Hoyt and similar methods in that diamond dust of appropriate grade is used for both grinding and polishing, and the grinding and polishing laps are covered with paper that may be discarded readily when dirty or contaminated.

The method consists of first obtaining a flat surface on the specimen of

---

[1] Hoyt, S. L.: Preparation of Microsections of Tungsten Carbide, *Trans. ASST*, Vol. 17, 1930.

[2] Unpublished research.

interest by cutting it with a diamond wheel fitted to a conventional cutoff machine.   The specimen may be cut in the unmounted condition or, if the specimen is small, it may be cut after appropriately mounting in bakelite or other synthetic resins.   A diamond wheel suitable for this purpose, and one that produces a fineness of surface scratches easily removed in the first grinding operation, is wheel No. 100-M, manufactured by the Carborundum Company.

After initial flattening of the specimen as described, the successive grinding operations are carried out on a series of three diamond laps, followed by the polishing operation on two additional laps, each unit rotating at about 175 rpm.   A design of lap found to be excellent for grinding and polishing consists of a metal disk, or preferably a micarta disk, over which

Fig. 17.   Micarta grinding and polishing disk covered with Keuffel and Esser tracing paper for preparing cemented tungsten carbide specimens.

is placed a circular sheet of high-quality tracing paper, *e.g.*, Keuffel and Esser paper No. 353R.   The paper is held onto the lap by means of a collared ring fitting over the edge of the paper and around the periphery of the lap, as illustrated in Fig. 17.

Each of the three grinding laps is charged respectively with diamond dust of decreasing particle size—Nos. 2, 4, and 7,[1] having respectively an average particle size of 74, 37, and 5 microns.

The diamond dust is applied to the lap as dry powder and in an amount equivalent to that which will adhere to the end of a moistened toothpick. Oil of lavender (USP XII quality) is ideally suited as a lubricant, and it is added to the lap before and during grinding by means of a medicine dropper. It is essential during grinding that the lap is never permitted to become

[1] As designated by J. K. Smit & Sons, Inc., New York, N.Y.

dry, nor should the amount of oil of lavender added be sufficiently great to make the lap surface excessively wet. During grinding, moderate pressure is applied to the specimen and the specimen is continuously moved from the center to the periphery of the lap. Grinding on any one lap requires only about 1 to 3 min, and in conformation to conventional practice of grinding, the specimen is turned about 90 deg when a change is made from one lap to another one that is charged with finer diamond dust.

The manipulations involved in polishing the ground specimen are similar to those described for grinding; the two grades of diamond dust used for polishing, however, are necessarily smaller in particle sizes. Preliminary polishing is best carried out with No. 8 diamond dust (average 2.5 microns)

Fig. 18. A cemented tungsten carbide compress, containing very large carbide particles, prepared for metallographic examination by grinding and polishing on paper-covered laps. 1500 ×.

and final polishing with No. 9 (0 to 2 microns). If the polishing operations are carefully executed, the recommended grades of diamond, particularly No. 9, will produce a polished surface comparable to that attained on softer metals and alloys when polished with levigated alumina.

The appearance of an etched specimen of cemented tungsten carbide prepared for metallographic examination by the method described is shown in Fig. 18.

Cast Iron. Because considerable difficulty is often encountered during the metallographic preparation of gray and malleable irons by the usual methods, particularly with respect to retention of graphite particles, it is necessary to prepare such specimens according to a special technique. Cast irons are best prepared by grinding the specimen on the usual three grades of emery papers (Nos. 0, 00, and 000), with prolonged grinding on a well-worn sheet of No. 00 paper.   Final grinding is generally carried out on a No. 000 emery paper which prior to use has been glazed with either graphite or soapstone.

Although dislodgment of the graphite particles may occur during grinding the specimen, it more frequently happens during the polishing operation particularly when the cloth is one that is deep piled.   To prevent such dislodgment and attending difficulties, final polishing is best conducted on a polishing lap fitted with a napless cloth such as fine silk or airplane wing-covering material.   Experience has further shown that a polishing abrasive of heavy magnesium oxide has less tendency to remove the graphite par-

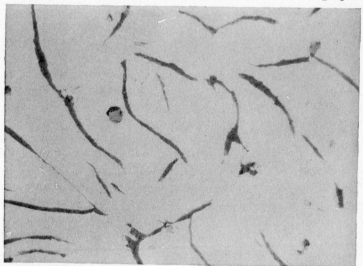

Fig. 19.   Illustrating the retention of graphite flakes in gray cast iron by the use of special metallographic preparatory techniques.   Specimen unetched.   500 ×.

ticles and will produce in this operation finer scratches and a higher luster on the polished surface than an abrasive of levigated alumina.

It is essential during the polishing operation that the cloth be kept damp but not wet and that polishing of the surface proceed in only one direction. Although the latter recommendation is contrary to the usual methods of polishing, it has been found that rotating the specimen while polishing will very quickly dislodge the graphite particles.   It will be noted during the first few minutes of polishing that a slight amount of dragging from the graphite will be evident—formation of comet tails—but this condition on the surface will disappear upon further polishing.   Polishing should not, under any circumstances, be continued in excess of that required to just remove the grinding scratches, and, in keeping with good technique, the polished surface should be examined microscopically at frequent intervals and the polishing operation discontinued when the surface first appears scratch free.

The retention of graphite flakes in gray cast iron, achieved through using the technique described, is shown in Fig. 19.

An alternate method of preparation for the retention of graphite particles in cast iron, as well as for nonmetallic inclusions in other metals and alloys, consists in carefully grinding the specimen on a series of three lead laps as already described, with fine and final polishing on Forstmann's cloth, using an abrasive of levigated alumina.

Preservation of Inclusions. The metallographic preparation of a specimen wherein it is essential to retain intact nonmetallic inclusions requires, for best results, a technique similar to that described for cast iron.

The hand method of preparation consists in grinding the specimen on the usual three grades of emery papers, with final grinding on a sheet of well-worn or glazed No. 000 paper. The last grinding operation is continued until the fine scratches are nearly invisible to the unaided eye.

Polishing is usually carried out on two lapping wheels. The first wheel is covered with a pileless cloth such as recommended for cast iron, and the second is generally covered with a more nappy cloth such as velvet or velveteen. The polishing abrasive may be either levigated alumina or magnesium oxide, with some preference for magnesium oxide in the preliminary polishing operation because it has less tendency to remove the inclusions. During polishing, the specimen is rotated counter to the rotation of the lap in order to prevent the formation of comet tails. The cloth is kept damp during polishing—never wet—and it should be almost dry when the polishing operation is completed.

The fine scratches introduced on the surface during preliminary lapping may be removed by carefully polishing the specimen on the second polishing wheel. During final polishing it is essential that the specimen be held in contact with the revolving lap under lightly applied pressure. Excessive contact pressure will cause ultimate dislodgment of the inclusions. A polishing abrasive of finely divided alumina has been found satisfactory for this operation.

The remarkable effectiveness in the special technique described for the retention of inclusions in steel is shown in Fig. 20.

STORAGE OF METALLOGRAPHIC SPECIMENS

The prepared surface of a metallographic specimen, in either the etched or unetched condition, may be temporarily preserved in a chemical desiccator or in a sterilizer cabinet. The atmosphere within the desiccator or cabinet may be dried with such satisfactory solid desiccants as Drierite

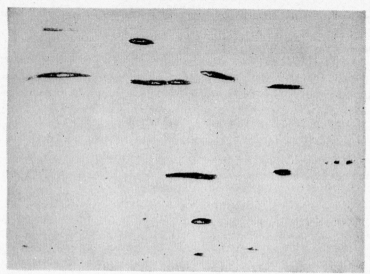

a. Inclusions removed during specimen preparation.   Note enlargement of voids as compared to actual size of inclusions shown below.   150 ×.

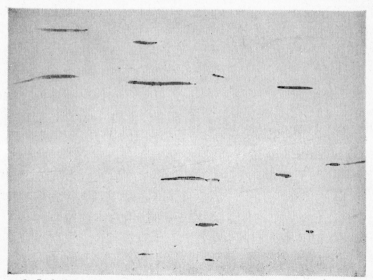

b. Inclusions preserved intact during specimen preparation.   150 ×.

Fig. 20.   Illustrating the effect of improper and proper metallographic preparatory techniques, particularly polishing, on the retention of nonmetallic inclusions.   (*Courtesy of U.S. Steel Corp. Research Laboratory.*)

(dehydrated calcium sulphate), anhydrous calcium or magnesium chloride, activated alumina, or Desicchlora.

Etched surfaces that are to be preserved for comparatively long periods of time may be coated thinly with a suitable lacquer, such as Aquanite A, Stonite 4100, or Murphy's Outside Lacquer. Lacquer coatings are best applied to prepared surfaces by partly immersing the specimen in a thinned solution of the lacquer, allowing the excess liquid to drain for a few seconds, and then permitting the lacquer to dry with the polished and etched surface in an up and horizontal position.

A thin coating of vaseline or neutral white mineral oil (Nujol) over the prepared surface has been found satisfactory in some cases for long-time preservation, particularly when storage is made in a desiccator.

MOUNTING OF SMALL SPECIMENS

Metallographic specimens that are either too small or too awkwardly-shaped to permit convenient handling during grinding and polishing, *e.g.*, chips, wires, small rods or tubing, sheet metal specimens, thin sections, etc., must be appropriately mounted in a suitable material or rigidly clamped in devices to facilitate in the preparational operations. The most common methods of mounting such specimens will be fully discussed.

Fusible Mounts. There are available a great many fusible mounting materials that have been found suitable, with some limitations, for mounting metallographic specimens. Such materials are sulphur, sealing wax, golaz wax, DeKhotinsky cement, dental plasters, and low-melting-point alloys. Since the melting points of these materials differ considerably one from the other, care must be taken in the selection of any particular one so that the temperature reached during melting will not alter the structure of the specimen being mounted. Furthermore, some of these materials, particularly the waxes, are soluble in alcoholic etching reagents, making such mounts undesirable.

Wood's metal and Lipowitz alloy are two common low-melting-point alloys used for mounting metallographic specimens. The composition of each alloy is as follows:

Wood's Alloy (M.p. 65.5°C)	Per Cent	Lipowitz Alloy (M.p. 70°C)	Per Cent
Bismuth	50.0	Bismuth	50.0
Lead	25.0	Lead	27.0
Tin	12.5	Tin	13.0
Cadmium	12.5	Cadmium	10.0

In etching a metallographic specimen previously mounted in a metallic-alloy mount, it often happens that the mounting alloy is at first etched preferentially to the specimen. This is due mainly to an electrolytic potential difference existing between the mount and the specimen, and the "protective" action of the mounting alloy in this respect may at times be so great that even upon prolonged etching it will be found impossible to etch the specimen satisfactorily.

Unless the specimen is properly mounted in any one of the materials described, at times considerable trouble arises during grinding due to clogging of the emery papers by the mounting material. This difficulty may be readily overcome by properly mounting the specimen so that the mounting material does not come into contact with the emery papers during subsequent grinding. A small brass ring approximately $\frac{1}{2}$ in. high and having an inside diameter sufficiently large so that the specimen may be contained therein is thoroughly roughened on the inside surface with a coarse file. The ring is then placed in a shallow pool of mercury about $\frac{1}{16}$ in. deep, and the specimen is held by means of tongs in the center of the ring. Both the ring and the specimen are then held in tight contact with the bottom of the mercury vessel and the molten mounting material is cast into the ring. After the mounting material has solidified, the ring and specimen are ground and polished as a single unit. It is obvious that this method of mounting will not prevent rounding of the edges of the specimen during metallographic preparation.

Synthetic Plastic Mounts. Mounting small metallographic specimens in synthetic plastic materials such as bakelite, Lucite, Tenite, and Transoptic is one of the most satisfactory methods now available to facilitate subsequent handling of such specimens. It affords a means of obtaining mounts of uniform size (usually 1 in. and 1.25 in. in diameter) and the method of mounting is applicable to specimens of almost any shape. The procedure of mounting is relatively simple but, because both heat and pressure must be applied simultaneously to secure a satisfactory mount, a special mounting press and auxiliary equipment are required, as shown in Fig. 21.

In general, metallographic molding plastics are an ideal medium for the purpose intended because, after appropriate molding, they are relatively hard (particularly the thermosetting plastics) and consequently do not readily clog emery papers with debris during subsequent grinding, and they are resistant to chemical attack by the usual metallographic etching reagents. Because they possess good electrical insulating properties, there occurs no differential etching between the specimen and the mounting material as so frequently happens when etching a specimen contained within a fusible alloy mount.

The necessity of applying both heat and pressure to synthetic plastic materials during molding makes the method inappropriate for mounting certain metals and alloys. Soft metals such as lead, tin, etc., are readily deformed during mounting owing to the relatively high pressures required, and hardened alloys may suffer structural changes due to the requisite temperature. Because of differences in hardness between the mount and the specimen, plastic mountings are not capable of preventing entirely the edges of a specimen from becoming rounded off during grinding and polishing. This circumstance is illustrated in Fig. 24a. If for any reason the edges of the specimen are the object of the examination, methods of mount-

Fig. 21. Mounting press, heating-unit assembly, and mold-cooling unit, for mounting metallographic specimens in thermosetting and thermoplastic resins. (*Courtesy of Buehler, Ltd.*)

ing other than in plastic materials will prove more satisfactory (see page 56).

Thermosetting Resins. The thermosetting plastic materials such as bakelite and aniline formaldehyde compounds are the most popular types of plastic used for mounting metallographic specimens. Bakelite molding powders are available in a variety of colors, which simplifies the identification and cataloging of mounted specimens. The mounts when properly made are very resistant to attack by the etching reagents ordinarily used, although they are decomposed by strong alkalies and concentrated oxidizing acids.

The thermosetting plastics, unlike the thermoplastic resins to be described, become more stable after molding owing to a chemical change, or curing, that occurs at the recommended molding temperature and pressure. This state of greater stability is not affected appreciably by temperature, even at temperatures approaching that which will cause the molded resin to char. For most grades of bakelite molding powders, a maximum temperature of 135 to 150°C (275 to 302°F) is required to assure proper curing with an associated molding pressure of 2500 to 3500 lb. per sq in. Since bakelite is set and cured within this temperature range, the specimen mount may be ejected from the molding die while it is still hot.

Schleicher and Everhart[1] have suggested a method whereby specimens may be mounted in bakelite without the use of the usual mounting press. Fresh bakelite resinoid (manufacturers' number BR-0014) is used, which at ordinary temperature is a thick and viscous liquid. The specimen is placed on the bottom of a previously greased porcelain crucible and enough bakelite resin is poured over the specimen to make a mount approximately $\frac{1}{2}$ in. high. The crucible and its contents are heated for about 12 hr at 85 to 90°C (185 to 194°F) and then the temperature is raised to 120 to 125°C (248 to 257°F) for an additional 2 hr. After cooling, the solid contents may be removed by gently tapping on the crucible.

An alternate method of mounting, requiring less time, consists of first coating the specimen to be mounted with bakelite varnish (S.D. 17) and allowing it to dry thoroughly. Drying may be hastened by placing the coated specimen in a drying oven at a temperature of 45°C (113°F). The specimen is then placed in a crucible and covered with a well-mixed solution of bakelite resin (No. XR-3220) containing 10 to 20 per cent by volume of bakelite hardener (No. XK-2997). The crucible and its contents are then heated for $\frac{1}{2}$ hr at 45°C (113°F), after which the temperature is raised to 65 to 75°C (150 to 167°F) for an additional $1\frac{1}{2}$ to 2 hr. This treatment fully hardens the mount.

Thermoplastic Resins. Mounting resins of the thermoplastic type such as polystyrene, methyl methacrylate compounds (Lucite), and cellulose-base materials are characterized by being completely transparent and crystal clear when properly molded, as illustrated in Fig. 22. This feature of transparency is a fortunate circumstance when it is necessary to observe critically the exact extent of sectioning of the mounted specimen during grinding and polishing or when it is desirable for any other reason to see the entire specimen within the mount. Thermoplastic materials are not affected by the usual etching reagents used nor by most concentrated alkalies or acids. They are, however, readily soluble in organic solvents such as acetone.

[1] Schleicher, H. M., and J. L. Everhart: Mounting of Small Metallographic Specimens and Metal Powders in Bakelite, *Metals & Alloys,* Vol. 5, 1934.

The thermoplastic resins do not undergo curing at the molding temperature as do the thermosetting plastics; rather they become soft and readily flow each time the appropriate combination of temperature and pressure is applied to the molded resin. Because of this inherent characteristic, thermoplastic mounts often become temporarily soft and gummy with attending difficulties when frictionally heated during initial flattening of the specimen and mount on a grinding belt, or even during too vigorous hand grinding on metallographic emery papers.

FIG. 22.   Metallographic specimens mounted in Lucite.   Note transparency of the mounts.
(Courtesy of Buehler, Ltd.)

Satisfactory mounts may be secured with thermoplastic resins by molding at a pressure of 2500 to 3500 lb per sq in. and a temperature between 140 and 165°C (284 and 329°F). To obtain a completely transparent mount the molded plastic must be cooled in about 6 or 7 min under maximum pressure from the molding temperature to a temperature of at least 75°C (167°F). This rate of cooling may be readily secured by chilling the molding die with wet cloths or by enclosing the die in a copper chilling block designed for this purpose and illustrated in Fig. 21. Ejecting the mount while still hot or allowing it to cool slowly in the molding die to ordinary temperature before ejection will cause the mount to be opaque.

Kehl and Church[1] have developed a transparent liquid resin for use in mounting metallographic specimens that may be cast at ordinary temperature and hardened at a temperature not exceeding 50°C (122°F). The plastic material is essentially a partially polymerized methyl methacrylate monomer that is allowed to polymerize to completion after it is appropriate-

[1] Kehl, G. L., and J. S. Church: Room Temperature Casting Resin for Metallographic Mounts, *Metal Progress*, Vol. 51, 1946.

ly cast about the specimen of interest. When completely hardened, the mount has characteristics similar to those of Lucite mounts secured by conventional heat-pressure molding techniques.

A methyl methacrylate monomer from which the casting resin may be prepared is obtainable from Rohm and Hass Company under the proprietary name of "Cement X." As supplied by the manufacturer, this monomer contains hydroquinone, which serves to inhibit premature polymerization during storage.

The first step in the preparation of the casting resin consists of removing completely the hydroquinone present. This is accomplished in a separatory funnel by treating at ordinary temperature approximately 1 liter of Cement X with a 5 per cent aqueous solution of sodium hydroxide. The reaction product formed is a red disodium salt of hydroquinone that is soluble in water but insoluble in methyl methacrylate. To insure complete elimination of the inhibitor, extractions are continued until the sodium hydroxide solution remains clear while in contact with the monomer. Approximately 850 ml of solution will be required for complete extraction.

The inhibitor-free monomer is then dried for approximately 15 hr over anhydrous sodium sulphate. The salt is removed by filtering and the monomer is again dried for another 15 hr with anhydrous sodium sulphate and again removed subsequently by filtration. Approximately 400 g of salt are required for each drying operation. It is essential after drying that the monomer be kept from intimate contact with moisture, since the presence of water will precipitate polymer in a granular form.

To serve as a catalyst in subsequent polymerization, 0.1 per cent by weight of benzoyl peroxide is dissolved in the water-free monomer. The solution is allowed to stand for about 15 hr and is then filtered to remove a slight precipitate.

The solution is then placed in a round-bottom flask fitted with a reflux condenser, a mercury-sealed stirrer, and a thermometer adjusted to read liquid temperature. Ten per cent by weight of Lucite molding powder (DuPont methyl methacrylate polymer) is added to the solution to initiate chain growth. The mixture is then heated in a water bath, and the temperature is allowed to rise to 80°C (176°F). The reaction mix is held at this temperature until the polymer is completely dissolved and the appropriate degree of polymerization has been secured, as indicated by the solution attaining a viscosity approximately equivalent to that of concentrated sulphuric acid.

When the appropriate viscosity of the solution has been obtained—after about 2 hr (or less) at the reaction temperature—the polymerization reaction is quickly arrested by immersing the flask into an ice bath. The partially polymerized resin may now be cast around a metallographic specimen and subsequently hardened; or the viscous solution may be preserved for approximately 30 days if it is stored under refrigeration, not exposed to sunlight, and the glass container is tightly stoppered.

It is essential during the partial polymerization process that the temperature does not exceed 80°C (176°F). If this should occur, it is likely that the reaction will suddenly go to completion with a marked evolution of heat and form a hard, foam-like brick.

The technique of mounting a metallographic specimen is similar to that followed in mounting small specimens in other casting materials such as sulphur, waxes, or low-melting-point alloys. To accelerate polymerization of the resin after casting, the mount is placed under an ultraviolet lamp and

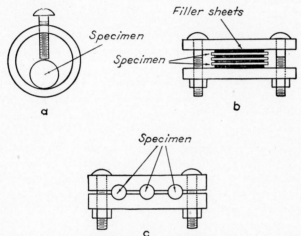

Fig. 23. Conventional types of clamping devices to facilitate the metallographic preparation of small specimens and thin sheets.

the temperature of the mount maintained at approximately 50°C (122°F) by adjustment of the lamp-to-mount distance. Higher temperatures are to be avoided because bubbles will form within the casting; temperatures lower than that recommended will increase considerably the time required for thorough hardening.

Mechanical Mounts. Small specimens may be conveniently mounted for metallographic preparation in laboratory-made clamping devices similar to those shown in Fig. 23. Thin sheet specimens, when mounted in the clamping device shown in Fig. 23b, are usually alternated with metal "filler" sheets which have approximately the same hardness as the specimens. The use of filler sheets will preserve surface irregularities of the specimen and will prevent to some extent the edges of the specimen from becoming rounded during grinding and polishing. In the preparation of ferrous sheet material, alternate filler strips of annealed copper are ideally suited for this purpose.

EDGE PRESERVATION

It is often desirable, and many times necessary, to examine microscopically a specimen out to its very edges. Unless the specimen has been properly mounted for this purpose, the edges may be rounded to such an extent during preparation that effective examination of this part of the specimen will be impossible.

The edges of sheet material or of a specimen having a square or rectangular cross section may be maintained reasonably flat by mounting the specimen in a clamping device, as shown in Fig. 23b. The material from which the clamp is made, as well as that of the filler sheets if any are used, should be approximately the same hardness as that of the specimen. This will prevent to some degree differential polishing and the ultimate rounding off of the edges of the specimen.

The edges of irregularly shaped specimens are best preserved by electroplating, after which they may be mounted in a conventional synthetic plastic to facilitate handling during subsequent grinding and polishing.

Ferrous specimens may be electroplated successfully with copper provided the specimen is thoroughly cleaned of grease, etc., before plating. A flash coating of copper is first applied to the specimen in order to prevent direct chemical attack of the specimen by the strongly acidic electrolyte employed for the second electroplating bath. The heavy edge-protecting coat of copper is electrodeposited on the specimen from a copper sulphate bath.

FLASH-COATING BATH

Cuprous cyanide.................. 22.5 g
Sodium cyanide.................. 34.0 g
Sodium carbonate............... 15.0 g
Water.......................... 1000 ml
Current density................ 0.2 amp per sq dm
Temperature.................... 30 to 40° C (86 to 104°F)
Specimen is cathode.
Copper anode.

COPPER SULPHATE BATH

Copper sulphate................. 250.0 g
Concentrated sulphuric acid........ 75.0 g
Water.......................... 1000 ml
Current density................ 2 to 4 amp per sq dm
Temperature.................... Room
Specimen is cathode.
Copper anode.

a. Specimen mounted in bakelite. Note excessive rounding off of the specimen edge. 1000 ×.

b. Specimen mounted in a steel clamp with alternate copper strips (see Fig. 23*b*). Note excellent preservation of specimen edge, as well as scale between specimen and copper strip. 1000 ×.

c. Specimen edge protected with an iron plate applied by electrodeposition. 1000 ×.

FIG. 24. The influence of mounting methods on the preservation of specimen edges during metallographic preparation.

Iron-plating of specimens for edge preservation is a recent development and has advantages over copper deposition in that only one electroplating bath is required and that satisfactory and tightly adhering deposits of iron can be made directly over mill scale or other oxide coatings. It is necessary however, that all grease, etc., be thoroughly removed before electroplating. This can be accomplished by electrolytic cleaning or more simply, in many cases, by washing the specimen in ether.

IRON-PLATING BATH

Ferrous chloride.................. 288.0 g
Sodium chloride.................. 57.0 g
Water........................... 1000 ml
Current density................. 5 to 20 amp per sq ft
Specimen is cathode.
Anode of low-carbon, low-metalloid iron sheet. It may be
    bound with linen tape to prevent the anodic sludge formed
    from entering the bath.

Experience has shown that operating the process at a temperature of 85°C (185°F) produces an iron coating that is very adherent, tough, and free from gas cavities. Operating at temperatures below 85°C results in a deposit that is porous, brittle, and more or less internally cracked, whereas at temperatures above 85°C the coating is somewhat coarse.

The remarkable effectiveness of an iron plate in preserving the edge of a specimen is shown in Fig. 24c.

References

Amberg, K.: Metallographic Grinding with Abrasive Fixed in Lead Disks, *Metal Progress*, Vol. 37, 1940.

ASM: "Metals Handbook," Cleveland, Ohio, 1948.

ASTM Standards, Part I-B, 1946.

Bassett, W. H., Jr., and C. J. Snyder: Method of Preparation of Lead and Lead Alloy Cable Sheath for Microscopic Examination, *Proc. ASTM*, Vol. 32, Part II, 1932.

Beregekoff, D., and W. D. Forgeng: Special Methods for Polishing Metal Specimens for Metallographic Examination, *Metals Tech.*, Vol. 6, 1939.

Ellinger, G. A., and J. S. Acken: A Method for the Preparation of Metallographic Specimens, *Trans. ASM*, Vol. 27, 1939.

Everhart, J. L.: A Note on the Mounting of Specimens in Bakelite, *Metals & Alloys*, Vol. 8, 1937.

Guillet, L., and A. Portevin: "Metallography and Macrography," George Bell & Sons, Ltd., London, 1922.

Hochschild, U. J.: Electropolishing of Microspecimens, *Metals & Alloys*, Vol. 22, 1945.

Holt, E. D.: Automatic Polishing of Metallographic Samples, *Metal Progress*, Vol. 54, 1948.

Jacquet, P.: Introduction to the Study and the Use of Electrolytic Polishing of Metals and Alloys, *Metaux Corrosion-Usure*, Vol. 18, 1943.

Jacquet, P., and P. Rocquet: *Compt. rend.*, Vol. 201, 1935.

Jeffries, Z., and R. S. Archer: "The Science of Metals," McGraw-Hill Book Company, Inc., New York, 1924.

Jenkinson, E. A.: The Iron-Plating of Specimens for Microscopic Examination, *J. Iron Steel Inst.* (London), Vol. 142, 1940.

Kehl, G. L., and J. S. Church: Room Temperature Casting Resin for Metallographic Mounts, *Metal Progress*, Vol. 51, 1946.

Lucas, F. F.: The Preparation of Iron and Steel for Microscopic Investigations, *Trans. ASM*, Vol. 24, 1936.

Meyer, G. F., G. D. Rahfer, and J. R. Vilella: Electrolytic Polishing of Steel Specimens, *Metals & Alloys*, Vol. 14, 1941.

Pellissier, G. E., H. Markus, and R. F. Mehl: Electrolytic Preparation of Iron and Steel Micro-Specimens, *Metal Progress*, January, 1940.

Rowland, D. H., and O. E. Romig: The Metallography of Galvanized Sheet Steel Using a Specially Prepared Polishing Medium with Controlled pH, *Trans. ASM*, Vol. 31, 1943.

Schleicher, H. M., and J. L. Everhart: Mounting of Small Metallographic Specimens and Metal Powders in Bakelite, *Metals & Alloys*, Vol. 5, 1934.

Tafts, H. J.: Mounting and Polishing of Tin Alloys, *J. Roy. Microscop. Soc.*, Vol. 56, 1936.

Vilella, J. R.: Improved Method of Polishing Metallographic Specimens of Cast Iron, *Metals & Alloys*, Vol. 3, 1932.

Vilella, J. R.: "Metallographic Technique for Steel," *ASM*, Cleveland, Ohio, 1938.

Vilella, J. R., and D. Beregekoff: Polishing and Etching Lead, Tin and Some of Their Alloys for Microscopic Examinations, *Ind. Eng. Chem.*, Vol. 19, 1927.

Waisman, J. L.: Metallographic Electropolishing, *Metal Progress*, Vol. 52, 1947.

Woodside, G. C., and H. H. Blackett: Polishing Metallographic Specimens with Diamond Dust, *Metal Progress*, Vol. 52, 1947.

CHAPTER 2

ETCHING OF SPECIMENS FOR MICROSCOPIC EXAMINATION

The preceding chapter has covered in detail the methods by which a metallographic specimen is prepared for microscopic examination. It now becomes of interest to discuss the principles and associated techniques by which such a prepared specimen is etched in order to make visible the many structural characteristics of the metal. Our experience indicates that microscopic examination of an unetched specimen will reveal few, if any, structural details, although the examination may be of considerable interest with regard to characteristics of the specimen not requiring an etch in order to be seen, such as surface defects, nonmetallic inclusions, etc.

Inasmuch as most metallographic examinations are made for the express purpose of determining the true structural characteristics of the specimen of interest, it is necessary that the various components of the microstructure be delineated with preciseness and extreme clarity. This circumstance is usually achieved by subjecting the polished surface of the specimen to the chemical action of some appropriate reagent under rather carefully controlled conditions.

In the duplex type of alloys, *i.e.*, alloys composed of two or more phases, structural components are revealed during etching by a preferential attack or staining of one or more of these constituents by the reagent, owing principally to existing differences in chemical composition of the phases and attending rates of solution. In the case of pure metals, however, and homogeneous single-phase alloys, structural contrast is established and grain boundaries are revealed mainly because of differences in the rate at which the various grains are attacked by the reagent, a circumstance directly associated with the orientation of the different grain sections with respect to the plane of the polished surface. A more detailed account of the etching mechanisms associated with single- and multiple-phase alloys and pure metals is discussed on page 64.

Formation of Disturbed Metal. During preparation of a metallographic specimen by grinding and polishing techniques, as mentioned heretofore, a layer of cold-worked metal is formed on the polished surface, commonly referred to as disturbed metal. This distorted formation is the natural consequence of grinding and polishing and it is inevitably formed to some degree regardless of the care exercised during metallographic preparation.

By virtue of intimate contact between the specimen surface and the finely divided abrasive particles used in grinding and polishing operations, the induced mechanical stresses and thermal effects arising therefrom are sufficiently great literally to smear the very topmost surface metal. A majority of electron-diffraction studies made on polished metallographic surfaces indicates that attending this temporary mobility of the free surface there is a considerable alteration in the periodicity of the surface atoms and, as a normal manifestation, complete destruction locally of the crystalline state. This disorganized surface metal, extending inwardly a distance equivalent to several interatomic distances, possesses a higher free energy than does the plastically deformed metal directly beneath it, and, because of this, is quickly and effectively removed by chemical dissolution during the initial stages of metallographic etching. As a consequence, the appearance of the metallographic structure of interest, after appropriate metallographic etching, will in no way be influenced by the initial presence of this highly disorganized free surface.

Directly beneath the free surface and its attendant state of disorganization, a layer of cold-worked or disturbed crystalline metal of some finite thickness is formed, which is distorted structurally most severely adjacent to the topmost surface and least near the unaffected base metal. This circumstance is associated with the deformation stresses, arising from grinding and polishing, diminishing in intensity over a relatively steep gradient as already defined. It is this disturbed metallic layer, whose presence so profoundly influences the appearance of the etched metallographic structure, that must be effectively removed from the surface of the prepared specimen by techniques to be described, before there is assurance that the true structure will be revealed unequivocally.

For reasons related heretofore, it is of the utmost importance to secure polished metallographic surfaces as free from distortion as is commensurate with the mechanical operations involved. The amount of disturbed metal formed during grinding and polishing is closely associated with a number of related circumstances, the more important being: (1) the chemical and structural composition of the specimen, (2) the grinding and polishing method used in preparation, (3) the care exercised during preparation, *i.e.*, whether heavy or light contact pressures were used, and (4) the nature of the polishing abrasive. With respect to the latter, a polishing medium such as iron oxide rouge very effectively flows surface metal during polishing with attending formation of considerable amounts of disturbed metal. In fact, the amount of surface flow in some cases may be so great that certain defects of the prepared surface, such as pits and scratches, will be completely covered over with flowed metal, giving to the polished surface a false appearance of complete homogeneity. Diamond dust, however—

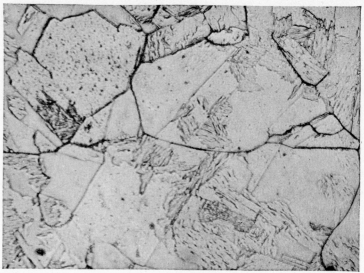

a. A pseudo structure revealed after a single etching treatment owing to the presence of disturbed metal.   500 ×.

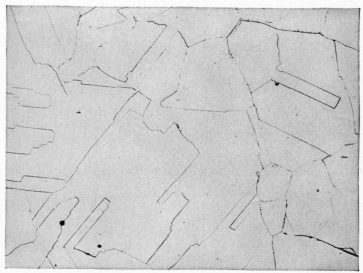

b. True structure revealed after complete removal of disturbed metal by alternately polishing and etching several times.   500 ×.

FIG. 25.   The influence of disturbed metal on the appearance of 18–8 stainless steel. *(Courtesy of U.S. Steel Corp. Research Laboratory.)*

which behaves more like a true cutting abrasive—polishes the surface of the specimen by removing metal rather than by flowing it and thus minimizes the amount of distortion produced.

Removal of Disturbed Metal. In the previous chapter, attention was directed to appropriate techniques of grinding and mechanical polishing whereby a minimum amount of disturbed metal would be produced. Because, as mentioned heretofore, it is practically impossible to secure a metallographic surface entirely free from distortion, it is therefore of the greatest importance that whatever disturbed metal is present be completely removed in order that the true structural characteristics of the specimen be revealed in final etching.

When a metallographic specimen is etched for the first time after polishing, a psuedo structure, as illustrated in Fig. 25a, is revealed that has an appearance entirely unlike that of the true structure. This circumstance arises from the presence of disturbed metal which, by a single etching treatment, is not completely removed. However, by alternately polishing and etching several times, the distorted surface is finally and completely eliminated and, as shown in Fig. 25b, the true structure of the specimen is unequivocally revealed. Usually three cycles of alternate polishing and etching are sufficient to remove the disturbed metal present, although in some cases, particularly those involving specimens of soft metals and alloys, many more cycles may be required to accomplish this purpose. Whether or not the distorted surface has been completely eliminated after any one trial may be readily ascertained by microscopic examination. For most metals and alloys, the alternate etching operations should not be in excess of the time required for final etching; and the polishing operations should be carried out with care, using light contact pressures in order to prevent the formation of additional disturbed metal on the specimen surface.

Etching Reagents. In general, reagents suitable for etching metallographic specimens are composed of organic or inorganic acids, alkalies of various kinds, and other complex substances, in solution with some appropriate solvent such as water, alcohol, glycerine, glycol, or mixtures of these solvents. The activity and general behavior of the usual metallographic etchants are related to one of the following characteristics: (1) the hydrogen ion concentration, (2) the hydroxyl ion concentration, or (3) the ability of the reagent to stain preferentially one or more of the structural components.

For a metal or an alloy to be satisfactorily etched and the desired details of the structure clearly revealed, requires that a reagent be used that is formulated with full cognizance of the composition of the specimen of interest and the structural phases present in it. For example, an etchant composed of ammonium hydroxide and hydrogen peroxide is ideally suited for etching copper and alpha brass, but it is completely unsuitable for etching iron and steel and other ferrous alloys. This same reagent, although performing well in etching single-phase brass alloys, is not too good compared to others for etching multiple-phase brass alloys such as the alpha-beta brasses.

From the many etching reagents suggested for etching a particular group of metals and alloys, a selection of any one must be made with careful consideration of the specific application for which the reagent is intended. As an example, both nital (a dilute solution of nitric acid and alcohol) and boiling sodium picrate are recommended for etching steel. The sodium picrate reagent is by no means a general-purpose etchant; rather it is one used specifically to distinguish between ferrite and iron carbide by preferential darkening of the carbide phase. Nital, however, is not at all suited for rendering a distinction between ferrite and iron carbide, but it is used principally as a general etching reagent for steel and for delineating grain boundaries in ferrite.

Metallographic etching reagents for iron and steel, alloyed steels, and the common nonferrous metals and alloys are given in Tables 2 to 13, pages 412 to 433.

Mechanism of Etching. When an etching reagent is applied to the polished surface of a metallographic specimen, structural details are revealed in part by a process of selective unbuilding of the structure from the surface downward. This mode of revealing the metallographic structure is possible only because the various constituents in a multiple-phase alloy, or the section planes of differently oriented grains in a pure metal or in a single-phase alloy, have inherently different rates of solution in the usual etching reagents. The operative characteristics of some reagents are such that, in addition to revealing structural details by preferential dissolution, certain phases of the structure are selectively discolored or stained. Reagents of picric acid and boiling sodium picrate behave in this manner, and discolor to some related degree the iron carbide phase existent in steel. Other reagents, such as Stead's (reagent No. 24, Table 2, page 412), react principally by selective deposition of reaction products on preferred structural constituents or on regions of limited chemical composition. When used for the purpose intended, Stead's reagent will dissociate and copper will be deposited preferentially on areas in steel relatively low in phosphorous content.

Duplex Alloys. The mechanism of etching multiple-phase alloys is essentially one of an electrochemical nature, owing to a difference in potential between the structural components when the specimen is brought into contact with the etching reagent. The phase at the higher potential is anodic, or electropositive, to the other and, as a consequence, tends to go into solution readily during normal etching. The cathodic, or electronegative, phase, being at a lower potential, is not significantly affected during normal etching or for a time of etching short of that which may bring about a change in the relative potential order of the phases. The difference in potential between the anodic and cathodic components is usually sufficiently great so that the electropositive phase is dissolutive at a fairly rapid rate in the usual etching reagents, and hence careful control is required during etching so as to prevent the specimen from becoming overetched. It is primarily because of this potential difference, which is absent in pure metals, that duplex alloys etch at a greater rate than either pure metals or single-phase alloys.

Owing to preferential dissolution of the anodic phase during etching, this component in duplex alloys is somewhat roughened and depressed from the plane of the polished surface, at least at the anode-cathode interface, and when observed microscopically it may appear darkened, due principally to shadow effects. The cathodic component, however, being unaffected during etching, will be left standing in relief with respect to the other phase and, particularly when existing structurally in somewhat massive form, it will appear bright and lustrous when examined microscopically.

By continuing the etching process beyond the time required to delineate clearly the various structural phases present, the component initially electropositive may become ennobled and thus possess with respect to etching all of the characteristics of an electronegative phase. This change in potential, when it occurs, may be attributed to selective deposition of reaction products, or the formation of other kinds of anodic protective layers. In any event, the original cathodic component may be rendered dissoluble, and upon further etching, the structure will become overetched and poorly defined. By arresting the etching process, however, before the anodic phase becomes ennobled, electrolytic action in the opposite direction may be avoided with attending better contrast established between the structural phases.

The process of etching duplex-type alloys as described is well illustrated in the etching of pearlite, shown in Fig. 26a. The lamellae of ferrite and iron carbide, of which pearlite is structurally composed, have inherently different mechanical and physical properties, and rather widely separated potentials in the usual metallographic reagents. The ferrite lamellae are relatively soft and ductile, and are electropositive to those of iron carbide. During normal etching, the ferrite platelets are partially dissolved and roughened, particularly along their edges, leaving sharply sloping valleys or troughs along the ferrite-carbide interfaces. This circumstance, together with the inherent thinness of the carbide lamellae compared to those of ferrite, gives rise to reflected light losses and shadow effects within the ferrite

a. Multiple-phase type—pearlite. 1000 ×.

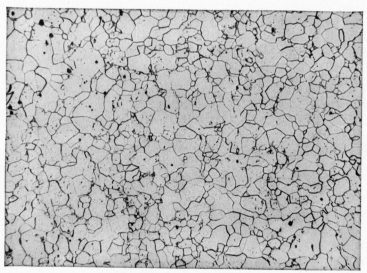

b. Single-phase type—ferrite. 100 ×.

FIG. 26.   Illustrating two types of metallographic structures that are etched by different mechanisms.

depressions, causing the carbide platelets to appear dark. However, when the carbide lamellae are relatively coarse as illustrated in the upper right hand portion of Fig. 26a, or when the iron carbide exists in more massive form as it might in hypereutectoid steels, the carbide phase will appear bright and lustrous, and it will be practically indistinguishable from massive ferrite unless etched specifically for purposes of distinction.

Pure Metals and Single-phase Alloys. The mechanism of etching pure metals and homogeneous single-phase alloys differs considerably from that associated with multiple-phase alloys. Inasmuch as only one structural phase exists in these metals, an operative etching mechanism governed by an electropotential phenomenon is hardly tenable. The difference in potential between the base metal and insoluble inclusions, and between grain boundaries and associated grain sections, is generally of such a small magnitude that it has very little, if any, influence on the etching process.

The etching of a pure metal or a single-phase alloy is a process of chemical solution of the metal by the reagent wherein each grain is dissolutioned at a rate dependent upon the orientation of the grain section with respect to the plane of the polished surface. Because the rate of solution of any one grain differs along different crystallographic planes, there is developed during etching a series of well-defined facets that are similarly oriented on any one grain section, but which as a group are differently oriented than those on neighboring grain sections. Although these etch-facets are usually not discernible after normal etching and at the customary magnifications employed for microscopic examination, their presence under some circumstances gives rise to what is known as oriented grain luster, described more fully on page 68. It is only after prolonged etching with certain highly active reagents that these facets will be sufficiently well developed and enlarged to be revealed microscopically as individual entities.

The grain boundary regions in single-phase polycrystalline metals are made evident in the microstructure because of circumstances associated with the mechanism by which such metals are etched. As shown in Fig. 26b, the regions of grain contact or abutment are manifested after etching by irregularly shaped, thin dark lines; and appear dark because of shadow effects and light losses in these areas. As a natural consequence of the different rates at which differently oriented grain sections are attacked during etching, it is possible that short sloping surfaces are formed between grains and thus provide appropriately oriented surfaces from which light will be reflected and lost outside of the microscope objective. It may be more tenable, however, that during normal etching there is developed at the grain boundary regions rather sharply defined valleys, due to these regions possessing higher free energy and consequential attending greater rates of solution than the grain sections proper.   Owing to scattered light and multiple reflections within these valleys, the grain boundaries as described before will appear as thin, dark lines.

Oriented Grain Luster.   After etching commercially pure metal or homogeneous single-phase alloys, it is often observed microscopically that some of the etched grain sections appear bright and lustrous whereas others appear rather dark and dull.   This phenomenon is known as oriented-grain luster, and is illustrated in Fig. 27.

By consideration of the nature of the metals and alloys which exhibit oriented-grain luster, it is obvious that the phenomenon cannot be explained satisfactorily on the basis of differences in composition of the various grains and attending differences in staining characteristics of the grain

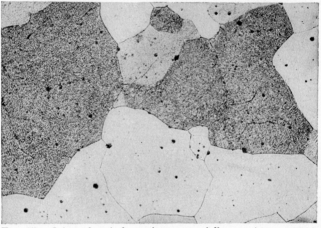

FIG. 27.   Oriented grain luster in commercially pure iron.   150 ×.

sections when etched.   It may be associated, however, with the presence of facets developed during normal etching which, as mentioned heretofore, are similarly oriented on any one grain section, but as a group are differently oriented from those on neighboring grains.

If a beam of illumination is made to impinge upon such an etched surface the direction of the reflected light from each grain section will depend upon the orientation of the etch facets.   When observed microscopically (or in some cases with the unaided eye if the grain size of the metal is relatively large), those grains whose facets are appropriately oriented to the incident light will appear bright, and those not properly oriented will appear dark. The precise degree of shading in luster between bright and dark will be dependent upon the amount of reflected light scattered from the various grain sections.

The origin of oriented-grain luster is illustrated in Fig. 28, which shows the vertical cross section of three etched grains and a schematic representation of the metallurgical microscope and plane-glass reflector. The facets developed on the etched sections of grains A and C are so oriented with respect to the incident illumination that all of the reflected light passes through the microscope to the observer's eye. These grains are thus rendered bright and lustrous in appearance. However, the facets produced on

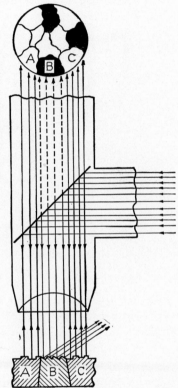

FIG. 28. Illustrating schematically the phenomenon of oriented grain luster in pure metals and homogeneous single-phase alloys.

the section of grain B, as illustrated in Fig. 28, are so oriented that the reflected light is lost outside of the microscope objective, and because none of this light reaches the observer's eye, grain B appears dark. By carefully tilting the specimen, if examined under vertical illumination, as in Fig. 28, or rotating it if observed under oblique illumination, so that the direction of the reflected light is changed, the grains which at first appeared light may be made to appear dark, and vice versa.

Etching Pits. If a pure metal or a homogeneous solid-solution-type alloy is etched with either very reactive reagents or with others for time in excess of that required for general microscopic examination, it is possible to produce etching pits in the various grain sections. The formation of etching pits is related to differences in etching rate along certain crystallographic planes, and is merely a continuation in development of the etch facets referred to above. The shape of the etching pits will be associated with the crystallographic system of the metal of interest; as for example, metals or alloys of the cubic system exhibit etching pits, as illustrated in Fig. 29, that correspond to geometric cavities derived from a cube.

Fig. 29.   Etching pits developed in pure copper by prolonged etching with ammonium persulphate reagent. 1000 ×. (*Courtesy of Arthur Williamson.*)

Etching Procedure. Although the manipulations involved in etching a metallographic specimen are relatively simple to carry out, a certain amount of skill is required on the part of the technician to secure a satisfactorily etched surface. Perhaps the most important preliminary consideration in the procedure is the selection of an appropriate etching reagent from the many that are recommended for any given metal or alloy. This selection, to be wisely made, requires judgment and knowledge of the behavior of the various reagents when used under recommended conditions. A selected etchant must be used for the specific purpose for which it is intended, and to secure the desired results, the directions pertaining to its use must be strictly adhered to. These directions usually specify the method by which the specimen is to be etched (immersion, swabbing, etc.), the temperature at which etching is to be carried out, and the approximate time required to produce a satisfactorily etched specimen. The time of etching, as will be shown later, depends in part upon the magnification at which the specimen is to be subsequently examined, and hence the time may vary considerably from that recommended.

Etching reagents are usually applied to the prepared surface of a specimen by either immersion or by swabbing; the appropriate method of application depending specifically upon the reagent to be used. By whatever method the specimen is etched, it is necessary that the specimen surface be clean and free from tarnish to ensure even and uniform wetting of the surface by the etching reagent. It is excellent technique to wash the specimen surface thoroughly in a stream of warm water before etching, carefully swabbing the surface with wet cotton, rinsing in ethyl alcohol or Petrohol, and finally drying in a blast of warm air.

Etching by methods of immersion, as the term implies, consists of suspending the specimen (polished surface downward) by means of tongs or the finger tips into a small vessel or Petri dish partly filled with the reagent. To eliminate clinging air bubbles and to ensure that fresh reagent is in contact with the surface at all times, the specimen is agitated moderately during etching, care being taken during the procedure so that the specimen surface is not scratched through contact with the bottom of the receptacle. Immediately upon immersion it will be noted that the polished surface, originally appearing bright and lustrous, becomes somewhat dull. This is an indication that etching is progressing and, after some experience, the technician can ascertain by observing the degree of dullness of the surface when the etching process should be stopped. Although the recommended time of etching prescribed for each reagent is only approximate, such data nevertheless serve as a guide to when the specimen should be removed from the etching solution.

When etching has progressed for a time believed appropriate to produce a satisfactorily etched surface, the specimen is removed from the etchant and quickly plunged into a stream of warm running water. This procedure immediately stops the action of the etching reagent and thoroughly removes from the specimen surface all traces of reagent. The specimen is then rinsed in ethyl or isopropyl alcohol to remove water droplets, and subsequently dried in a stream of warm air.

The etched specimen is now ready for microscopic examination. If the structure has not been completely and satisfactorily revealed at the magnification of interest, the specimen may be etched for a longer time. If upon further etching no improvement is noted in the appearance of the structure, a different etching reagent may produce more satisfactory results. It is important, however, that before a new reagent is used, the surface of the specimen is repolished to remove the effects of the first etch.

The general procedure of etching a metallographic specimen by swabbing techniques is similar in part to that already described. The polished surface to be etched, after being cleansed as described before, is vigorously swabbed with a tuft of soft cotton thoroughly saturated with the selected etching reagent.   If the etching time is comparatively long, as indicated by the time recommendation for the reagent used, the solution may be replenished on the cotton by frequently dipping the tuft into a dish of fresh reagent.   To prevent soiling of the fingertips when using reagents that stain, the cotton tuft may be held by means of tongs or by wrapping the cotton around a wooden applicator stick.

Time of Etching.   The time of etching a specimen is next in importance to the proper selection of the etching reagent used.   The visual appearance of an etched structure or a photomicrograph of that structure considered to be of high quality is always characterized in part by a certain degree of crispness and brilliancy.   These characteristics depend to a large extent upon how precisely delicate contrasts in the structure have been developed during etching, and is a circumstance that is directly associated with the time of etching.

Depending upon the metal to be etched and the etching reagent selected, the etching time may vary from a few seconds to 30 min or longer.   Many reagents have been purposely devised with low activity so that the time of etching is relatively long, thus permitting the degree of etching to be carefully controlled.   Other reagents are much quicker in their action and these, which are the more common, require skill on the part of the technician during use to produce a satisfactory etch.

As mentioned heretofore, the time of etching is dependent in part upon the magnification at which the structure is to be examined, and upon the circumstance of whether the structure is to be photographed or only visually examined.   As illustrated in Fig. 30, etching for a time appropriate to reveal fully the structure at low magnification is much in excess of that required for proper delineation of structural details at high magnification. Conversely, etching for a time so as to reveal maximum details at high magnification is much too short a time to delineate satisfactorily the structure at low magnification.   In cases where the structure is to be photographed, and particularly so if at high magnification, the time of etching should be slightly less than that required to produce maximum and optimum contrast for visual microscopic examination.   It is usually better technique to secure contrast between the structural components in the photomicrograph by manipulations in printing and developing than to risk obliteration of delicate structural contrasts by too long an etching time.

With regard to precise etching time, cognizance must be taken of the magnification at which the structure is to be examined.   A specimen that is underetched for examination at 150×, as illustrated in Fig. 31a, lacks considerably in structural details, and particularly in this case, lacks in complete development of the grain boundaries.   Overetching the specimen

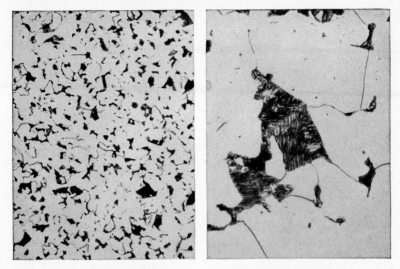

a. Magnification, 150 ×.   b. Magnification, 1000 ×.
Specimen appropriately etched for examination at 150 ×.   Time of etching, 10 sec.

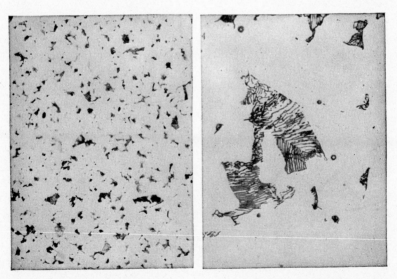

c. Magnification, 150 ×.   d. Magnification, 1000 ×.
Specimen appropriately etched for examination at 1000 ×.   Time of etching, 3 sec.

FIG. 30.   The influence of time of etching on delineation of the microstructure at relatively high and low magnifications.

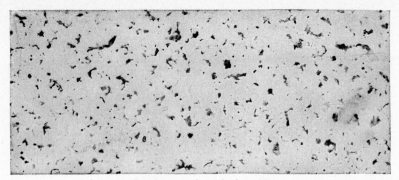

a. Specimen underetched.   150 ×.

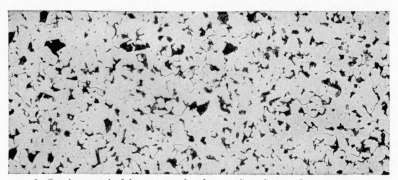

b. Specimen etched for a somewhat longer time than *a* above.   150 ×.

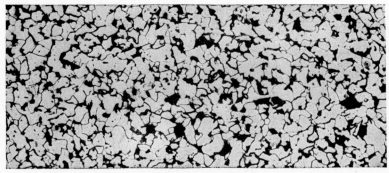

c. Specimen overetched.   150 ×.

Fig. 31.   The influence of time of etching on the appearance of the metallographic structure at the same magnification.

for examination at the same magnification, however, is just as objectionable as underetching. As shown in Fig. 31c, many structural details are obliterated by overetching, the grain boundary regions are widened considerably, and in general the structure as revealed is not intrinsically characteristic of the specimen.

When the behavior of a selected etching reagent is not known with certainty, it is generally more desirable to underetch, as in Fig. 31a, than to overetch the specimen, as illustrated in Fig. 31c. If the specimen after a first attempt is found to be insufficiently etched for satisfactory examination at the desired magnification, the etching process can usually be repeated without further preparation of the surface. Some reagents, however, are prone toward staining if attempts are made to repeat the etching treatment. It is only necessary in such cases to remove the effects of the first etch by repolishing, and then subsequently etching the specimen for a longer time.

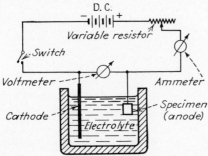

FIG. 32. Electrical circuit and arrangement of apparatus suitable for most electrolytic etching operations.

A specimen that is overetched can be corrected only by repolishing on the final polishing lap and then re-etching for a shorter time. In some cases, if the specimen is severely overetched, it may be necessary to regrind the specimen on either No. 00 or No. 000 emery paper and subsequently polish the specimen on one or more polishing laps. To repolish such a specimen without first regrinding, would tend to round off the structural constituents in high etch relief and in general, would produce unsatisfactory development of the structure in the re-etching treatment.

Electrolytic Etching. Electrolytic etching is a particularly useful etching method for revealing the microstructure in such materials as thermocouple alloys, severely cold-worked metals, heat- and corrosion-resistant alloys, and alloys that exhibit surface passivity during conventional etching. The electrolytic method of etching consists of passing a direct current varying between a fraction of an ampere and several amperes, through a suitable electrolyte wherein the specimen to be etched is usually made the anode and some insoluble material, such as platinum or graphite, is made the cathode. Although direct-current electrolysis is the more common procedure, alternating current is sometimes used for electrolytic etching of certain metals, such as platinum and platinum alloys. An electrolytic-cell arrangement, suitable for electrolytic etching, is shown in Fig. 32. A commercial electrolytic polishing and etching unit, designed specially for treating metallographic specimens, is illustrated in Fig. 15, page 31.

The selection of an appropriate electrolyte depends primarily upon the composition of the metal or alloy to be etched, and upon what constituents in the structure are to be revealed by etching. Electrolytes suitable for etching some of the more common metals and alloys, together with approxi·

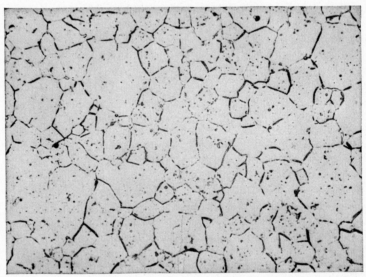

FIG. 33.   The metallographic structure of Alumel revealed by electrolytic etching.   500 ×.
Technical data: electrolyte, 5 per cent aqueous solution of hydrochloric acid; current density, 4.5 amp per sq in.; time of etching, 10 sec.

mate current densities to be used and other pertinent data, are given in Table 13, page 433.

In cases where mounted specimens are to be etched electrolytically, it is essential that the mounting medium be a nonconductor of electricity (the thermosetting and thermoplastic materials are ideal for this purpose), and that provisions be made to secure an electrical connection between the positive-cell terminal and the specimen. The connection may be accomplished by a touch-wire arrangement or by a flexible, insulated lead wire attached to the specimen and passed out from the back side of the mount.

Figure 33 illustrates the microstructure of a thermocouple-element alloy as revealed by electrolytic etching.

Methods Alternate to Etching. Although the etching methods as described are the most important and widely used procedures for revealing the metallographic structure of a metal or an alloy, it is sometimes possible and often desirable to reveal the structure, in part at least, by methods known as relief polishing, etch-relief polishing, heat etching, and heat tinting.

Relief Polishing. The methods of relief and etch-relief polishing may be employed successfully in cases where it is known that an alloy is composed of relatively hard and soft structural constituents. By appropriate techniques[1] of hand polishing, and using at times an abrasive suspended in some dilute acid solution, it is possible to abrade the softer constituent at a greater rate than the harder one, thus leaving the harder phase in bold relief at the end of the polishing operation. When such a polished specimen is examined microscopically, and particularly when slightly oblique illumination is used, a distinction may readily be made between the different structural phases.

As far as routine metallographic preparation is concerned, these methods of partially revealing the microstructure are now obsolete, and even in special cases the techniques are rarely employed.

Heat Etching. Heat-etching methods are rather specialized procedures for revealing certain structural characteristics of an alloy, and they have been used with success in revealing the prior austenitic grain size in low-carbon steels, as described on page 277. The technique of heat etching consists essentially of preferential vaporization of different structural phases or grain boundary regions under rather precisely controlled conditions of temperature and inertness of the atmosphere surrounding the specimen of interest.

Heat Tinting. When a properly prepared metallographic specimen is uniformly heated in air to some comparatively low temperature, the various structural components are oxidized to a degree commensurate with the precise composition of the phases involved. Because the various structural components have inherently different tendencies toward oxidization at the same temperature, the amount of oxide or thickness of the oxide layer formed on each separate phase will be different and of course will be of different composition. These attending circumstances influence the absorption limit of the oxide layers with respect to visible radiation, thereby affording a means of distinguishing between the structural components solely on the basis of differences in color or tone of the oxide films. An example of heat-tinted manganese bronze is shown in Fig. 34.

Heat-tinting methods are particularly useful for partially revealing the microstructure in nonferrous alloys, particularly those in the as-cast condition, where considerable porosity may be present. There is no danger associated with this method, as is so often encountered in conventional etching, of the specimen becoming stained by seepage of the etching reagent

---

[1] For a more detailed account, see Heyn, E., and M. S. Grossman: "Physical Metallography," John Wiley & Sons, Inc., New York, 1925.

from such porous areas.   The techniques of heat tinting are also used successfully for revealing certain significant features in the structure of plain-
carbon and alloy steels (see Table 2, page 412, Nos. 5 and 23), and for
identifying various constituents in cast iron, as designated in Table 14,
page 436.   The method is not applicable, however, to those metals and
alloys that may suffer structural changes at the temperatures required to
produce satisfactory oxidation.

The manipulations involved in heat tinting are relatively simple and, as
in the case of conventional etching methods, some skill is required on the
part of the technician.   The specimen to be heat tinted is first prepared for

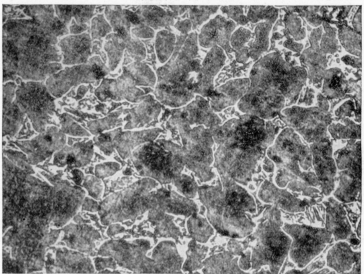

Fig. 34.   The metallographic structure of manganese bronze revealed by heat tinting
methods.   150 ×.

metallographic examination by techniques already described and, after
final polishing, the specimen is thoroughly cleansed with carbon tetrachloride or petroleum ether.   This cleansing step is necessary to remove all
foreign material from the surface so that during subsequent heat tinting
oxidation will proceed uniformly.   Depending upon the metal or alloy to be
oxidized, the specimen is either heat-tinted directly after cleansing, or
etched prior to heat tinting in order to remove disturbed metal formed during polishing, whose presence might otherwise affect the outline of the oxidized patterns.   In the case of cast iron, the specimen is immersed for a
few seconds in a 50 per cent aqueous solution of ammonium hydroxide to
prevent the etching reagent subsequently used from being absorbed by the
graphite and exuded when the specimen is heated; and finally etched in
either a 2 per cent phosphoric acid solution or 2 per cent nital.

The etched or unetched specimen, as the case might be, is gently heated on either an electric hot plate or in a sand bath, or sometimes by floating it in a bath of molten tin. The maximum temperature to which the specimen is heated depends upon the composition of the alloy and is one best determined by trial. In most cases, however, the temperature rarely exceeds 400 to 500°C (752 to 932°F). When by visual observation it appears that satisfactory contrast might be developed between the structural components as evidenced by a change in color of the surface undergoing oxidation, the specimen is removed from the heating medium and quenched in mercury. To prevent objectionable staining of the heat-tinted structure, care must be exercised during quenching so that the oxidized surface does not come into contact with the mercury bath.

If it is determined upon microscopic examination that the structure is not satisfactorily revealed, the heat-tinting procedure may be repeated until the desired structural contrast is secured. This may require either a longer time at temperature or a change in the temperature to somewhat higher or lower limits. A specimen that has been oxidized in excess of that required for appropriate delineation of the phases of interest must be repolished and subsequently heat-tinted with more attention paid to the maximum temperature reached.

References

ASM: "Metals Handbook," Cleveland, Ohio, 1948.

ASTM Standards, Part I-B, 1946.

Corson, M. G.: Etching Technique, *Iron Age*, August, 1941.

Evans, V. R.: "Metals and Metallic Compounds," Vol. 1, Longmans, Green & Co., Inc., New York, 1923.

Greaves, R. N., and H. Wrighton: "Practical Microscopical Metallography," D. Van Nostrand Company, Inc., New York, 1925.

Heyn, E., and M. A. Grossmann: "Physical Metallography," John Wiley & Sons, Inc., New York, 1925.

Houghton Research Staff: "Pickling Steel with Modern Inhibitors," 1937.

Materials & Methods: Data sheet No. 113, April, 1946.

Mechel, R.: An Electrolytic Etching Method for Wrought Magnesium Alloys, *Z. Metallkunde*, Vol. 33, 1941.

Raub, E., and G. Buss: Etching of the Platinum Metals and Their Alloys with A. C. Electrolysis, *Z. Elektrochem*, Vol. 46, 1940.

Sauveur, A.: "The Metallography and Heat-treatment of Iron and Steel," University Press, Cambridge, Mass., 1935.

Shepherd, H. H.: Heat Tinted Microsection of Non-ferrous Metals, *Metal Progress*, Vol. 38, 1940.

Vilella, J. R.: "Metallographic Technique for Steel," ASM, Cleveland, Ohio, 1938.

Watkins, S. P.: Practical Metallography of the Stainless Steels, *Metals & Alloys*, Vol. 14, 1941.

Williams, G. C., and G. Rieger: The Electrolytic Etching of Brass, *Trans. Electrochem. Soc.*, Vol. 77, 1940.

CHAPTER 3

METALLURGICAL MICROSCOPES AND PHOTOMICROGRAPHY

PRINCIPLES AND OPTICAL PARTS

Having already described in some detail the methods by which a metallographic specimen is best prepared and subsequently etched for microscopic examination, it is now appropriate to discuss the principles of the metallurgical microscope and the associated techniques of photomicrography. The metallurgical microscope is composed of two distinct and separate optical systems—the objective and the eyepiece—whose primary purpose, when appropriately used together, is to reveal details in an object that are too small in size to be normally seen with the unaided eye. The extent to which such details are revealed and the clarity with which the microscopic image can be observed will depend largely upon the degree to which the objective and eyepiece are corrected for aberrational and other optical errors. It is important that the metallographer possess a rational understanding of simple optical systems and the principles involved with the skillful handling of a microscopic unit. Only under such conditions will it be possible for the instrument to perform to the degree of perfection of which it is capable.

Principle of the Microscope. A metallurgical microscope, in comparison to a biological type, differs in the manner by which the specimen of interest is illuminated, owing to the inability of visible radiation to propagate through a metallographic specimen, however thin it may be. As illustrated in Fig. 35,[1] a horizontal beam of light from some appropriate source is diverted, by means of a plane-glass reflector, downward and through the microscope objective onto the surface of the specimen. A certain amount of this incident light will be reflected from the specimen surface, and that reflected light which again passes through the objective will form, in a manner to be described, an enlarged image of the illuminated area.

A microscope objective is composed of a number of separate lens elements which, as a compounded group, behave as a positive and converging type of lens system. If an illuminated object, such as the surface of a metallographic specimen, is placed just outside of the equivalent front focus point

[1] For a detailed account of the methods by which metallographic specimens are illuminated for microscopic examination, see p. 120.

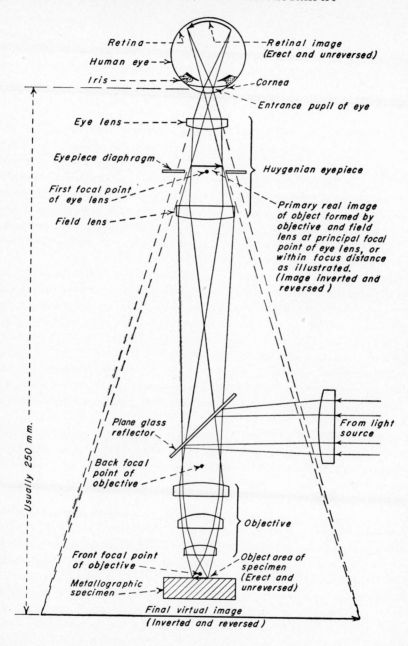

Fɪɢ. 35. Illustrating the principle of the metallurgical compound microscope, and the trace of rays through the optical system from the object field to the final virtual image.

of the objective, as illustrated in Fig. 35, a primary real image[1] of greater dimensions than those of the object field will be formed at some distance beyond the rear lens element of the objective. The size of the primary image with respect to that of the object field will depend upon the relative distances at which the object and image exist from the objective; and the precise distance beyond the objective where the image is positioned will depend upon the focal length of the objective and the distance existing between the plane of the object and the front focus point of the objective.

By appropriately positioning the primary image with respect to a second optical system (the eyepiece), the primary image may be further enlarged by an amount related to the initial magnifying power of the eyepiece. Inasmuch as the separation between the objective and the eyepiece is fixed at some definite distance, equivalent to the mechanical tube length of the microscope, the primary image may be properly positioned with respect to the eyepiece by merely focusing the microscope, *i.e.*, increasing or decreasing the distance between the object plane and the front lens of the objective.

Although an objective is capable of forming a real image by itself, as already mentioned, some eyepieces, such as Huygenians, assist in forming this image. As illustrated in Fig. 35, the primary image is formed by the objective in conjunction with the field lens of the eyepiece, and the microscope is so focused that the primary image is located at the focal point, or within the focus distance as in Fig. 35, of the eye lens of the eyepiece. Such precise positioning of the primary image is essential in order that the final image can be formed and rendered visible to the observer when looking into the eyepiece. If now the entrance pupil of the eye is made to coincide with the exit pupil of the eyepiece, the eye lens in conjunction with the cornea lens of the eye will form a second real image on the eye retina. This retinal image will be erect and unreversed, and owing to the manner of response of the human brain to excitation of the retina, the image will appear to the observer to be existing in space at some distance in front of the eye. This space image, since it has no real existence, is known as a virtual image and it will appear to be inverted and reversed with respect to the object field.

Because of physiological and psychological eccentricities associated with human behavior, most normal-sighted persons will invariably focus a microscope so that the primary image is formed at such a location within the focus distance of the eye lens that the virtual image will appear to exist at

---

[1] A real image is one that is formed by intersecting image-forming light rays, as, for example, the image formed on a reflecting or ground-glass screen by a lantern-slide projector. The real image, as formed by a microscope objective, is inverted and reversed with respect to orientation of the object.

approximately the near point of the eye.[1] Far-sighted observers, however, will generally focus so that the virtual image will appear farther away than 250 mm; and others may focus the primary image to coincide with the focal point of the eye lens and thus render the virtual image at infinity.

When a positive type of eyepiece (see page 100) is used in conjunction with an objective, the manner by which the final virtual image is formed is similar to that already described. Although not illustrated, the primary image is formed only by the objective (the field lens does not assist in forming the image); and by appropriately focusing the microscope, the primary image is located at or within the equivalent focus point of the field-and-eye lens combination. Thus the eyepiece combination, behaving as a simple positive lens to which it may be referred, forms in conjunction with the eye a final virtual image in precisely the same manner as does the eye in conjunction with the eye lens of a Huygenian eyepiece.

Microscope Objectives. Of the three essential optical systems incorporated in the metallurgical microscope—objective, eyepiece, and illuminating system—the objective is probably the optical part most critical with regard to influencing the quality of the observed image. Although the clarity and definition of the image will suffer if any of these systems are inferior or inefficiently used, the image quality is nevertheless most adversely affected by departure of an objective from near optical perfection.

Metallurgical-microscope objectives may be divided into four general groups, namely, achromats, semiapochromats, apochromats, and monochromats, which are a special class of objectives for use with ultraviolet light. The distinction between the first three classes is based upon the design of the objective, the degree to which optical errors such as coma, astigmatism, and other aberrational effects have been corrected, and the extent to which desirable properties in the objective have been developed. Typical microscope objectives of different focal lengths, numerical apertures, and initial magnifying powers are illustrated in Fig. 36.

Chromatic Aberration. When white light is passed through a simple positive lens from some source outside of its principal focus point, the light will be dispersed and a series of color images of the source will be focused at different points along the principal axis of the lens. Because the index of refraction of a medium, such as optical glass, is greater the shorter the wave length of visible radiation passing through it, the violet or blue image of the source will lie closer to the emergent side of the lens than will the red image;

---

[1] The near point of the eye refers to the closest point in front of the entrance pupil at which eye focus can be attained with complete comfort and with maximum eye accommodation. Although the near point varies with age, for a normal-sighted observer who has reached maturity, it is at a distance of approximately 250 mm (10 in.).

between these extremes of the visible spectrum other colored images will be appropriately positioned, corresponding to intermediate wave lengths.

The inability of a simple lens to focus sharply at the same point all of the colored images arising from the dispersion of white light, and the attending unequal refraction of the various wave lengths, is known as longitudinal chromatic aberration. Associated with this longitudinal error is a second aberrational effect, known as lateral chromatic aberration. This optical error arises from the variation in focal length of a lens with different wave lengths of light, giving rise to the formation of colored images of unequal sizes.

For practically all microscopic work, the existence of chromatic errors to any appreciable extent is objectionable, since the image formed will be

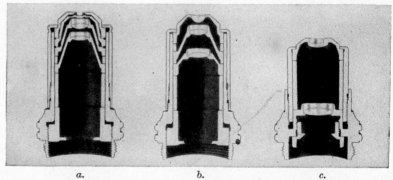

a.   *b.*   *c.*

a. 97 × apochromat—oil immersion. F. L.—1.8 mm; N. A.—1.25.
b. 43 × achromat—dry. F. L.—4 mm; N. A.—0.65.
c. 10 × achromat—dry. F. L.—16 mm; N. A.—0.25.

Fig. 36.   Cut-away sections of three typical microscope objectives. (*Courtesy of Bausch and Lomb Optical Co.*)

surrounded by color halos, and in general will lack in definition and clarity. Although inherently present in simple lenses, chromatic aberration may be limited to various degrees in compound lens systems such as objectives by forming the different lens elements from optical glasses and minerals having different refractive and dispersive characteristics.

Achromatic objectives, described more fully on page 86, are corrected chromatically for two selected color regions, involving usually the red and the green, as illustrated in Fig. 37. Apochromats, however, are corrected for three principal color regions—usually the red, green, and violet portions of the spectrum—and partly because of this better chromatic correction, these objectives are superior to achromats with regard to performance.

Spherical Aberration.   It is known that when monochromatic light, *i.e.*, light of a definite wave length, is passed through a simple positive lens from a source outside of its principal focus point, a series of images of the source will be formed along the principal axis of the lens. The light in passing through the outermost margins of the lens will be refracted to a greater degree and the image formed thereby will be at a point closer to the emergent side of the lens than in the corresponding case of the same wave length of light passing through the lens nearer the principal axis. This inherent optical error in an uncorrected lens is known as spherical aberration, and it, as well as other aberration effects related to it, are attributable mainly to the curved surfaces characteristic of lens elements.

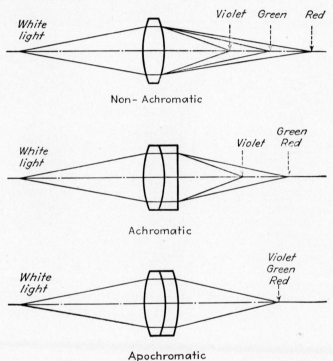

FIG. 37. Illustrating schematically the phenomenon of longitudinal chromatic aberration, and the degree of correction usually applied to achromatic and apochromatic objectives.

When white light, instead of monochromatic radiation, is passed through a simple lens, the manifestation of spherical aberrational errors becomes somewhat more complex. The white light is dispersed into its color components as described for chromatic aberration, and this circumstance, together with the variation in refraction of light in the different zones of the lens, gives rise to the formation of a multiple series of successive color images along the principal axis, as illustrated in Fig. 38.

When such an adverse condition exists to any great extent in a microscope objective, the attending image will appear fuzzy and indistinct.

However, by applying appropriate optical correction to an objective through proper formulation of the various lens elements with regard to their radii of surface curvature, and through the use of different optical glasses and minerals, spherical-aberrational errors may be reduced to a minimum. It is impossible to achieve perfect spherical correction over the entire aperture of an objective, and although correction may be pre-

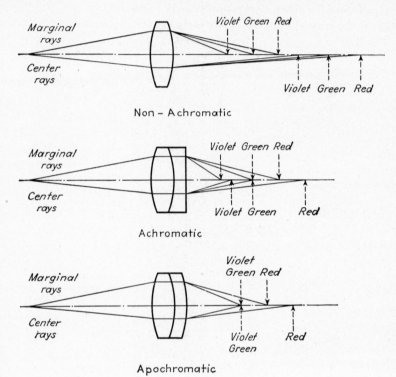

Fig. 38.   Illustrating schematically the phenomenon of spherical aberration, and the degree of correction usually applied to achromatic and apochromatic objectives.

cisely applied to axial and peripheral rays of light, there will always remain some intermediate portion of the lens system that is not fully corrected.

When an optical system is corrected as perfectly as possible for spherical errors, such as in apochromats and to a lesser extent in high-powered achromats, it is practically impossible to correct fully for lateral chromatic errors. Fortunately, however, this residual color error may be rendered insignificant in its effect on the quality of the image by using in conjunction with such objectives an eyepiece that is overcorrected, as described on page 101.

Achromatic Objectives. Achromatic objectives, illustrated in Fig. 36, are probably the most popular class of microscope objectives used, owing principally to their relative freedom from aberrational and other optical errors and their relatively low cost. They are available for metallographic use (optically corrected to be used without a specimen cover glass) in a wide range of focal lengths, varying from about 32 to 1.8 mm, with associated initial magnifications of 4 to 97× respectively. Their optical performance is particularly good when those of low or medium power are used in conjunction with Huygenian or intermediately corrected eyepieces, and those of high power (the semiapochromatic objectives to be described) are used with compensating eyepieces to correct for residual color errors.

As illustrated in Fig. 38, achromatic objectives are spherically corrected for only one designated region of the spectrum (usually the green-yellow region), and chromatically for two color regions, generally the green and the red. Because correction for spherical aberration is not extended to the violet and red, or for chromatic aberration to the violet region, achromatic objectives are incapable of rendering an image that possesses true color relationships. Although the image will be free from prominent color halos, some residual color will be present that is particularly noticeable when the focus of the microscope is slightly changed. For most low-powered microscopic work, however, these remaining color fringes are not too objectional.

Achromatic objectives, because of the usual extent to which they are corrected for aberrational errors, perform best when used in conjunction with illumination that is appropriately filtered to render wave lengths of light corresponding to about the middle region of the visible spectrum. Images of inferior quality will be produced if illumination of longer or shorter wave lengths is used, and hence the use of red, blue, or violet light filters should be avoided.

Apochromatic Objectives. Apochromatic objectives represent the finest microscope objectives obtainable, and owing to the high degree of perfection to which they are corrected for optical errors, their cost is relatively great. They are available in fewer focal lengths than other types of objectives, but in comparision to equivalent achromatic objectives, they possess greater numerical apertures and generally higher initial magnifications. With due regard to the excellent optical qualities inherent in apochromatic objectives, they nevertheless render an image that lacks decidedly in flatness of field, being in this respect inferior to achromats. In order to secure an image of the high quality which an apochromatic objective is capable of forming, it must be used under the conditions for which it was designed and intended, and all adjustments pertaining to illumination and the microscope in general must be made with preciseness and the highest degree of criticality.

Apochromatic objectives, owing to the greater number of lens elements incorporated in their design, are more fully corrected for aberrational errors than are achromats. Chromatically they are corrected for three principal color regions (red, green, and violet) as illustrated in Fig. 37, and spherically corrected for two, violet and green. Because of the high degree to which apochromats are corrected for spherical aberration, a certain amount of lateral chromatic errors persist, as mentioned heretofore. This residual color error gives rise to an image field that is slightly fringed with color when the objective is used with a Huygenian or other simple-type eyepiece. However, by using a compensating eyepiece that is designed specifically to correct for the variation in size of the color images formed by the objective, a final image of superb quality may be produced. Because of this, it is essential that apochromatic objectives, in order to perform efficiently, must be used with eyepieces having compensating characteristics.

Owing to the nearly complete freedom from aberrational errors, apochromatic objectives are highly desirable for visual use at high magnifications or for purposes of photomicrography. They perform efficiently when used with white light, light of daylight quality, and green-yellow or blue illumination secured by appropriate filtering.

Semiapochromatic Objectives. Semiapochromatic objectives, often referred to as Fluorites, are essentially achromats of high quality. With respect to their correction for aberrational errors, they represent a compromise between achromatic and apochromatic objectives, and with regard to other optical qualities and general performance, they approach those of apochromats. As in the case of apochromats, semiapochromatic objectives of high power render the best image when used in conjunction with compensating eyepieces.

Special Objectives. As mentioned heretofore, microscope objectives are usually designed to perform most efficiently when used in conjunction with an eyepiece at some definite mechanical tube length. Owing to the manner by which metallographic specimens are illuminated (pages 81 and 125), the converging image-forming rays from the objective may suffer distortion in passing through the plane-glass reflector, and thus render a final image that lacks in definition. This objectionable circumstance may be eliminated by the use of objectives corrected for an infinitely long tube, the principles of which are more fully discussed on page 126. Most manufacturers of microscopic equipment now supply such objectives for metallographic work.

To eliminate the inconvenience associated with changing objectives on a microscope, some table-type metallurgical microscopes are equipped with a revolving nose piece (holding up to four objectives) and parfocal objectives. With respect to equivalent conventional types of objectives, parfocals of different focal lengths are designed so that their mechanical body mounts are of equivalent length. This offers the advantage that when the microscope is once brought into focus with one of the objectives on the nose piece, other objectives may be rapidly rotated into position above the object without seriously affecting the focus of the instrument.

Properties of Objectives. A microscope objective possesses a number of inherent properties and characteristics that arise as a natural consequence of combining together two or more lens elements to form a compound optical system. Many of these properties are desirable and are enhanced by appropriate design and correction of the lenses, whereas others are objectionable and in many cases cannot be eliminated completely without sacrificing, in part at least, more desirable qualities.

Some of the important interrelated properties of a microscope objective are (1) magnifying power, (2) aperture or numerical aperture, (3) resolving power, (4) vertical resolution, (5) illuminating power, and (6) curvature of image field. The most important of these characteristics will be discussed separately and in detail.

Magnifying Power. All objectives possess initial magnifying power, *i.e.*, the ability of the objective to magnify the real object a definite number of times without the aid of an eyepiece. Usually the initial magnification of an objective is engraved on the objective mount, although in the case of objectives of less recent manufacture, the initial magnification may be designated by a letter or a number. To obtain magnification data of such objectives reference must be made to the manufacturer's catalogue.

In order to complete the optical system of a microscope and to make visible the primary image formed by the objective, the introduction of a second optical system—the eyepiece—is required at the upper end of the microscope tube. Eyepieces, like objectives, possess initial magnifying power, and such data are usually engraved on the eye lens mount.

The total visual magnification secured by the combination of a given objective and eyepiece depends not only upon their respective initial magnifying abilities but also upon the distance by which these optical systems are separated in the microscope. All objectives, with the exception of those corrected for infinity, are designed for use at some very definite mechanical tube length, usually 160 to 250 mm. If the tube length is altered appreciably from that for which the objective is corrected, the apparent magnification of the final virtual image will be changed and, as already mentioned, the image will be somewhat distorted.

Total magnification data of a microscope usually refer to the image as observed visually, and are based upon the size of the virtual image with respect to the object field when the microscope is so focused that the virtual image appears to exist at 250 mm. When a given combination of objective and eyepiece is used at the appropriate tube length, the total magnification is equal to the product of the initial magnifications of the two optical systems. If an objective has an initial magnifying power of 54.0×, for example, and the eyepiece used in conjunction with it has an initial magnification of 5.0×, the total magnifying ability of the combination will be 270×.

In photomicrography or in other cases where the image formed by the objective is projected by the eyepiece onto a screen, the total magnification of the real image is related to the initial magnifying powers of the optical parts, the tube length of the microscope, and the projection distance, *i.e.*, the distance from the exit pupil of the eyepiece to the plane of the screen. To approximate the magnification of a projected image, the following equation may be used:

$$M = \frac{DM_1M_2}{250} \tag{1}$$

where M = magnification to be determined
M_1 = initial magnification of the objective
M_2 = initial magnification of the eyepiece
D = projection distance, mm, usually determined with sufficient accuracy by measuring the distance from the eye lens of the eyepiece to the focusing screen

The denominator factor of 250 in Eq. (1) refers to the distance in millimeters for which the total magnification of the objective and eyepiece was computed. That is, the total magnification of the virtual image as visually observed at a distance in space of 250 mm is equivalent to the total magnification of the real image, formed by the same objective and eyepiece, at a projection distance of 250 mm, as measured above.

From the above equation it is evident that the magnification of a projected image, other factors remaining constant, is directly proportioned to the projection distance.

In cases where it is necessary to determine the total magnification of a projected image with greater accuracy than can be secured by use of Eq. (1), a stage micrometer may be used. As illustrated in Fig. 39, a stage micrometer consists of a flat metal plate (for use with reflected light) or a glass plate (for use with transmitted light) that is accurately engraved with lines spaced 0.01 mm apart over a distance of about 1.0 mm. To determine the total magnification under a given set of conditions, the stage micrometer is used as the object, and the image of the engraved lines is appropriately projected onto the focusing screen. The distance between the lines of the image is then carefully measured, and by simple calculations the exact magnification of the projected image can be determined.

Numerical Aperture. The numerical aperture, or light-gathering ability of an objective, is probably one of its most important constants, but unfortunately it is not always given the consideration that is warranted. It is because of numerical aperture, which for any objective is a function of design, that fine details in an object may, within limits, be completely and clearly resolved. This resolving power of an objective is nearly proportional to the value of its numerical aperture, with other factors, such as wave length of illumination, microscope adjustments, etc., remaining constant.

The purpose of all objectives is to receive and to combine into an image a larger solid cone of light than can normally be received by the unaided eye. The clearness of the image and the resolution of fine detail will depend in part upon the numerical value of this solid cone of light. For any particular objective, this cone of light is best expressed as the angular

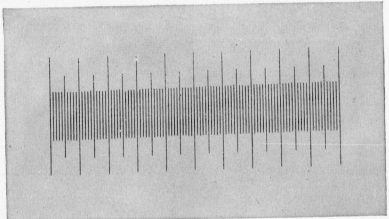

Fig. 39. Image of a stage micrometer at a magnification of 75×. Each division is equivalent to 0.01 mm.

aperture wherein reference is made to that solid light cone whose apex lies at the object, or at a point on the object transmitting light, and whose base corresponds to the light opening of the system of lenses formed by the objective mount.

Since the resolution of fine detail by an objective is essentially governed by the amount of light entering the objective, it is evident that the larger the angular aperture, the greater will be the resolving power of the objective. However, the amount of light received by the objective is further influenced by the index of refraction of the medium between the front lens of the objective and the surface of the object—the medium through which the image-forming rays must pass. This is diagrammatically illustrated in Fig. 40.

In considering a dry objective, *i.e.*, one wherein the medium between the objective and object is air (index of refraction $n = 1.0$), it will be noted in

Fig. 40a that a ray of light, designated as R_1, leaves the object at an angle of 30 deg to the principal axis of the objective and just enters it at the extreme edges of the front lens element. Light ray R_2, however, being of greater angularity than R_1, is not received by the objective, but is lost outside of it.

If the medium of air between the front lens of the objective and the object is replaced by one of greater refractive index, such as cedar oil (index of refraction $n = 1.5$), it will be observed as shown in Fig. 40b that light ray R_2 is refracted or bent in passing through the oil so that it is now received by the objective where formerly it was not. For the same reason, light ray R_1, in the case of oil immersion, passes through the objective closer to the principal axis than it did formerly. Thus a wider angle of light rays or a greater amount of light is received by an oil-immersion objective than by a dry objective.

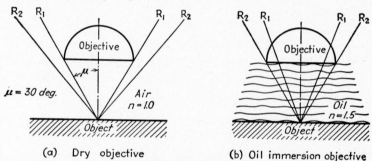

(a) Dry objective   (b) Oil immersion objective

FIG. 40.   Illustrating schematically the principle of numerical aperture of dry and oil-immersion objectives.

It is evident, therefore, that a direct measure of the light-gathering power of an objective cannot be made solely upon the angular aperture of the cone of light entering the objective or solely upon the index of refraction of the medium between the objective and object. It has been shown, however, that the numerical aperture of an objective (N.A.) may be expressed mathematically as the product of the sine of one-half the angle of light aperture (μ in Fig. 40a) and the index of refraction (n) of the medium between the front lens of the objective and the object; that is,

$$\text{N.A.} = n \sin \mu \qquad (2)$$

Thus, with reference to Fig. 40a, the numerical aperture of the objective in air will be

$$1.00 \times \sin 30 \text{ deg} = 1.00 \times 0.5 = 0.5 \qquad (3)$$

In the case of oil immersion with cedar oil, the numerical aperture will be

$$1.50 \times \sin 30 \text{ deg} = 1.50 \times 0.5 = 0.75 \qquad (4)$$

The importance of numerical aperture will become apparent as the other properties of objectives are discussed.

Resolving Power. The resolving power, or the ability of an objective to produce sharply defined separate images of closely spaced detail in the object (sometimes referred to as fineness of detail), is fundamentally dependent upon the numerical aperture of the objective and the wave length of light used to illuminate the object.

If a very narrow central pencil of light is passed into an objective, the finest separate images of detail that can be observed with sufficient eyepiece magnification and such a narrow pencil of light may be expressed mathematically as

$$\text{Fineness of detail} = \frac{\lambda}{\text{N.A.}} \tag{5}$$

where λ = wave length of illumination
  N.A. = numerical aperture of the objective
If, however, this narrow pencil of light is increased in diameter, or the original narrow beam is made oblique, so that in either case the aperture of the objective is completely flooded with light, the resolution so obtained with that particular objective will be a maximum, and it may be expressed as

$$\text{Fineness of detail} = \frac{\lambda}{2\,\text{N.A.}} \tag{6}$$

If an objective with a numerical aperture of 1.00 and an initial magnification of $100\times$ is used with green light ($\lambda = 0.00053$ mm) under such conditions of illumination that the resolving power equals $\lambda/2$ N.A., then the maximum resolution that can be obtained is equivalent to lines spaced $0.00053/(2 \times 1)$, or 0.000265 mm, apart. With proper eyepiece magnification, these lines will appear as distinct and separate images, whereas lines that are spaced closer together than 0.000265 mm will not be resolved, reregardless of the extent to which the total magnification is increased (see "empty magnification," page 95).

The human eye at a distance of 250 mm from an object can resolve lines or other detail that are spaced about 0.11 mm apart. Therefore, in order that the eye may just resolve the lines that were originally spaced 0.000265 mm apart and initially magnified to 100 diameters by the objective, the minimum initial magnification of eyepiece necessary will be $0.11/(0.000265 \times 100)$, or approximately $4\times$. The minimum total magnification will therefore be 100×4, or 400 diameters.

If this same reasoning is applied to objectives of different numerical apertures, it will be found that theoretically the minimum total magnifica-

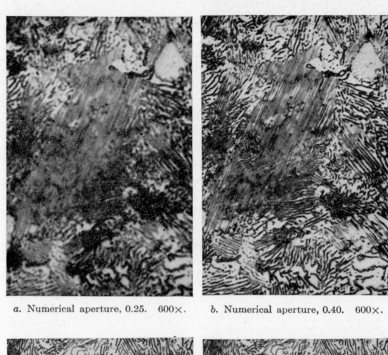

a. Numerical aperture, 0.25.   600×.   *b.* Numerical aperture, 0.40.   600×.

c. Numerical aperture, 0.65.   600×.   *d.* Numerical aperture, 0.95.   600×.

FIG. 41.   The influence of numerical aperture of an objective on resolution at the same magnification and wave length of illumination.

tion necessary to observe clearly the lines or details resolved by that particular objective will be equivalent to approximately 400 times the objective's numerical aperture. It has been found by experience, however, that at this minimum total magnification the image is not magnified sufficiently to be studied comfortably by the eye. This condition may be alleviated without seriously affecting the quality of the image by increasing the total magnification to some value within an approximate range of 500 to 1000 times the numerical aperture of the objective. For most microscopic work this upper limit of 1000 should not be exceeded, for beyond this limit the image will usually lack sharpness and clear definition. However, under ideal working conditions not usually encountered in ordinary microscopic work, it is possible with oil-immersion apochromatic objectives to obtain, at magnifications of 2000 times the numerical aperture, photomicrographs that appear reasonably sharp and distinct at a distance of 250 mm from the eye.

It is evident from the past discussion that, for any given wave length of illumination, the resolution of fine detail in an object is greater the higher the numerical aperture of the objective, and for any given objective (constant numerical aperture) the resolution is greater the shorter the wave length of light used to illuminate the object field.

The influence of numerical aperture on the resolution of fine details (using a given wave length of illumination) is illustrated in Fig. 41. Without doubt, the details of the object are more fully resolved at 600× by an objective having a numerical aperture of 0.95 than one possessing a numerical aperture of only 0.25. Since all microscopic examinations are for the purpose of observing details in an object that are normally too small to be resolved by the unaided eye, it is evident that the potential resolving power of an objective is the most important consideration, and not total observed magnification, as is commonly believed. When details of an object are resolved to an extent commensurate to the full resolving power of the objective, and are magnified beyond a total magnification of about 1000 times the numerical aperture of the objective, such images are referred to as possessing "empty magnification," *i.e.*, nothing whatsoever has been revealed in the way of additional details in the object that could not have been clearly seen at some lower magnification.

Figure 42 illustrates the effect of the wave length of illumination on resolution for an objective of given numerical aperture. It will be noted that there is a small but significant increase in the resolution obtained with blue illumination as compared to that obtained with nearly monochromatic red. However, the improvement in resolution brought about by using objectives of greater numerical aperture is far better than that achieved by decreasing the wave length of visible light used. The influence of shorter wave lengths

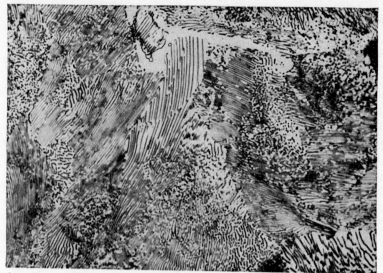

a. Specimen illuminated with nearly monochromatic red radiation.   Wave length approximately 6500 Å.   1200×.

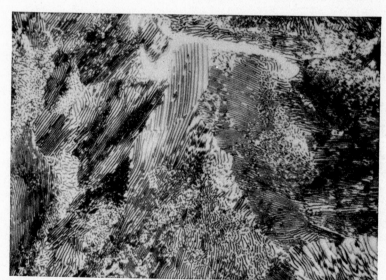

b. Specimen illuminated with nearly monochromatic blue radiation.   Wave length approximately 4500 Å.   1200×.

Fig. 42.   The influence of wave length of illumination on resolution at the same magnification and objective numerical aperture.

is not apparent to any great extent until wave lengths in the region of ultraviolet are used.

Vertical Resolution. Vertical resolution, sometimes referred to as depth of focus or penetration, is the ability of an objective to produce a sharply focused image when the surface of the object is not truly plane. Such a surface is encountered, for example, when a metallographic specimen has been severely overetched, or when certain constituents of the structure are depressed or elevated from the etched surface.

The vertical resolution is inversely proportional to the numerical aperture and initial magnification of the objective; unfortunately this relationship cannot be changed by any known method of lens correction. Practical attempts have been made to improve the field-depth property of an objective by decreasing the diameter of the illuminant entering it or by inserting in the rear of the back lens of the objective an appropriate diaphragm. Although such methods of improvement are effective, they decidedly reduce the effective numerical aperture of the objective and hence impair the resolution. Methods such as these should therefore never be used unless depth of focus, for some particular reason, is of greater importance than resolution.

In general, microscope objectives that have greater depth of focus than other objectives of the same numerical aperture are usually not so well corrected for other qualities.

Curvature of Field. All objectives inherently produce to some degree a curved image field, *i.e.*, a condition wherein the sharpness of the centrally focused image declines toward the outer edges of the field of view. This condition varies with numerical aperture, and becomes of serious and objectionable proportions in apochromatic objectives or others of high numerical aperture. With these objectives, however, some improvement in flatness of field may be secured, particularly if the image is projected, by using appropriate eyepieces as described below.

There is no means of obtaining from an objective full resolving power and a perfectly flat field at the same time. The often-used method of improving flatness of field by closing down the aperture diaphragm is not to be recommended because by so doing resolution is sacrificed at the gain of usually less important field flatness.

Microscope Eyepieces. An eyepiece, or ocular, as it is often called, is an optical system in the microscope whose primary purpose is to enlarge the primary image formed by the objective and render it visible as a virtual image, or to project the primary image as a real image, such as in photomicrography. In some cases, it also compensates for residual aberrational errors in the objective. Eyepieces possess an initial magnifying power, the value of which refers to the extent to which the primary image will be enlarged when a final virtual image is formed at a viewing distance of 250 mm, or a real image at a projection distance of 250 mm.   The initial magnification value is usually engraved on the eye-lens mount of the ocular, such as 5×, 7.5×, 10×, etc., but in the case of eyepieces of less recent manufacture, the magnification is designated by a single number or letter. The initial magnifying power of such eyepieces may be determined by reference to the manufacturer's literature.

Eyepieces may be divided into three general groups, namely, (1) negative types, (2) positive types, and (3) photographic or amplifying types.   With regard to the first two types, the lens element nearest the eye, when the ocular is appropriately positioned in the microscope, is referred to as the eye lens, and the element nearest the objective is known as the field lens.

Negative Types.   A negative type of eyepiece, of which the Huygenian is probably the most representative, consists, in its simplest form, of two nonachromatic, plano-convex lens elements that are mounted in the eyepiece tube with the convex side of both elements toward the objective. Between the two lens elements is a diaphragm, precisely located at the first focus point of the eye lens, as illustrated in Fig. 43a.   The term "negative" as applied to this group of oculars is not appropriate because the eyepieces do not in any way behave as a true negative lens system, such as amplifying eyepieces to be described.   The term merely implies that these oculars behave differently than positive types, and that the equivalent focus point of the eye-and-field-lens combination lies between the two lens elements.   Because of the location of the equivalent focus point, negative-type eyepieces cannot be used as an ordinary magnifier.

A Huygenian eyepiece does not render an image completely free from distortion, owing to the lack of optical correction applied.   Because of this, they are recommended for use only with low- and intermediate-powered achromatic objectives, where a slight amount of image distortion at the outer edges of the field can be tolerated.   Negative eyepieces, such as Hyperplanes and Periplans, which are partly corrected for aberrational errors, are recommended for use with high-powered achromats and semi-apochromatic objectives.   These intermediately corrected oculars produce a somewhat larger field of view than Huygenians, and render a flatter image field than compensating eyepieces, to be described.

As mentioned heretofore, the field lens of the Huygenian eyepiece assists the objective in forming a primary image, and by appropriately focusing the microscope, this image is positioned at or within the focus of the eye lens.   When it is located at some distance within the focus point, as illustrated in Fig. 35, the eye lens in conjunction with the human eye forms a virtual image at a viewing distance of about 250 mm.   Owing to the influence of the field lens, the primary image will be formed at a point closer to the objective than otherwise and hence produce an image that is slightly

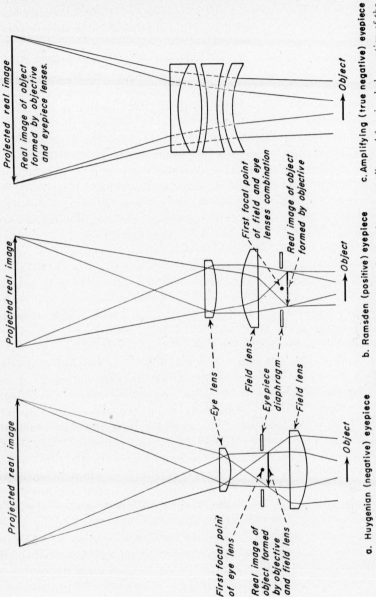

FIG. 43. The principle of image projection (as in photomicrography) by eyepieces normally used for visual observation of the image, and by an amplifying type of eyepiece.

a. Huygenian (negative) eyepiece   b. Ramsden (positive) eyepiece   c. Amplifying (true negative) eyepiece reduced in size. This circumstance, however, is not objectionable since it permits a larger and more brightly illuminated area to be included in the field of view.

The Huygenian eyepiece may be used for either visual examination, as described above, or it may be used for projecting an image, as in photomicrography. In principle, projection of the image by the eyepiece involves the formation of a second real image, rather than one that is virtual as for visual observation. This is brought about by changing the focus of the microscope (increasing slightly the distance between the front lens of the objective and the object) so that the primary image is shifted from the focus point, or from within the focus distance of the eye lens, to some appropriate position outside of the focus point, as illustrated in Fig. 43a. With the primary image in this new location, the eye lens, behaving as any simple positive lens, will form a second real image when a focusing screen is appropriately placed to intersect the image-forming rays.

Positive Types. A positive type of ocular is one consisting of two or more lens elements so computed and combined together that the optical system as a whole behaves as an equivalent positive lens. The first principal focus point of the system lies in front of the field lens, and hence under appropriate conditions the eyepiece may be used as a simple magnifier. This circumstance is one that usually enables a distinction to be made between a positive and negative eyepiece.

The Ramsden ocular, illustrated in Fig. 43b, is the simplest of positive-type eyepieces, and it probably is the one most commonly used. It consists of two plano-convex lens elements—the eye lens and the field lens—that are so positioned in the ocular body mount that the convex side of the lenses are toward one another. To define the image field an ocular diaphragm is located in front of the field lens at the equivalent focal plane of the lens combination. With regard to chromatic aberration, the Ramsden eyepiece is more in error than the Huygenian, but it is definitely better with respect to spherical errors. For equivalent initial magnifying powers, the Ramsden renders a slightly smaller field of view than negative-type eyepieces, which under most circumstances is not objectionable.

When used in a microscope for either visual examination or image projection, the field lens of the Ramsden eyepiece in no way assists in the formation of the primary real image. Rather, the image is formed solely by the objective, and by appropriately focusing the microscope, the primary image is positioned at, or slightly within, the equivalent focus point of the eyepiece system. The lens elements of the eyepiece, functioning as a single unit, then form in conjunction with the eye a virtual image in the same manner as the virtual image is formed by the eye and the eye lens of a negative-type eyepiece. By slightly refocusing the microscope so that the primary image is positioned outside of the equivalent focus point of the ocular, a second real image may be projected, as illustrated in Fig. 43b.

Compensating Types. Compensating oculars are a group of specially designed eyepieces that are chromatically overcorrected in order to compensate for the residual color errors present in semiapochromatic and apochromatic objectives. Compensators are designed as either positive or negative types, and owing to the extent to which they are corrected for optical errors, they are available with higher initial magnifying powers than either Ramsden or Huygenian eyepieces. Usually the higher magnifying eyepieces are of positive design, and those of lower power are negative.

To secure an image as free from chromatic errors as it is possible to attain, semiapochromatic and apochromatic objectives of high power should always be used in conjunction with compensating eyepieces. The curvature of image field, however, will be decidedly more pronounced than that produced by intermediately corrected oculars, or that resulting from a simple type of eyepiece used in conjunction with low- to moderately-powered objectives. Compensators are not suitable for use with achromatic objectives of low power because the overcorrected characteristics of these eyepieces may introduce adverse chromatic effects into the final image.

Amplifying Types. Amplifying lenses, of which Ampliplans and Homals are representative, are a group of so-called eyepieces designed specially for use in photomicrography or for purposes of image projection over a relatively short distance. An amplifier consists of a number of lens elements which, as a group, function as a true negative lens system (not to be confused with the term as applied to Huygenian eyepieces). Since the exit pupil of an amplifier lies between the lens elements and the objective, the ocular cannot be used for visual examination nor can it be used as a simple magnifying system.

Because amplifiers are corrected for aberrational errors to about the same extent as compensating eyepieces, they are particularly well suited to be used with apochromatic and semiapochromatic objectives. The image field resulting therefrom will be exceptionally flat, provided the initial magnifying power of the objective used is one for which the amplifier is designed.

As illustrated in Fig. 43c, the amplifying lens functions as a part of the objective system to form the primary real image. The amplifier is always used at a reduced mechanical tube length, secured by an appropriate eyepiece adapter, and because of its location within the microscope system, the image-forming rays pass into the amplifier before the primary image can be formed by the objective as is normally the case. The first real image, therefore, is formed by the objective-amplifier combination, and can be viewed when a focusing screen is appropriately placed to intersect the image-forming rays.

Measuring and Reticle Types. It is often necessary to measure accurately the dimensions of small surface defects on metallographic specimens, microhardness impressions, etc., or to superimpose a micrometer measuring scale or other type of reticle on a projected image and, perhaps, incorporating it on a finished photomicrograph. This may be readily accomplished by means of a measuring-type eyepiece which consists essentially of a negative- or positive-type ocular so designed that an appropriate reticle may be placed in the focal plane of the primary real image, and thus be superimposed on the final image. In the positive type, the reticle is placed coincident with the plane of the ocular diaphragm, and by adjustment of the lens combination with respect to the fixed reticle, the reticle markings can

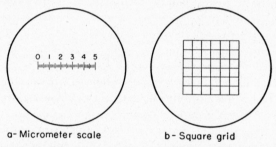

a- Micrometer scale   b- Square grid

Fig. 44. Typical reticles of the fixed type useful for microscopic measurements.

be sharply focused. Positive eyepieces are particularly appropriate for use with either movable or stationary reticles, because any aberrational errors present will affect both the image and the reticle markings equally. In the case of negative-type measuring eyepieces, the reticle must be placed between the field and eye lenses in the plane of the ocular diaphragm. This circumstance gives rise to serious design problems if the reticle is of a lateral movable type, and also introduces parallax between the image and the reticle markings owing to some change in position of the primary image when the reticle markings are sharply focused by adjusting the eye lens.

A reticle consists of a reference mark, a micrometer scale, or other geometric markings, inscribed on a glass disk or formed by fine threads, wire, or spider-web strands. The reticle may be of a fixed type, as those illustrated in Fig. 44, or it may be movable for purposes of length measurements, such as in the Filar eyepiece to be described. All micrometer-type reticles must be calibrated for a particular objective-and-measuring-eyepiece combination, if precise and direct measurements are desired. This may be readily achieved by calibrating the reticle scale against the spacings on a stage micrometer (see page 91).

The Filar eyepiece is probably the most useful micrometer ocular used for metallographic measurements. It consists of a Ramsden type of eyepiece wherein the field and eye lenses are of equivalent focal lengths. The composite reticle of the eyepiece consists of a fixed micrometer scale ruled at intervals of 0.5 mm; a vertical hairline that can be moved across the field of view by means of a micrometer screw; and a fixed horizontal hairline running through the center of the field in a direction parallel to the axis of the screw. The horizontal hairline is for the purpose of orienting the specimen of interest with respect to the direction of movement of the vertical line. Sharp focus of the scale and hairlines may be secured by adjusting the eyepiece lenses mounted in a focusing sleeve. The micrometer screw which actuates the vertical hairline is fitted with a graduate drum divided into 100 equal divisions, each division corresponding to 0.01 mm lateral movement of the hairline across the field. One-tenth of each division can be accurately estimated, thus enabling measurements to be made to 0.001 mm.

It is essential in the use of a Filar type of eyepiece that the vertical hairline be moved in only one direction when a measurement is being taken between two points. Movement in only one direction will eliminate errors in measurement that would otherwise be introduced owing to mechanical backlash of the screw.

Low-reflectance Coatings. It is well known that when light is passed through an objective or other lens system, the internal flare, reflections, and light losses occurring at air-glass surfaces render an image inferior to that which the optical system is capable of forming under more favorable circumstances. To minimize the effects of internal flare and other adverse conditions on image quality, optical parts for metallurgical microscopes, including objectives, eyepieces, prisms, plane-glass reflectors, etc., are now available with the various polished surfaces coated with low-reflectance films.

Low-reflectance films are extremely thin layers of special materials, having appropriate optical qualities, that function by reducing light losses caused by reflection from optical surfaces and redistributing this energy so that a greater percentage of the incident light is transmitted. Because of this redistribution of radiant energy, the over-all efficiency of the optical system is increased, the image quality and contrast is improved, the effective aperture of the system is increased, and with respect to photomicrography, photographic exposure is decreased. The materials of which low-reflectance coatings are composed have an index of refraction, when existing as very thin films, as near as possible to the harmonic mean between that of optical glass (1.52 to 1.66) and that of air (1.0). When such materials are appropriately applied to polished optical surfaces, any incident light will be reflected in about equal amounts by the air-film surface and the film-glass surface. By adjusting the thickness of the film with very great precision during application, the waves of the two reflected beams of light, for any selected wave length, can be rendered out of phase and thereby obliterate or cancel out one another. Other wave lengths whose dimensions are not too greatly different from that for which the thickness of the coating was

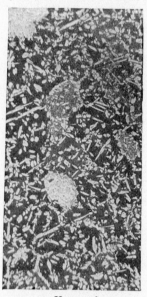 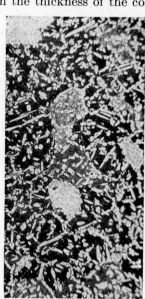

a. Uncoated.   *b.* Balcoated.

FIG. 45.   Illustrating the improvement in image quality of a metallographic structure secured through the use of coated objectives.   50×.   (*Courtesy of Bausch and Lomb Optical Co.*)

adjusted will be partially canceled out, with the effect becoming less pronounced as either end of the visible spectrum is approached.

The methods by which antireflectance coatings are applied to polished optical surfaces vary, depending upon the material that is used. The so-called "cold process" involves film formation by air oxidation (not commercially feasible), controlled etching with hydrofluoric acid fumes, spraying of the material by means of an air gun, or spreading droplets of the coating material over the optical surface by means of centrifuging. The most satisfactory method, however, and the one currently used by most manufacturers of optical equipment, is the "hot process." This process involves the vaporization of either calcium fluoride (n = about 1.30 when in the form of thin films) or magnesium fluoride (n = approximately 1.40) in a highly evacuated system and the deposition or condensation of these vapors on the optical surfaces to be treated.   The precise thickness of the film is controlled by visual observation, and when appropriate colors of reflected light from the treated surface are observed, the process is discontinued.   Calcium and magnesium fluorides are ideal for the purpose intended because of their individual indices of refraction, as noted above, and because even in the form of very thin films they have a resistance to mechanical damage and chemical deterioration equivalent to that of most optical glasses to which the coatings are applied.

The improvement in the appearance of metallographic structures through the use of coated objectives is illustrated for manganese bronze in Fig. 45. For this particular metallographic specimen, the amount of flare present when using a conventional objective (Fig. 45a) was 29 per cent,[1] as compared to only 9 per cent when using a filmed objective.   The amount of flare in both cases was determined by photoelectric methods.

Care of Optical Parts.   In all phases of microscopy it is essential that optical parts be maintained scrupulously clean and free from finger prints, dust, oil films, etc., so that the quality of the image obtained will be equal to that which the equipment is capable of producing.   Without question, and as so strikingly illustrated in Fig. 46, objectives and oculars cannot function efficiently if the lens surfaces are in a dirty and dusty condition.

Particles of loose dust and dirt can usually be removed from optical surfaces by blowing air over such surfaces with an all-rubber ear syringe.   If this method is not effective, brushing the surfaces with a soft camel's-hair brush, or wiping them with lens tissue, will undoubtedly prove more successful.   Care must be exercised in these operations, however, so that the optical surfaces are not scratched or otherwise damaged by abrasion. Oil and finger prints are best removed by wiping the marred surfaces with lens tissue moistened with xylol (never alcohol or other organic solvents); drying with fresh lens paper; and finally removing any adhering paper lint with an air syringe.   This procedure is ideally suited for removing oil adhering to the front lens of an oil-immersion objective which, by standards of good microscopic technique, should be done immediately after the objective has been used.

Under no circumstances should optical parts, particularly objectives, be taken apart for cleaning.   This work should always be done by the manufacturer, who is well equipped with facilities to ensure proper reassembling.

Optical parts should always be handled with a reasonable amount of care. Although they are not fragile, they may be seriously damaged if accidentally dropped or maltreated in any way; and if improperly or carelessly han-

[1] Ratio of the unwanted light (flare) to the total light reaching the image plane, expressed as a percentage of the total light.

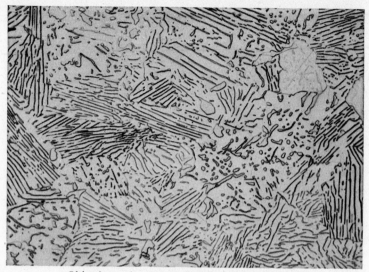

a. Objective and eyepiece scrupulously clean. 1000×.

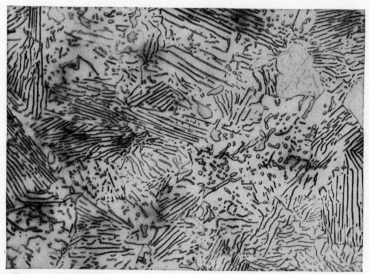

b. Exposed lens surfaces of objective and eyepiece covered with dust. 1000×.

FIG. 46. Illustrating the effect of unclean optical parts on the quality of the projected image.

dled, the optical surfaces may become scratched and thus be rendered use-less. During storage, optical pieces should not be subjected to extreme and sudden changes of temperature. The unequal expansion and contraction of the various parts resulting therefrom may be sufficiently great to cause permanent damage and separation of the lens elements comprising the optical system.

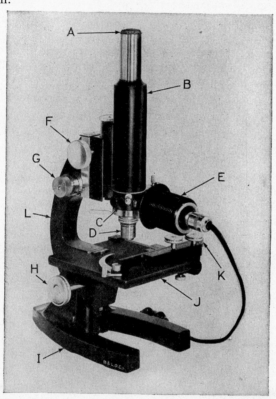

Fig. 47. Table-type metallurgical microscope. (*Courtesy of Bausch and Lomb Optical Co.*)

A. Eyepiece.
B. Microscope tube.
C. Vertical illuminator containing a plane glass reflector.
D. Objective.
E. Illuminating source, consisting of a lamp housing, a 6-volt electric light bulb, a diffusion screen, and a green-yellow light filter.

F. Coarse-focusing adjustment.
G. Fine-focusing adjustment.
H. Adjustment for vertical positioning of the object stage.
I. Microscope base.
J. Object or specimen stage.
K. Adjustment for lateral positioning of the specimen.
L. Microscope limb.

PRINCIPLES OF MICROSCOPIC EQUIPMENT. PHOTOMICROGRAPHY

In the preceding section of this chapter the principles of the metallurgical microscope and the optical systems associated therewith have been de-scribed in detail. It is now of importance to consider the design and the appropriate use of a microscope and a photomicrographic instrument, and to discuss the functioning principles of the many parts and adjustments that so profoundly influence the quality of the final image.

Table-type Microscopes. A conventional table-type metallurgical microscope, illustrated in Fig. 47, consists of a base or stand to which is attached a microscope arm, an object stage, and a movable body tube. The microscope tube supports the vertical illuminator (see page 120), the objective and, at the upper end, the eyepiece. The illuminating source—usually consisting of a 6- to 8-volt filament lamp, a simple condensing lens to collimate the light beam, and an appropriate light filter—is contained within a suitable housing and, as a unit, attached to the vertical illuminator. An aperture diaphragm, the principles of which will be discussed later, is generally provided to afford a means of regulating the amount of light from

Fig. 48. Specimen holder designed for use with a table-type metallurgical microscope to maintain the plane of the specimen surface normal to the principal axis of the microscope. (*Courtesy of Buehler, Ltd.*)

the source entering the microscope system. The object stage is either permanently fixed below the objective or, as illustrated in Fig. 47, it is movable in a vertical direction to allow accommodation of metallographic specimens normally too large to be otherwise handled.

To provide an image field that is in focus over the entire field of view and to permit the specimen to be moved laterally on the stage from center to edge without the necessity of appreciable refocusing of the microscope, it is essential that the prepared surface of the specimen lie in a plane normal to the optical axis of the microscope. Unless the prepared surface is plane and parallel to the one opposite it, it will be necessary to mount the specimen in order to meet the above requirement. A simple method of so doing consists of placing a small mass of plasticene on a glass plate, around which is placed an accurately machined ring with edges lying in parallel planes, approximately 1 in. apart.   The diameter of the ring is of no significance, except that it must be large enough to contain the specimen.   The specimen is then forced into the plasticene (prepared surface upward and protected with lens tissue) by another glass plate until the specimen surface is in the same plane as the upper edge of the ring.   The top glass plate and ring are then removed and the specimen is ready for examination.

FIG. 49.   Typical arrangement of apparatus for securing photomicrographs with a table-type metallurgical microscope.   (*Courtesy of Bausch and Lomb Optical Co.*)

A. Camera extension bellows.
B. Support rod for camera bellows.
C. Lamp housing containing a ribbon-filament electric lamp.
D. Water-cooling cell.
E. Metallurgical microscope as in Fig. 47.
F. Compur type of camera shutter.
G. Cable release for actuating shutter.
H. Hand magnifier for critically focusing the projected image.

An alternate method of mounting consists of embedding the specimen in a small mass of plasticene on a glass plate and making the surface of the specimen parallel to the base plate by means of a Leitz leveling device or a similar unit.   More elaborate specimen mounts may be obtained with special holders, one type of which is shown in Fig. 48.

Although a table-type metallurgical microscope is designed principally for convenience in visual examination of metallographic specimens, it nevertheless can be used with good success as a photomicrographic instrument. As illustrated in Fig. 49, an arrangement suitable for this purpose can be secured through the use of certain auxiliary equipment, consisting of a vertically mounted camera bellows, a Compur type of shutter, and an illuminating source of somewhat greater intensity than the conventional low-voltage lamp described. A suitable source of illumination may be secured from a high-intensity ribbon-filament lamp, a portable carbon-arc arrangement, or a concentrated-arc lamp more fully described on page 114.

Fig. 50. Bausch and Lomb metallograph equipped with a carbon-arc illuminating source. Model ILS. (*Courtesy of Bausch and Lomb Optical Co.*)

It is claimed by some metallographers that the quality of photomicrographs secured from an arrangement similar to that illustrated in Fig. 49 is comparable or even superior to that obtained on a regular photomicrographic instrument. This may be true if more or less ideal working conditions are prevalent and if very careful attention is directed toward critically aligning the microscope and related equipment. Under less ideal circumstances, however, it is reasonable to expect that a photomicrographic unit, being designed specifically for the purpose intended, will yield more satisfactory image reproductions than the less versatile table-type-microscope arrangement described.

Photomicrographic Instruments.   A photomicrographic unit, more often referred to as a metallograph, is essentially a high-precision microscopic instrument designed for both visual examining and permanent recording of metallographic structures by photographic methods.   As illustrated in Fig. 50, a metallograph is basically composed of four separate systems, namely, (1) a table stand and optical bed, (2) an inverted Le Châtelier-type microscope, (3) a horizontally mounted camera bellows, and (4) an illuminating source of sufficient intensity to render a clear and bright image after the light has passed through the optical systems as typically illustrated in Fig. 51.   Compared to a bench-type metallurgical microscope, a metallograph is usually equipped with more elaborate instrument adjustments and refinements in design, so as to render an image of the highest possible quality commensurate with the performance of the optical parts used.

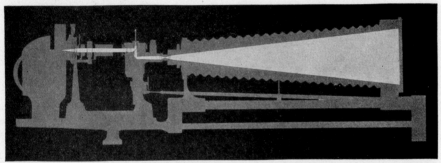

Fig. 51.   Path of light through the Bausch and Lomb metallograph, illustrated in Fig. 50. (*Courtesy of Bausch and Lomb Optical Co.*)

The metallographs most widely used in this country are those manufactured by the Bausch and Lomb, Carl Zeiss, and Ernest Leitz optical companies.   Basically their instruments are the same in principle but differ somewhat in design and in special, proprietary features.

The essential parts comprising any of the metallographs mentioned above are (1) an illuminating source; (2) a condensing lens or system of lenses to collimate the light beam from the source; (3) a cooling cell to abstract heat from the illuminating beam before it enters the microscope proper; (4) an aperture diaphragm to control the amount of light entering the vertical illuminator and objective; (5) a field diaphragm to minimize internal glare within the microscope and to enhance contrast in the image; (6) a vertical illuminator to direct the incident light from the source through the objective and onto the surface of the specimen; (7) an inverted type of microscope; (8) a camera bellows for photomicrography; and (9) a complete set of objectives and eyepieces.   Some of these component parts are illustrated in Fig. 52.

Illuminating Source.   Owing to the tortuous optical path that the illuminating beam is required to follow when passing through a metallograph, as shown in Fig. 51, and the inevitable light losses arising therefrom, it is essential that the illuminating source be of rather high intensity so as to

Fɪɢ. 52.   Illustrating some of the essential parts of a metallograph.   (*Courtesy of Bausch and Lomb Optical Co.*)

A. Vertical and horizontal adjustments for the light source.
B. Condensing lens for focusing the illumination.
C. Adjustable field-of-view diaphragm.
D. Water-cooling cell.
E. Rack to support light filters other than *H*.
F. Light shield.
G. Light shield to prevent impingement of stray illumination on the vertical illuminator housing and objective.
H. Green-yellow light filter.
I. Shaft supporting plane glass and prism reflectors.
J. Adjustments for lateral positioning of the microscope stage.

K. Adjustable-aperture diaphragm.
L. Stage aperture over which specimen is placed for examination.
M. Specimen holder.
N. Eyepiece for visual observation of image.
O. Eyepiece for projection of image onto focusing screen.
P. Clutch onto which rod is attached to permit focusing of the microscope from a position behind the focusing screen.
Q. Fine-focusing adjustment.
R. Coarse-focusing adjustment.
S. Vertical illuminator housing and objective seat.

render a final projected image that is reasonably brilliant and clear.   This is most usually achieved by means of a carbon-arc source which is capable of rendering a beam of illumination of considerable brightness.   Depending upon the current and voltage applied to maintain the arc, a crater brightness of 175 to 200 candles per sq mm can be developed at the positive electrode. This is a rather high order of brightness for an illuminating source, considering that the measured brightness of the sun is about 1600 candles per sq mm.

The mechanical features of most carbon-arc arrangements permit the two carbon electrodes to be moved simultaneously by a single shaft, so as to start the arc and subsequently maintain the arc through appropriate adjustment of the electrodes as they burn away. Usually, however, the hot ends of the electrodes are automatically maintained at a more or less uniform distance apart (usually about $3/16$ in.) by means of a clockwork mechanism or a reversing-type electric motor. This automatic-adjusting feature ensures a more uniform and continuous beam of illumination than could be secured by periodically adjusting the position of the electrodes by manual manipulations. As illustrated in Fig. 50, the carbon electrodes and feeding mechanism are enclosed within an appropriately ventilated lamp housing which, as a unit, can be adjusted by means of centering screws so that the illuminate may be well centered to the condensing system and to the entrance port of the vertical illuminator.

The arc across the electrode ends may be sustained by either direct or alternating current of appropriate voltage, with preference for the former if either source is available. Owing to the inherent behavior of alternating current, the illumination arising therefrom is not as steady nor as uniform in intensity as is possible to secure with direct current. The illumination is usually very flickery and to an extent considered objectionable for either visual examination or photomicrography.

Depending upon whether direct or alternating current is used, the size of the carbon electrodes varies. In the case of alternating current, the two electrodes are of the same diameter, usually 6 mm. With direct current of 5 to 10 amp, the horizontal electrode (positive terminal) is usually 8 mm in diameter and the vertical electrode (negative terminal) is generally 6 mm in diameter. In the Zeiss metallograph, however, the size of the horizontal and vertical carbons is reversed with respect to their positions. In any event, it is essential that the horizontal electrode be connected to the positive side of the direct current source in order to ensure complete utilization of the light from the arc. Whether or not the electrodes are appropriately connected can be readily ascertained by observing through the colored-glass window of the lamp housing which of the carbons is the brighter. The one appearing brighter and exhibiting a definite crater formation is the positive electrode.

In some metallographs the conventional carbon-arc source is replaced by either a low-voltage, high-amperage photomicrographic lamp or a conventional ribbon-filament lamp of high illuminating intensity. One such lamp of recent development, known as the tungsten-arc source,[1] is essentially a combination of a mercury-vapor discharge and an incandescent tungsten electrode. The elements of the lamp are enclosed within an evacuated glass bulb and operate on low voltages supplied by a special power-pack unit. When in operation, the tungsten electrode (anode), consisting of a small cylinder about ⅛ in. in diameter, is heated to a temperature approaching that at which tungsten melts. One of the circularly-shaped ends of the anode is normal to the axis of the instrument, and when appropriately heated this surface serves as the radiating area.

Although the tungsten arc is not capable of producing as bright illumination as is the carbon-arc source, it nevertheless is more intense than that derived from a ribbon-filament lamp. In contrast to the carbon arc, any of the incandescent sources mentioned have the advantage that once they are aligned in the optical system of the instrument they require little further attention. It is only when the filament or the electrode supports droop, owing to continuous use, that the lamp requires repositioning and adjusting.

Perhaps the finest source of illumination for general microscopic work, and particularly so for photomicrography with a table-type microscope as described on page 110, is the concentrated-arc lamp.[2] These lamps are available in four sizes, rated at 2, 10, 25, and 100 watts. They are capable of developing a brightness of from 100 candles per sq mm in the 2-watt size to about 50 candles per sq mm in the 100-watt lamps. The brightness factor is considerably higher than that of the ordinary tungsten-filament lamps (average about 10 candles per sq mm), but somewhat lower than that of the carbon-arc source. The circular luminous area to be described, serving as the radiating source, varies from about 0.003 in. in diameter for the 2-watt lamp to 0.059 in. in the 100-watt size.

The concentrated arc operates on direct current from a special power-pack system that supplies the appropriate high voltage required for starting the arc and the subsequent lower voltage necessary for continuous operation. As illustrated in Fig. 53, the principal elements of the concentrated arc are two permanently fixed electrodes. The electrodes are mounted within a glass bulb which, after thorough evacuation, is filled with some inert gas such as argon. The cathodic element consists of a tantalum, molybdenum, or tungsten tube of appropriate dimensions, filled with a specially prepared zirconium oxide. The anode consists of a circular plate

[1] Manufactured by the General Electric Company. The tungsten-arc lamp is an optional illuminating source supplied with the Bausch and Lomb Research metallograph.

[2] Developed by the Western Union Telegraph Company. A complete portable unit, consisting of a concentrated-arc lamp, condensing system, lamp housing, and power pack, is available from the Fish-Schurman Corporation, New York, N.Y.

of one of these high-melting-point metals, with a centrally cut hole that is slightly larger in diameter than the cathode tube. The two elements, as illustrated, are mounted so that the exposed surface of oxide of the cathode is but a few hundredths of an inch behind, and directly in back of, the center hole of the anode plate.

The high intensity of visible radiation derived from the concentrated arc is associated with the presence of a very thin film of molten zirconium metal at the forward end of the cathode element. To produce this layer of metallic zirconium, the cathode is subjected to a "forming" process during manufacture of the lamp. The process consists of forming an arc between the anode plate and the cathode tube by means of a high-potential direct-current source. The rise in temperature resulting therefrom renders the

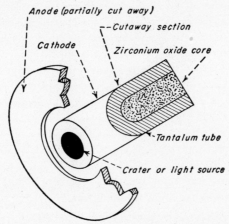

FIG. 53. The principle component elements of the concentrated-arc microscope lamp. *(Courtesy of Society of Motion Picture Engineers.)*

zirconium oxide electrically conductive so that the arc is then maintained between the anode and the oxide filling of the cathode. The temperature of the cathode soon reaches the melting point of zirconium oxide, and when this occurs, some of the oxide is reduced to metallic zirconium in the form of a thin molten film spread over the surface of the cathode. As soon as the zirconium surface film is formed, the temperature of the cathode drops below the freezing point of the oxide—a circumstance associated with the better electron-emitting characteristics of zirconium metal at high temperatures compared to that of its oxide. As a consequence the zirconium oxide within the cathode tube undergoes solidification and serves to support the film of liquid metal on its surface. This, then, completes the forming process, and when the lamp is permitted to cool, the zirconium surface film solidifies and remains on the cathode surface to be again heated and to serve as the radiating source whenever the lamp is relighted.

Light Collimator and Cooling Cell. The visible radiation emitted at the light source is collimated by a condenser lens of appropriate focal length and directed into the entrance opening of the vertical illuminator. The condensing system is mounted just in front of the lamp housing (about 3 to 5 in. from the illuminating source) and provision is made so that the condenser can be conveniently adjusted to focus the illumination. Between the collimating lens and the vertical illuminator a glass water cell is interposed for the purpose of abstracting heat from the illumination before it passes into the microscope proper. The light beam from the source, particularly if the source is a carbon arc, carries sufficient heat energy to cause erratic thermal expansion of the microscope system. Unless the light beam is adequately cooled before it enters the microscope system, the physical distortion arising therefrom may be sufficiently great to render a final image that is also slightly distorted. Furthermore, optical parts which have been joined by balsam or similar cements may suffer permanent damage if heated for long periods of time at even slightly elevated temperatures.

Light Filters. A rack support is usually provided on the emergent side of the cooling cell into which colored light filters of the proper transmission may be mounted. Light filters, as used in metallographic work, are essentially of three types: (1) dyed gelatin sheets cemented between two plates of clear glass (Wratten filters); (2) solid-colored glass filters; and (3) liquid dye solutions, which may replace the water in the cooling cell and thus serve both as a coolant and as a color filter. Solid-glass filters are preferable as they are more durable than Wratten-type filters and more convenient to handle and interchange than liquid filters.

In general the purpose of a light filter in microscopy is to absorb as completely as possible, from the incident white light passing through it, all undesirable wave lengths of radiation and to transmit a narrow band of the visible spectrum that for one or more reasons may be desired. By appropriate choice of filters with regard to their transmission characteristics, their use may serve to enhance contrast in the image and to improve the definition and resolution of fine detail in the specimen. Usually for metallographic applications, light filters are employed principally to render a quality of illumination comparable to that for which objectives have been corrected and, as a consequence, to secure a degree of resolution commensurate to the ability of the optical system.

The use of light filters for the primary purpose of enhancing structural contrast, such as might be desirable in highly colored nonferrous alloys, may result in some sacrifice of definition in the image owing to the wave lengths required for establishing good contrast differing from those for which the objective was corrected. However, in cases where it is desirable to render structural detail as dark as possible in contrast to surrounding elements, a filter should be used which is complementary to the color of the detail, *i.e.*, one that transmits wave lengths within the absorption band of the detail of interest. If it is desired to render maximum contrast or to secure better rendering of detail within a single-colored phase, a supplementary filter should be used which transmits wave lengths that are essentially the same as those reflected from the phase of interest.

Inasmuch as achromatic objectives are corrected for aberrational errors in the green-yellow region of the spectrum, they perform most efficiently when used in conjunction with filters transmitting these wave lengths of radiation. A blue filter, although desirable for use with apochromatic objectives, cannot be used with achromats without causing some distortion in the final image.

Apochromatic objectives, being more highly color-corrected than achromats, will function efficiently with either green-yellow or blue illumination. Blue illumination is desirable when these objectives are used since, as mentioned heretofore, the resolution is greater for any given objective (provided proper optical correction has been made) the shorter the wave length of light used to illuminate the object. In many cases, the resolution obtained with blue illumination may be 25 per cent greater than that obtained with green-yellow light. Since the correction of apochromatic objectives has not been extended to the region of ultraviolet, the use of blue filters that do not absorb entirely the ultraviolet radiation may cause some loss of sharpness in the image. Ultraviolet radiation may be removed in these cases by using, in conjunction with the blue filter, a special solid-glass ultraviolet filter, or by replacing the water in the cooling cell with the following solution:

Distilled water.......................... 1000 cc
Sulphuric acid.......................... 1.0 cc
Quinine sulphate....................... 3.0 g

Aperture Diaphragm. In the optical system of all metallographs there are appropriately placed two adjustable iris diaphragms—the aperture and the field-of-view diaphragms—whose primary purpose is to enhance the quality of the final image.

The physical location of the aperture diaphragm varies with the model and design of the instrument, but usually it is placed in the path of the illuminating beam somewhere between the collimating condenser lens and the vertical illuminator, as illustrated in Fig. 52. Regardless of its physical location, the aperture diaphragm is always positioned in the optical system so that by means of an interposed positive lens system, the diaphragm opening is imaged in the rear focal plane of the objective. Owing to the position of the diaphragm image, it is therefore possible to control the amount of light entering the objective by merely regulating the diameter of the diaphragm opening.   This control over the illumination is the principal function of the aperture diaphragm.

The amount of light from the source that enters the rear lens of the objective, as governed by the opening of the aperture diaphragm, is of considerable importance with regard to final-image quality.   As shown heretofore, the resolution of fine detail in a metallographic specimen is dependent in part upon the amount of light entering the objective.   As the aperture diaphragm is stopped down, the image of the diaphragm edge cuts into the objective aperture and thus reduces the effective numerical aperture of the system.   Under such circumstances the resolution is seriously impaired and the general quality of the image is rather poor owing to the presence of pronounced interference fringes.   The loss in resolution associated with a nearly closed aperture diaphragm is well illustrated in Fig. 54.

When the opening of the aperture diaphragm is increased, the resolving power of the system is likewise increased within limits.   Theoretically at least, maximum resolution for any objective will be secured when the diaphragm opening is so adjusted that the rear lens of the objective is completely flooded with light.   Actually, however, at full aperture the marginal aberrational errors of the objective become so pronounced, and the internal glare and reflections within the microscope system are so serious, that the improvement in image quality and resolution falls short of that which should theoretically result.

It is evident, therefore, that for any objective there will be some optimum opening of the aperture diaphragm that will yield the best image.   The appropriate diaphragm setting is best determined by visual examination of the image, and so adjusting the diaphragm that the best possible resolution is obtained with a minimum of image glare.   When by this procedure the diaphragm is correctly adjusted but the image is too bright to be comfortably viewed, the light intensity may be reduced by inserting ground-glass screens in the path of the light beam.   Under no circumstances should the aperture diaphragm be adjusted merely to increase or decrease the intensity of image brightness.   Although the practice of using ground-glass screens will destroy "critical" illumination, as will be noted later, it is better to sacrifice quality of illumination than to impair resolution.

Field-of-view Diaphragm.   The principal function of the illuminated field, or field-of-view, diaphragm is to minimize the internal glare and multiple light reflections within the microscope system which, when achieved, will markedly improve contrast in the image.

The field diaphragm, like that of the aperture diaphragm, is physically located at some appropriate place between the vertical illuminator and the collimating lens.   As shown in Fig. 52, it is located in some metallographs

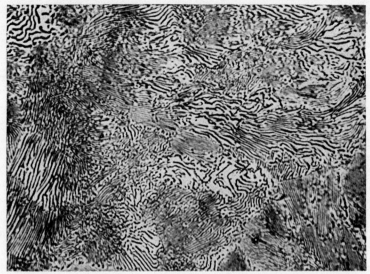

a. Diameter of diaphragm opening, 3 mm. Note poor resolution. 1000✕.

b. Diameter of diaphragm opening, 9 mm. Note excellent resolution. 1000✕.

FIG. 54. Illustrating the influence of the aperture diaphragm opening on resolution of the objective.

just in front of the adjustable light condenser. With regard to the optical system of the instrument, the diaphragm is so positioned that an image of the diaphragm opening is projected into the plane of the specimen surface, thus rendering the edge of the diaphragm visible in the image field when the microscope is appropriately focused. Because of the location of the diaphragm image plane, as contrasted to that of the aperture diaphragm, adjustment in the opening of the field diaphragm will have no real effect on resolution. In some cases, however, when image glare is reduced to a minimum by appropriate adjustment of the field diaphragm, the resolution may appear to be improved. This improvement is only apparent, and not to be considered real.

As illustrated in Fig. 55, when the field diaphragm is stopped down, the size of the image field is reduced but at a gain in image contrast, owing essentially to partial elimination of internal light flare within the microscope system. This illustration further shows that neither resolution nor image brightness is affected by the aperture of the field diaphragm, since the three photomicrographs were secured under identical conditions, except for the relative setting of the field diaphragm.

To contribute to good image quality in visual examination, the field diaphragm should be stopped down to such an extent that the image of the diaphragm edge just begins to enter the image field as defined by the limiting diaphragm of the eyepiece. In photomicrography, the diaphragm should be adjusted so that the size of the visible image on the focusing screen of the instrument corresponds approximately to, or is slightly smaller than, the larger dimension of the photographic plate or film.

Vertical Illuminator. Bright-field Illumination. The conventional form of illumination by which metallographic specimens are illuminated for microscopic examination is known as bright-field illumination—a condition of lighting that renders a dark image on a bright, well-lit background field. Bright-field illumination is obtained by means of a vertical illuminator and appropriate light reflectors contained therein that direct the incident light beam from the source into the microscope objective. The light emerging from the objective is then utilized in illuminating a small area of the specimen surface. Deviation of the incident light beam is achieved by either a total reflecting prism or more commonly by a plane-glass reflector inclined 45 deg to the axis of the incident illuminate. The vertical illuminator and reflectors are mounted directly in back of the microscope objective, as illustrated in Figs. 47 and 52. Usually the prism and plane-glass reflectors are mounted on the same support shaft within the vertical illuminator housing to provide a convenient means of changing from one type of reflector to another.

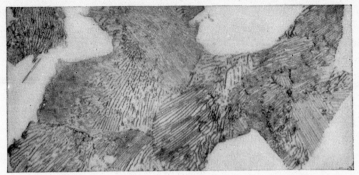

a. Diaphragm wide open.   Note poor contrast.   1000×.

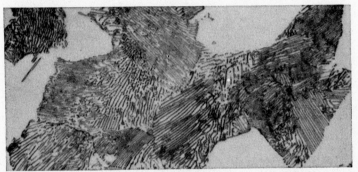

b. Diaphragm partially closed.   Note improved contrast in image over *a*.   1000×.

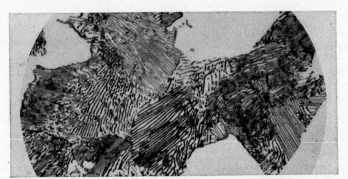

c. Diaphragm closed to position shown.   Note excellent contrast.   1000×.

FIG. 55.   Illustrating the influence of the field-of-view diaphragm opening on contrast in the image.   (*Courtesy of U.S. Steel Corp. Research Laboratory.*)

In bright-field illumination, regardless of the type of reflector used, the microscope objective first serves as a condensing system to the incident light beam, and then as an image-forming system to the light reflected into it from the surface of the specimen.   The fact that the microscope objective first functions as a light-condensing system is one of the features distinguishing bright-field from dark-field illumination to be described.

Prism Reflector.   The prism reflector, when appropriately positioned in the microscope system, as shown in Fig. 56, serves to reflect all of the light incident upon it into the microscope objective and onto the surface of the specimen.   To enable the image-forming rays to reach the eyepiece, the

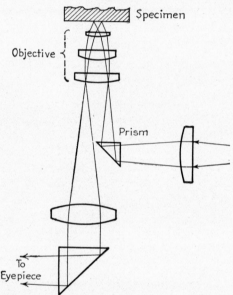

Fig. 56.   Illustrating the functioning principle of the total reflecting prism in the vertical illuminator.

prism reflector is so positioned that only about one-half of the area of the objective aperture is obstructed.

As was previously demonstrated, the resolving power of an objective is decreased by any factors or set of circumstances that effectively reduce its numerical aperture.   As illustrated in Fig. 56, the position of the prism reflector reduces the effective numerical aperture of the objective by a factor of about one-half in one direction, and as a consequence, the resolving power of the objective is proportionally decreased.   At relatively low magnifications, such as 100× or less, this reduction in resolution is not too significant; but at high magnifications the loss in resolving power through the use of the prism reflector is indeed very great.   Figure 57 conclusively

a. Specimen illuminated by means of a plane-glass reflector.   Note excellent resolution.
1000×.

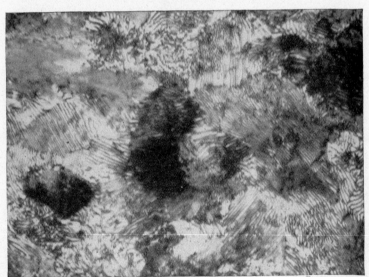

b. Specimen illuminated by means of a total reflecting prism.   Note poor resolution.
1000×.

Fig. 57.   Illustrating the influence of the method of illuminating the specimen on resolution of fine detail.

shows this to be the case, and serves to illustrate that at relatively high magnifications it is impossible to secure full resolving power and good image quality by use of the prism reflector.

The principal use of the prism reflector, and only at low magnifications, is to render a more brilliant image than can be secured through the use of

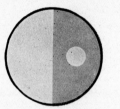 

a. Correct.   *b.* Incorrect.

FIG. 58.   Illustrating the appropriate positioning of the total reflecting prism relative to the incident beam of illumination, as determined by sighting through the eyepiece tube.

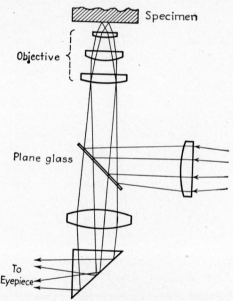

FIG. 59.   Illustrating the functioning principle of the plane-glass reflector in the vertical illuminator.

the plane-glass reflector; to enhance contrast in the image; and to obtain, through appropriate setting of the prism, slight obliquity of the illumination incident on the specimen surface.

To secure an evenly illuminated image field, which is a primary requisite in microscopy, it is essential that the prism reflector be properly oriented and aligned with respect to the incident illumination. This is best achieved by merely observing the image and rotating the prism slightly until the entire field is uniformly brilliant. On most metallographs the shaft supporting the prism reflector can be rotated through a small angle. An alternate orienting procedure consists of focusing the microscope on a highly polished surface, such as an unetched metallographic specimen, stopping down the aperture diaphragm; and then adjusting the prism until the spot of light defined by the aperture diaphragm (as seen when looking into the eyepiece tube through a pinhole cap) is positioned as illustrated in Fig. 58a.

Plane-glass Reflector. The plane-glass reflector, as illustrated in Fig. 59, is inclined 45 deg to the principal axis of the microscope objective (or axis of the illuminate) when appropriately positioned in the vertical illuminator. On some metallographs proper orientation of the plane glass is assured by a mechanical lock arrangement on the supporting shaft, on others the plane-glass reflector can be rotated through a small angle, hence requiring care in adjusting to secure an evenly illuminated image field. As in the procedure

a. Correct. *b.* Incorrect.

Fig. 60. Illustrating the appropriate positioning of the plane-glass reflector relative to the incident beam of illumination, as determined by sighting through the eyepiece tube.

followed for orienting the prism reflector, the plane glass may be adjusted while observing the image field, or by rotating the plane glass through a small angle until the spot of light as seen through a pinhole cap appears as in Fig. 60a.

The plane-glass reflector is not capable of producing an image field so brilliant or with so much contrast as is the prism reflector. This lower brilliancy is associated with the greater light losses encountered each time the light rays are reflected from, or passed through, the plane-glass reflector. Although the total light losses arising therefrom may be as high as 80 per cent for uncoated reflectors, the image field will generally be sufficiently bright for comfortable observation and the use of reasonably short photographic-exposure times in photomicrography. Because the plane-glass reflector in no way reduces the effective numerical aperture of the objective, the resolving power of the system is therefore not adversely affected. As illustrated in Fig. 57a, best image quality is obtained at moderately high and high magnifications only when the plane-glass reflector is used.

The use of the plane-glass reflector in conjunction with conventional type of microscope objectives, *i.e.*, objectives corrected for some definite tube length, may produce some astigmatism and other forms of distortion in the final image. This is particularly true if the plane glass is mechanically strained in any way or if the polished surfaces of the reflector are not optically plane and parallel. In principle, image distortion under these circumstances is associated with unequal refraction of the converging image-forming rays as they pass through the plane glass to the eyepiece of the system.

To correct this source of distortion, some metallographs are now equipped with a thicker plane glass than used formerly, to ensure greater rigidity of the reflector and to afford greater accuracy during manufacture in producing plane-parallel surfaces. In conjunction with the use of this reflector of greater thickness, special microscope objectives are employed that are corrected for an infinitely long tube length, *i.e.*, the image-forming rays leaving the objective are parallel. Hence, these parallel rays in passing through the plane-glass reflector suffer only parallel displacement and are converged to a point of focus only after they pass through a positive lens system located on the emergent side of the reflector.

Calcite Prism. The Bausch and Lomb Research metallograph has incorporated in its design an ingenious vertical illuminator consisting of a special calcite prism, which may be used either as a bright-field illuminator (equivalent to a plane-glass reflector) or as a polarizing illuminator. In contrast to the ordinary plane-glass reflector, the calcite prism has a number of distinct advantages.

When used as either a bright-field or a polarizing illuminator, the light losses arising from internal reflections are considerably less, making the calcite prism approximately 4.5 times as efficient as the plane-glass reflector. As related heretofore, a certain amount of astigmatism is introduced in the image (unless infinity-corrected objectives are used) as the converging rays from the objective pass through the plane glass. This objectionable circumstance does not exist when the calcite prism is used, owing to the functioning principles of the calcite illuminator.

The examination of a metallographic specimen in plane-polarized light, using a plane-glass reflector, requires the use of a polarizing prism, which is introduced in the incident light beam usually between the light source and the reflector, and the use of an analyzer, which is placed either on top of the eyepiece or between the eyepiece and the objective. The latter position, from a mechanical point of view, is usually difficult to achieve. Perhaps the most serious objection to the use of a plane-glass reflector with plane-polarized light, in contrast to the calcite prism, is the unavoidable introduction into the system of elliptically polarized light by the reflector. This causes only a small portion of the observed field to become perfectly extinguished when the polarizer and the analyzer are 90 deg out of phase.

The calcite-prism reflector, shown schematically in Fig. 61, consists of two pieces of calcite, *A* and *B*, cut from the same crystal. The two parts are cemented together along the line *ab*, and the orientation of the prism is such that the optical axis is parallel to the end of the prism and in the plane of the cemented surface.

The incident light from the illuminating source enters the prism from below, where it is divided into two characteristic rays, one the ordinary and the other the extraordinary ray. The ordinary ray is reflected at the cemented surface and directed toward the blackened surface of the prism, where it is absorbed. The extraordinary ray, however, passes through the prism and into the objective as plane-polarized light, the plane of polarization being in the plane of the drawing. The plane-polarized light, falling

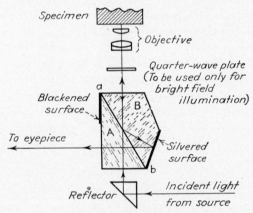

Fig. 61. Illustrating the functioning principle of the calcite prism for bright-field and polarizing illumination.

on the etched surface of the specimen, will be partly or fully depolarized, depending upon the condition of the surface. Any specular areas on the surface of the specimen will reflect the plane-polarized light back through the objective and the prism to the light source. None of the light in passing through the prism will be reflected at the cemented surface, and hence none will be directed to the eyepiece. Such areas of the specimen will therefore appear dark.

The plane-polarized light falling on areas of the specimen surface that are less specular will upon reflection be depolarized, *i.e.*, the plane of vibration of the reflected light will be rotated from that of the incident beam. This depolarized light, in passing back through the prism, will be partly reflected at the cemented surface and directed onto the silvered surface of the prism, where it will again be reflected into the eyepiece. Precisely what has been brought about by the calcite prism when examining a metallographic specimen is equivalent to that of crossed Nicol prisms when examining transparent materials or thin sections with transmitted light.

Bright-field illumination may be secured from the calcite prism by inserting between the prism and the microscope objective a quarter-wave plate with its slow-ray direction making an angle of 45 deg with the vibration plane of the prism. This plate may be inserted by rotating a disk mounted on the metallograph directly below the objective seat.

The plane-polarized light, emerging from the prism, is passed through the quarter-wave plate and becomes circularly polarized. The circularly polarized light, after reflection from the specimen surface, is again passed through the quarter-wave plate and becomes once again plane-polarized, but the plane of vibration is in a plane perpendicular to the plane of the drawing. Thus a 90-deg change of phase has occurred, which is equivalent to a change brought about by the introduction of a half-wave plate between a polarizer and an analyzer.

The polarized light reflected from the specimen surface through the quarter-wave plate (now 90 deg out of phase with the incident light) is passed into the prism and is totally reflected at the cemented surface and again at the silvered surface to the eyepiece.

Vertical Illuminator. Dark-field Illumination. It has been shown that in bright-field illumination the microscope objective serves a double purpose—first as a condensing system to the illumination and, second, as an image-forming system to those rays of light passing into it from the surface of the specimen.

When visible radiation is passed through an objective, the amount of light reflection and scattering that occurs at various glass-air interfaces may be sufficiently great to reduce image brightness and, to some extent, image quality. These adverse effects are most pronounced when the objective first serves as a condensing system and when the specimen surface, or areas thereof, are roughened and reflect little, if any, of the incident light. In part, these objectionable effects on the final image are reduced in dark-field illumination, principally because in this form of illumination the objective functions only as an image-forming system.

The principles of dark-field illumination are illustrated in Fig. 62. By means of a circular light stop, appropriately interposed in the path of the incident light beam, the illumination enters the vertical illuminator in the form of a hollow cylinder or cone of light. The incident light reflected from the plane-glass reflector, as illustrated, is then passed upward along the outside of the microscope objective and is reflected onto the specimen surface by a concave metal reflector whose inside surface is of a contour appropriate for the purpose. Because of the light stop blocking out the central portion of the illuminating beam, no light whatsoever passes into the objective. This form of illumination renders light incident on the specimen surface of greater angularity and of higher numerical aperture than that secured through bright-field illumination.

Under conditions of dark-field illumination as described, a first-surface ideal mirror serving as the specimen will appear dark in the image field, since none of the reflected light will enter the objective owing to the high obliquity of the reflected radiation. A metallographic specimen, however, having surface properties unlike those of an ideal mirror, will scatter some of the incident light, the amount depending upon the precise condition of the specimen surface. That light reflected from the highly specular areas

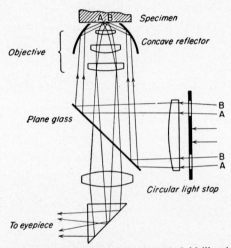

Fɪɢ. 62. Illustrating the principle of dark-field illumination.

of the surface will be lost outside of the microscope objective and, as a consequence, these areas will appear dark in the image. The less specular regions, however, will appear bright in the image field, owing to a large percentage of the light reflected from these areas passing into the objective. In conformance with these principles, the image rendered by dark-field illumination will be complementary, with respect to dark and light areas, to that obtained by bright-field illumination, as illustrated in Fig. 63.

Dark-field illumination is distinct in that the contrast established in the image is of excellent quality; the rendition of color in the image is quite natural and appropriately balanced; and the resolution is improved over that secured by bright-field illumination, other factors remaining constant, due to the relatively high numerical aperture of the incident illumination. Notwithstanding these desirable features, dark-field illumination is rarely

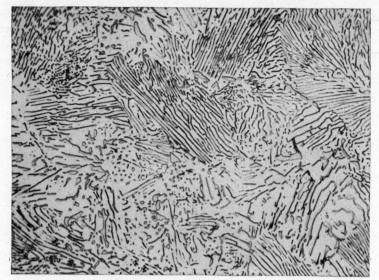

a. Bright-field illumination. 1000×.

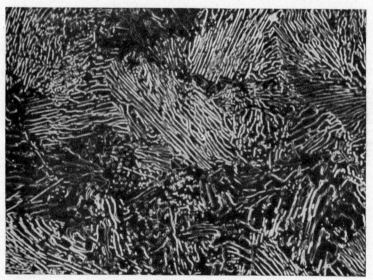

b. Dark-field illumination. 1000×.

FIG. 63.   The appearance of pearlite under bright- and dark-field illumination.

used in metallographic work.   This circumstance is probably related to the general lack of dark-field equipment on most instruments and to the fact that metallographers are more accustomed to viewing bright-field images.

Oblique Illumination.   Oblique illumination may be used to good advantage in cases where there is little contrast produced between the structural components of the specimen by axial illumination and where it is desirable to accentuate certain of the constituents of the structure which are depressed or standing in relief.   Oblique lighting, primarily because of cast shadows, imparts an appearance to the structure decidedly more natural than that produced by plane axial illumination.   This is well illustrated in Fig. 64.   It should be noted, however, that surface defects, such as pits, scratches, etc., are emphasized by oblique illumination and in many cases this circumstance may distract from the over-all appearance of the structure.

Oblique illumination may be readily obtained by stopping down the aperture diaphragm by an appropriate amount and then decentering it with respect to the normal optical axis.   The obliquity of the light rays may be easily controlled by this method, and any desired angle of inclination, within limits, may be obtained.

At low magnifications, when only slightly obliqued illumination is desired, the aperture diaphragm may be left undisturbed and obliquity of the illumination brought about by turning the prism reflector through a very small angle with respect to its normal position in the vertical illuminator.

Conical-stop Illumination.   It is possible with some metallographs to obtain increased contrast and pleasing relief effects in the image by a form of oblique lighting known as conical-stop illumination.   This type of illumination is distinct in that the surface of the specimen is illuminated by a hollow cone of light rather than by a solid cone as in the case of plane axial or vertical illumination.

To obtain conical-stop illumination, a small opaque disk or stop, varying in diameter from about 1 to 5 mm, is placed in the path of the light beam, the plane of the stop being normal to the axis of the illuminant.   With respect to the optical system of the metallograph the stop is so placed, usually nearly coincident with the aperture diaphragm, that it is imaged approximately in the rear focal plane of the objective.   On all metallographs incorporating the feature of conical-stop illumination a holder is provided so that the stop may be appropriately placed.

The light beam from the illuminating source, as shown in Fig. 65, does not illuminate the central portion of the rear lens of the objective because of the circular stop, and hence through this locally stopped area the in-

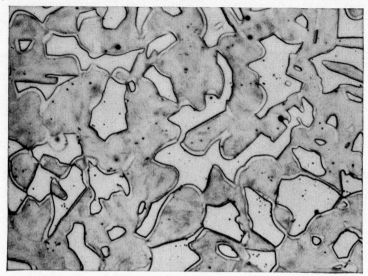

a. Plain axial illumination. 1000×.

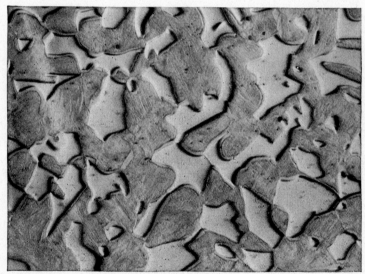

b. Oblique illumination secured by decentering the aperture diaphragm. Note striking relief effects. 1000×.

FIG. 64.   The influence of different methods of illuminating the specimen on the appearance of the image.   (*Courtesy of U.S. Steel Corp. Research Laboratory.*)

cident light is unable to pass into the objective and aid in illuminating the specimen. However, the light passing into the objective around the image of the stop may do so unhindered, and as the light rays pass through the objective toward the surface of the specimen they are refracted by the various lens elements so that they come closer and closer to the principal axis of the objective. Thus, in effect, the light emerging from the objective and striking the surface of the specimen is uniformly obliqued and is in the form of a hollow cone.

The image-forming rays reflected from the surface of the specimen are passed through the entire objective aperture (the stop does not interfere with the passage of the image-forming rays as it is only imaged in the rear of the objective), thereby utilizing the full numerical aperture of the

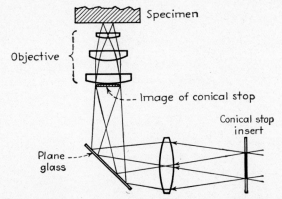

Fig. 65. Illustrating the principle of conical-stop illumination.

objective to form the image. Thus, the illumination is obliqued uniformly with no deleterious effect on the resolution.

If the conical stop is placed eccentrically in the path of the light beam, and this position is sometimes preferred to one more symmetrical, the specimen will then be illuminated more obliquely from one side than from the other. Usually this eccentricity may be effected by decentering the stop and the aperture diaphragm.

"Critical" and Köhler Illumination. Inasmuch as the manner of illuminating metallographic specimens influences to some measure the quality of the final image, the illuminating systems in table metallurgical microscopes and metallographs are precisely and carefully designed to render either so-called "critical" or Köhler illumination. These forms of illumination are distinct in that a comparatively aberrational-free image of the primary or some secondary light source is projected into the plane of the specimen surface, and as a consequence, an image field is rendered that is fully illuminated and uniformly brilliant. Unless this is achieved, the rays of light entering the objective from the specimen surface will be insufficient to render an image of the quality that the microscope system is capable of forming, and the image quality will then fall short of perfection, within limits, by an amount roughly proportional to the deficiency involved.

Critical illumination is theoretically obtained when the illuminating source and the optical system of the microscope are so arranged that the primary light source, *e.g.*, the luminous crater of the positive electrode in a carbon-arc system, is imaged in the plane of the specimen surface, and the rear lens of the objective is uniformly flooded with light. In practice, such conditions are never fully achieved, and purposely so, because of objectionable circumstances associated with defects in the illuminating source, such as gas bubbles in the carbon arc, being imaged simultaneously with the object field of the specimen. The departure, however, from strictly

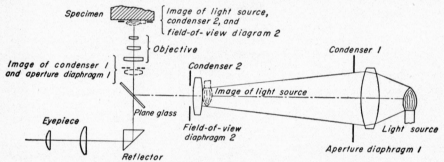

FIG. 66. Illustrating schematically the principle of critical illumination.

critical illumination to avoid this is of small order and has little effect on image quality.

By means of an adjustable aspheric condenser lens 1, as illustrated in Fig. 66, an image of the luminous area of the light source is formed at the front side of condenser lens 2. The illuminated surface of condenser 1 and the aperture diaphragm 1 located just in front of it are imaged by condenser 2 into the rear focal plane of the objective. The rear lens of the objective is thus evenly illuminated, and owing to the position of the image of the aperture diaphragm, the amount of light entering the objective can be readily controlled.

Because of the focal length of condenser 2—which is appropriate for the purpose intended—and owing to its location in the optical system, the illuminating source, the evenly illuminated surface of condenser 2, and field-of-view diaphragm 2 are brought into nearly coincident focus on the surface of the specimen when the objective system is correctly focused. To avoid imaging of source defects in the plane of the specimen surface, condenser 1 is adjusted so that the image of the source is formed just ahead of condenser 2.

Köhler illumination will achieve the same desirable lighting qualities as critical illumination, but through principles of imaging in the plane of the specimen surface a uniformly illuminated secondary source instead of the primary source. With reference to Fig. 66, Köhler illumination may be secured by an arrangement of the system which will image the evenly illuminated surface of condenser 1 in the plane of the specimen surface, and the primary light source in the rear focal plane of the objective. Owing to the location of the condenser-lens image, its surface must be free of defects such as fingerprints, scratches, etc., which otherwise would be visible in the final image of the specimen.

Methods of Focusing. Critical focus of the projected image prior to photographing is of primary importance. Generally at low magnifications focusing may be accomplished by mere visual examination of the image on the ground-glass focusing screen. A focusing rod is provided so that the image may be focused from a position in back of the ground glass where the operator would normally stand. Because of field curvature accompanying the use of high-numerical-aperture objectives, the center portion of the projected image should always be brought into sharp focus as this is the region of the image field that will normally be registered on the photographic plate.

Images of fine structures may at times be difficult to focus perfectly on the ground glass because of the inherent haziness associated with this type of screen. Furthermore, difficulty may be experienced in this respect when the illumination is not very intense. Focusing under these conditions may be readily carried out with the aid of a special focusing magnifier. To use such a magnifier, the ground-glass screen is removed from the metallograph, and the magnifier is held in position at the back of the bellows either by hand or supported by a clear plate of glass, which replaces the ground glass. A critical focus may then be made while viewing the image through the magnifier.

Method of Determining Correct Plate Exposure. An important factor that contributes to securing good photomicrographs is the length of time to which the photographic plate or film is exposed to the projected image. This exposure time varies inversely with the brightness of the image as observed on the focusing screen and with the speed of the photographic emulsion on the plate or film. For any given objective and ocular combination, a great many things will affect the brightness of the image—among them, the aperture-diaphragm opening, intensity of the light source, cleanliness of the optical parts, and projection distance. Unless the technician has had considerable experience in estimating exposure times by examining the brightness of the image field, a method must be employed whereby this time may be determined accurately.

The correct exposure time for any given arrangement of the metallograph is best determined by examination of a trial exposure plate. Such a plate is relatively simple to obtain, provided the opaque slide cover of the plate-holder is ruled with lines spaced approximately ¾ in. apart. The purpose of the ruled lines is to facilitate exposing strips of the photographic plate for different lengths of time. The entire plate is first exposed for a pre-determined length of time, for example, 2 sec, after which the slide cover is

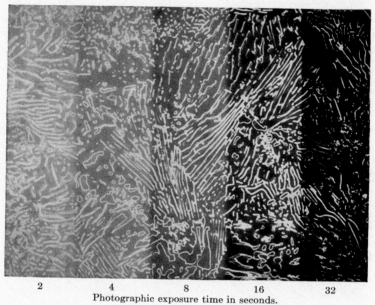

2   4   8   16   32

Photographic exposure time in seconds.

Fig. 67.   The appearance of a trial-exposure negative.

moved inward to shield the first strip of plate as indicated by the first ruled division on the cover. The remaining uncovered portion of the photo-graphic plate is then exposed for another 2 sec. After this exposure, the slide cover is moved to the second ruled division (the plateholder cover now shields two differently exposed strips of the plate) and the uncovered por-tion of the plate is exposed for an interval of 4 sec. This procedure is repeated until the last remaining strip of the plate has been fully exposed. Each time the slide cover is moved in to the next ruled division, after the first, the uncovered portion of the plate is exposed for a time double that of the previous exposure.

This method of obtaining exposure data produces a wide range of ex-posure times on the same photographic plate, and, as shown in Fig. 67, the exposure times as related to the example cited are in the order of 2, 4, 8, 16, and 32 sec. After the plate has been developed and processed, it may be examined visually and the exposure time noted that renders an image of appropriate photographic density. Substantially the same exposure time should be used for securing a full-size negative under the same conditions.

Effects of Vibration. To eliminate as far as possible the transmission of shock or other forms of vibration from the surroundings to the micro-scope and camera, all metallographs are equipped with shock-absorbing devices. Shock absorbers are essentially mechanical-spring or rubber devices which support the entire metallographic system as a single unit. This arrangement prevents any one part of the instrument from moving relative to another when subjected to external vibrations, and hence will prevent the photographic image from becoming blurred during photo-graphic exposure.

Experience has shown that the deleterious effects of vibrations during photographic exposure is roughly proportional to the magnification at which the photomicrograph is secured. For example, vibrations of rather small magnitude have little if any effect on image quality at a magnifica-tion of 50×, whereas at 1000 or 1500× the effects are usually ruinous.

The beneficial effects arising from the use of shock absorbers is illustrated in Fig. 68. In both instances (*a* and *b*) vibrations were set up in the stand of the metallograph by a person walking heavily across the floor at about 5 ft from the instrument.

Suggestions for Successful Photomicrography. To aid in securing photomicrographs of truly high quality, the following suggestions are made, with particular reference to manipulations attending the proper use of a metallograph. The processes of photographic development and negative printing, which in part determine the quality of a photomicrograph, will be discussed in the chapter to follow.

1. One of the most important factors contributing to successful photo-micrography is the final prepared condition of the metallographic specimen of interest. By the very nature of the photographic process it is impossible for a photomicrograph to be of higher quality than that which is character-istic of the specimen surface and the attending degree of perfection attained in preparation.

2. The selection of optical parts will depend mainly upon the total magnification desired (see item 9). To assist in proper selection, most instruction manuals pertaining to the instrument will list appropriate com-binations of objectives, eyepieces, and bellow extensions that will give standard total magnifications. Lacking such data, however, objectives should be selected on the basis of the best compromise between resolving power, flatness of field, and vertical resolution. Because resolving power

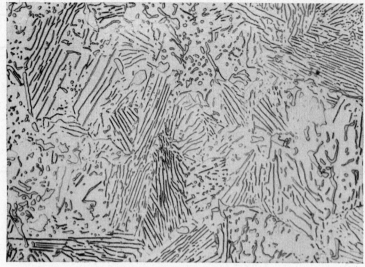

a. Shock absorbers in action.   1000×.

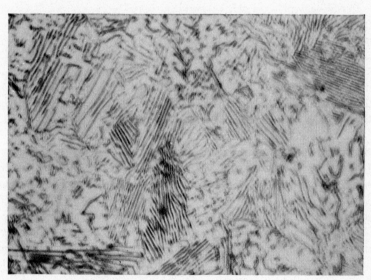

b. Shock absorbers not in action.   1000×.

Fig. 68.   The influence of vibrations imposed upon the metallograph during photographic
exposure on quality of the image.

is usually most significant, the total magnification should not be greater than about 1000 times the numerical aperture of the objective selected (see page 95).

3. Unless specifically recommended, the selection of an eyepiece appropriate for projection should be made on the basis of the size of image field, the extent to which the ocular is corrected, and the magnifying power desired. The initial magnifying power of the ocular should be of such a value that the desired total magnification can be secured at relatively long bellow extensions rather than short. This will aid in rendering a comparatively flatter image field. Whenever possible, an amplifier should be used provided it is appropriately matched to the objective used. A compensating eyepiece should always be used in conjunction with high-powered semiapochromatic and apochromatic objectives.

4. The source of illumination should be carefully adjusted so that the illuminating beam is well centered to the entrance port of the vertical illuminator. If a carbon arc constitutes the source, it is essential that the two electrodes are properly connected, with regard to polarity, to the source of direct current (see page 113).

5. The type of light filter used should be checked with regard to its transmission characteristics. The principal function of the filter in most cases is to render illumination of a quality comparable to that for which the microscope objective is corrected (see page 116).

6. The reflector contained within the vertical illuminator housing should be aligned as described on page 125. Under no circumstances should the prism reflector be used at magnifications over about 100 diameters, unless a gain in image contrast and brilliancy is warranted at the sacrifice of resolution.

7. With the specimen properly positioned on the microscope stage, focus the microscope, and locate in the center of the field that portion of the specimen which is to be photographed.

8. While viewing the image through the eyepiece, adjust the aperture diaphragm to render the best possible appearing image. The diaphragm should be centrally located, unless slightly obliqued lighting of the specimen is desired.

As discussed on page 118, the aperture diaphragm is never to be stopped down merely to decrease the brightness of the image field. If the image is too brilliant to be comfortably observed, ground-glass screens may be interposed in the illuminating beam and removed, if necessary, when the image is finally projected onto the focusing screen.

9. Set the camera bellows at a projection distance corresponding to that required to render the desired total magnification. Standard total magni-

fications, as recommended by the ASTM, are 25, 50, 75, 100, 150, 200, 250, 500, 750, 1000, 1500, and 2000 diameters. The projection distance may be determined by the use of equation (1), page 90; by reference to the instrument maker's instructional literature; or by the use of a stage micrometer, as described on pages 90 and 91.

10. The image should now be projected onto the focusing screen. This is accomplished in some instruments by withdrawing the visual-examining eyepiece tube from the path of the light rays or in others by rotating out of position a total reflecting prism, and by opening the camera shutter.

11. After refocusing the image on the focusing screen, the field-of-view diaphragm should be adjusted so that the visible image field is about equal to or is slightly smaller in diameter than the greater dimension of the photographic plate or film.

12. Final selection and orientation of the image field should now be made.

13. Carefully put shock absorbers in action if the instrument is not permanently suspended.

14. Critically focus the projected image with strict attention paid to the center portion of the image field. When necessary a focusing magnifier should be used to assure critical focus.

15. The camera shutter should now be closed; the focusing screen removed; and a loaded photographic plateholder appropriately inserted. All of these operations must be carried out with care, so that the adjustments already made will not be disturbed from their critical settings.

16. If the photographic exposure time for the conditions involved are not known with certainty, a trial exposure plate should be secured (see page 135).

17. When the proper exposure time has been determined, a full-size photographic reproduction of the image should be obtained. Adjustments in the metallographic system, other than refocusing the projected image or selecting another image field, should never be made after the trial exposure plate has been obtained. To do so would invalidate the determined time of exposure.

SPECIAL METALLOGRAPHIC MICROSCOPY

Photomicrography with Ultraviolet Radiation. It was shown heretofore that the resolution of fine detail in a metallographic specimen could be effectively increased by using microscope objectives of relatively higher numerical apertures or by decreasing the wave length of visible radiation employed to illuminate the specimen of interest. With high-quality apochromatic objectives, the shortest wave lengths of light that can be feasibly employed are those of about 4000 to 5000 Å. Owing to the difficulties encountered in the design of an objective with a numerical aperture greater than about 1.60, the successful approach to greater resolution is therefore through the use of shorter wave lengths of radiation, even though such radiation extends beyond the lower end of the visible spectrum.

Ultraviolet radiation of wave lengths less than 4000 Å has been successfully employed for improving the resolving power of the microscope system. Because the conventional kinds of optical glasses from which optical parts are constructed absorb ultraviolet down to wave lengths of about 3000 Å, microscopy with ultraviolet radiation necessitates the use of special optical parts. The lens elements of such special objectives and eyepieces are constructed from quartz, fluorite, calcite, or combinations of these minerals, so that practically all wave lengths of ultraviolet down to about 2000 Å can be transmitted.

A source of radiation suitable for ultraviolet microscopy is one consisting of a quartz mercury arc that emits wave lengths of 3650 Å. With radiation of this dimension, used in conjunction with optical parts appropriately corrected, the resolution will be increased about 33 per cent over that secured with an objective of the same numerical aperture and green illumination of about 5400 Å.

The increase in sharpness of the image and the greater resolution of detail that should theoretically result through the use of ultraviolet may at times be offset by the highly developed technique that is required for its successful application. Metallographic specimens must be prepared to a high degree of perfection, as ultraviolet radiation will reveal surface defects, scratches, etc., that normally would not be apparent with visible radiation.

Focusing of the projected image in ultraviolet photomicrography is an extremely difficult procedure. In a limited way, this difficulty is met by using special objectives that are corrected for visible mercury green as well as for ultraviolet of 3650 Å. Under these circumstances, the image may be focused with visible green light and the photographic exposure made with ultraviolet by interposing in the radiation beam (after focusing) an appropriate filter transmitting radiation of only 3650 Å of the mercury spectrum.

Lucas[1] has employed fluorescent screens and a carbon method to focus the image directly with ultraviolet radiation. Carbon is applied in the form of a thin line on the surface of the specimen by means of typewriting carbon paper and a stylus. Since carbon absorbs complete ultraviolet radiation, the carbon line is visible on the focusing screen and may be focused.

[1] Lucas, F. F.: An Introduction to Ultraviolet Metallography, *Trans. AIME*, Vol. 73, 1926.

a. Specimen illuminated with green-yellow light.   1500×.

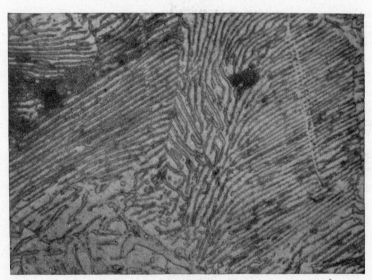

b. Specimen illuminated with ultraviolet radiation of wave length 3650 Å.   1500×.

FIG. 69.   The appearance of pearlite as photographed under visible and ultraviolet radiation.
(Courtesy of Bausch and Lomb Optical Co.)

As illustrated in Fig. 69, the resolution secured with ultraviolet radiation is superior to that obtained with visible illumination, but the image lacks in good contrast and high brilliancy. These deficiencies may be attributed to the fact that ultraviolet is strongly absorbed by metals and that the

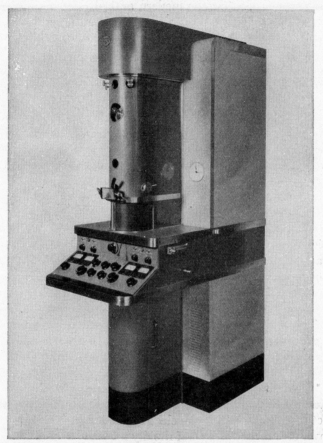

Fig. 70. The RCA electronic microscope. Universal model EMU. (*Courtesy of Radio Corporation of America.*)

image may be slightly out of focus, owing to the extreme difficulties associated with direct focusing.

The Electronic Microscope. Without doubt the greatest contribution made during recent years to the general field of microscopy was the conception, development, and refinement of the electronic microscope system, and the technical perfection of a usable, compact, and versatile commercial instrument. The electronic microscope represents the latest achievement in the relentless search for systems capable of greater resolving power than that obtained from either conventional-light or ultraviolet microscopes. This improvement in resolution has been secured through the use of extremely short wave lengths of radiation—an approach that was the only feasible one considering that the modern optical light microscope has already reached the theoretical limit of resolution as defined mathematically by Abbe, and that fundamental considerations of lens design prevent the successful application of very short wave lengths of ultraviolet radiation.

The modern electronic microscope, one model of which is shown in Fig. 70, is capable of direct total magnifications of 1000 to 30,000 diameters, and by the use of auxiliary optical equipment, the magnification may be extended to about 100,000 diameters. The resolution attainable with the electron microscope is of the order of 0.002 micron, corresponding to a useful total magnification of about 10,000 diameters, as compared to only about 0.2 to 0.3 micron secured with the most modern light microscope and slightly less than 0.1 micron with the ultraviolet instrument. Further improvement in resolution is not practical in the present electronic instrument, owing to the inherently ineffective numerical aperture of the microscope system. Actually, the effective numerical aperture of a high-quality, oil-immersion glass objective is something of the order of 100 to 1000 times greater than that of the electronic microscope "lenses." It should be remembered, however, that the tremendous resolving power of the electronic microscope is due to the very short wave lengths of radiation used and not to effective improvements in the numerical aperture of the system. The low order of numerical aperture in some respects is advantageous, since it achieves great depth of focus, or vertical resolution. The electron microscope is capable of vertical resolution to about 2.0 micron, as compared to an oil-immersion objective which is only capable of about 0.06 micron. This and other desirable characteristics of the electronic microscope image are illustrated in Fig. 71.

The functioning principle of the electronic microscope is one of forming an image by converging a stream of rapidly moving electrons, after passing through a thin-section specimen (about 10^{-5} in.) or transparent replica thereof, to a point of focus. The beam of electrons made incident on the specimen originate from a cathode-ray discharge tube operating at 50 to 60 kv, and is collimated to the specimen and converged after passing through it by a series of electrostatic- or, more commonly, magnetic-field "lenses." The formation of the image is geometrically identical with that of the optical microscope but differs somewhat in the fundamental mechanism by which the image is formed. In the optical microscope, image formation is dependent upon differences in the nature and the amount of

a. Optical microscope.   1500×.

b. Electronic microscope, employing a polystyrene-silica replica.   11,000×.

FIG. 71.   The structural appearance of high-purity nickel as observed under the optical and electronic microscopes.   (*Courtesy of R. H. Heidenreich.*)

interaction between the illumination and various points of the specimen—the differences arising from differences in selective absorption, index of refraction, scattering power for the illumination used, reflectivity characteristics of various oriented surfaces, and the size of the smaller particles. In the case of the electronic microscope, however, these differences arise from variation in scattering power and mass density of various specimen points, interference conditions for crystalline particles, and the size of the particles.

The rapidly propelled electrons passing through the specimen may be considered, according to physical theory, to possess a definite wave length. This wave length is variable, being shorter the higher the velocity of the electrons. The electronic wave length normally employed in the electronic microscope is of very small dimension; for example, when electrons are accelerated by about 50 kv, they behave as though they have a wave length of approximately 0.05 Å. This wave length is about 100,000 times shorter than the visible radiation (4000 to 5000 Å) normally used with the conventional light-optical microscope.

The high-velocity stream of electrons serving as the source of "illumination" in the electron microscope and the magnetic or electrostatic field coils serving as the "lenses" may be defined in analytical terms equivalent to the light-microscope system. The "microscope tube" enclosing the "lenses" is evacuated to about 0.0001 mm Hg pressure to prevent scattering of the electrons and retardation of their motion through collisions with air molecules. As illustrated in Fig. 72, the electron stream from the source is collimated by an electromagnetic coil of appropriate shape and directed through the specimen of interest. This collimating coil functions in a manner similar to the substage condenser system in a biological microscope of the optical type, and may be used to control the angular aperture and the intensity of the electron stream at the specimen.

The electrons, after passing through the specimen of interest, are focused by a second magnetic coil—the "objective"—to form an image directly above the projection, or "eyepiece," coil. This intermediate image is at a magnification of 50 to 100 diameters, depending upon the precise position of the specimen in the system and the field strength of the coil. That portion of the intermediate image formed by the "objective" coil, which passes through the "eyepiece" coil, is remagnified between 20 and 300 diameters and projected onto an appropriately positioned fluorescent screen where the final image can be observed, or photographed by replacing the screen with a photographic plate. By controlling the magnetic intensities of the "objective" and "eyepiece" coils, the total magnification of the final image may be adjusted to between 1000 and 30,000 diameters.

Unfortunately the standard electronic microscope cannot be used for the direct observation of opaque surfaces, such as metallographic specimens, without a great loss in resolving power. However, a scanning type of electron microscope is now under development[1] which, when perfected, will in all probability extend the range of direct observation in metallographic microscopy to the extent that the standard instrument has already extended it in the bacteriological and associated fields.

At the present time metallic structures can be observed with the standard electronic microscope only through the use of transparent replicas that faithfully define the contour of the etched surface of a metallographic specimen. Various methods of preparing such replicas have been devised,

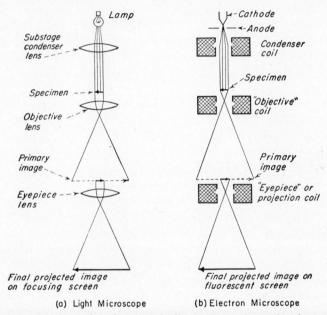

FIG. 72. Illustrating the analogy between the light and the electronic microscope systems.

such as the collodion, silver-collodion, Formvar, and attending shadow-casting techniques, which are fully described elsewhere. The polystyrene-silica, or two-step, process, however, developed by Heidenreich, is probably the most satisfactory replica method available at the present time.

The polystyrene-silica method consists of first securing an impression of the specimen surface in polystyrene by appropriately molding polystyrene plastic in a conventional metallographic mounting press, as illustrated on page 51. The specimen is then removed from the polystyrene casting by either mechanical means or, if necessary, by dissolution of the specimen in an appropriate solvent. The replica surface of the polystyrene is then

[1] Zworykin, V. K., and J. Hillier: Electronic Microscopy, *Sci. Monthly*, Vol. 59, 1944.

coated with silica, applied by vaporization techniques, to a thickness of about 200 Å.   By partially dissolving the polystyrene base with ethyl bromide, the silica replica may be loosened and removed to serve as the specimen in the electronic microscope.

Photomicrography in Color.   The high degree of perfection attained in the development and manufacture of color photographic materials, and the opportunity now afforded the metallographer to process some of these materials in his own darkroom, has prompted considerable interest in color photomicrography as related to metallography.   Although excellent color photomicrographs have been secured of highly colored nonferrous alloys, it is questionable as to whether or not color photomicrography will ever be used so extensively as ordinary photographic techniques.   In part this is attributable to the difficulties encountered in securing faithful color reproductions, and aside from the aesthetic value and novelty effects, color photomicrographs rarely serve a purpose more useful than conventional black-and-white prints.

Without question, the techniques associated with color photomicrography are somewhat involved and complicated, and require expert attention to render proper color balance in the final positive transparency or in the paper print.   Excellent techniques have been developed by Loveland[1] and Butterfield[2] and need not be repeated here.

Perhaps the greatest difficulty encountered in securing faithful and well-balanced reproductions in color is in correcting the illumination of the microscope to a quality precisely comparable to that for which the color film is balanced.   In principle, this is accomplished by carefully filtering the illumination from a reproducible source (preferably a carbon-arc or a ribbon-filament lamp) so as to either raise or lower the color temperature of the illuminate to match that of the film being used.   There are available various color films which are balanced to some given color temperature. Kodachrome regular, for example, is matched to daylight (5400 to 6000° K color temperature), Kodachrome type A is adjusted to 3400° K, which matches the color temperature of photoflood lamps, and Kodachrome type B and Ansco Color film are balanced to 3200° K, which matches the color temperature of a series of photographic lamps manufactured by the General Electric Company.

Because of complications introduced by the microscope system, the color temperature of the image-forming rays reaching the color film may be different from that of the source, even though the color temperature of the source is appropriately adjusted.   This color imbalance may be attributed to selective chromatic absorption for certain colors by the glass optical

[1] Loveland, R. P.: Metallography in Color, *ASTM Bull.* 128, May, 1944.
[2] Butterfield, J. V.: Color Photomicrography, *Educational Focus*, Vol. 17, No. 2, 1946.

parts of the microscope, and to the introduction of false colors in the image because of residual chromatic errors in the system. At low magnifications of about 100 diameters the color imbalance is not significant, but at moderately high and high magnifications the color deficiency in the image must be compensated for by additional filtering. This is best accomplished by the use of specially prepared color-filter solutions available from the Eastman Kodak Company.

The latitude of photographic exposure of color film is rather narrow and, hence, to secure proper color rendition in the final photomicrograph, the correct exposure time must be known with some degree of accuracy. Since conventional exposure meters are not very useful in photomicrography, comparative methods for determining the exposure time under a given set of conditions are generally used. One such method consists of securing a series of trial exposures on conventional black-and-white film and then calculating the exposure time for a given color film from the known photographic speeds of the two materials. An alternate method consists of securing two identical series of trial exposures, as described on page 135, one series on the color film of interest and the other on Kodak Super Speed Direct Positive paper. These two exposure series serve as standards so that under a new set of conditions it is only necessary to secure a black-and-white exposure, and then by comparison to the standard, an estimation can be made as to how the color reproduction will appear.

Replica Reproduction. Various techniques have been suggested, other than photographic, for permanently recording metallographic structures for observation at low magnifications and precise surface conditions of metallic specimens and certain nonmetallic materials. Of these methods the transparent-replica method is probably the most suitable for metallographic work, because the techniques involved are relatively simple, the recording is permanent, and the surface condition of the specimen is faithfully and precisely reproduced, as illustrated in Fig. 73.

The replica method[1] involves the use of a transparent plastic film about 0.0055 in. in thickness which, when used, is softened on one side by an appropriate solvent. The softened surface of the film is then brought into intimate and close contact with the specimen surface of interest and, after the solvent has completely evaporated, the film is stripped off. Resulting therefrom, the contact side of the film is an exact reproduction of the contour of the specimen surface, being thinner in areas corresponding to elevated points of the specimen surface and thicker in regions associated with depressed areas of the surface such as grain boundaries, scratches, pits, etc. Owing to variations in thickness of the transparent replica and at-

---

[1] Materials suitable for this purpose, consisting of plastic films, solvent, mounting cards, etc., are available in kit form from The Faxfilm Co., Cleveland, Ohio.

tending differences in film densities, an image of the replica (and secondarily, an image of the original specimen surface) will be formed when the replica is examined microscopically with transmitted light, or projected onto a screen by a conventional type of lantern-slide projector. Special projectors, however, are available for this purpose, which are capable of producing a total image magnification of about 100 diameters at relatively short projection distances.

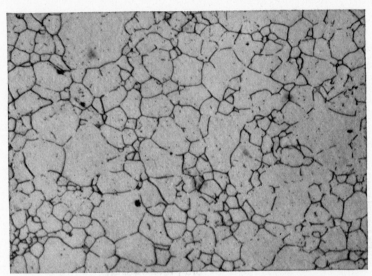

Fig. 73.   The metallographic structure of 18–8 stainless steel recorded by the transparent plastic-replica method.   100×.   (*Courtesy of the Faxfilm Co.*)

References

Allen, R. M.: "Photomicrography," D. Van Nostrand Company, Inc., New York, 1941.

Benford, J. R.: Anti-reflection Films for Metallographic Objectives, *Trans. ASM*, Vol. 36, 1946.

Brown, E. B.: "Optical Instruments," Chemical Publishing Company, Inc., Brooklyn, N. Y., 1945.

Buckingham, W. D., and C. R. Deibert: Characteristics and Applications of Concentrated-Arc Lamps, *J. Soc. Motion Picture Engrs.*, Vol. 47, 1946.

Butterfield, J. V.: Color Photomicrography, *Educational Focus*, Vol. 17, No. 2, 1946.

Eastman Kodak Co., "Photomicrography," 14th ed., Rochester, N. Y., 1944.

Foster, L. V.: A Polarizing Vertical Illuminator, *J. Optical Soc. Am.*, Vol. 28, 1938.

Gage, S. H.: "The Microscope," 17th ed., Comstock Publishing Company, Inc., Ithaca, N. Y., 1941.

George, H. S.: Conical Illumination in Metallography, *Trans. ASST*, Vol. 4, 1923.

Hall, C. E.: Electron Microscope, *Complete Photographer*, Vol. 4, 1942.

Heidenreich, R. D.: Interpretation of Electron Micrographs of Silica Surface Replicas, *J. Applied Physics*, Vol. 14, 1943.

Heindenreich, R. D.: Techniques in Applied Electron Microscopy, *J. Optical Soc. Am.,* Vol. 35, 1945.

Loveland, R. P.: Metallography in Color, *ASTM Bull.* 128, May, 1944.

Lucas, F. F.: An Introduction to Ultraviolet Metallography, *Trans. AIME,* Vol. 73, 1926.

Milne, L. J.: Coated Lenses for Biological Photography, *J. Biol. Phot. Assoc.,* Vol. 14, 1945.

Patterson, W. L.: The Optics of Metallography, *Trans. ASST,* Vol. 2, 1921.

Picard, R. D.: New Development in Electron Microscopy, *J. Franklin Inst.,* Vol. 239, 1945.

Prebus, A., and J. Hillier: Construction of a Magnetic Electron Microscope of High Resolving Power, *Can. J. Research,* Vol. 49, 1939.

Shillaber, C. P.: "Photomicrograph in Theory and Practice," John Wiley & Sons, Inc., New York, 1944.

Vilella, J. R.: "Metallographic Technique for Steel," ASM, Cleveland, Ohio, 1938.

Williams, R. C., and R. W. G. Wyckoff: Applications of Metallic Shadow-Casting to Microscopy, *J. Applied Phys.,* Vol. 17, 1946.

Zworykin, V. K., and J. Hillier: Electronic Microscopy, *Sci. Monthly,* Vol. 59, 1944.

CHAPTER 4

THE PRINCIPLES OF PHOTOGRAPHY

The production of a photomicrograph of good quality, *i.e.*, one which is brilliant, clear cut, and marked by good contrast between the various constituents in the structure, depends upon the care with which the following manipulations have been carried out: (1) preparation of the specimen, (2) the adjustment of the metallograph as described in the preceding chapter, (3) the selection of an appropriate plate or film, (4) the exposure of the photographic plate to the projected image, (5) the development of the plate or film to produce a negative of the image, and (6) the printing of the negative to obtain a positive, representing the finished photomicrograph.

The making of high-quality photomicrographs involves not only a knowledge of specimen preparation and the use of a photomicrographic apparatus, but a knowledge of photography as well, for it is only by proper attention to the techniques of photography that a photograph which is truly representative of the subject can be obtained.

In the discussion to follow no attempt will be made to embrace the entire field of photography. Such a discussion would, of necessity, require the presentation of special techniques, such as negative retouching, print toning, etc., which are of no importance in photomicrography. It is much more important that a good photomicrograph be obtained by a correct exposure and proper processing than by correcting the faults of the negative and print by special corrective methods.

PHOTOGRAPHIC PLATES AND FILMS

A photographic plate or film is essentially a transparent base material, such as glass or celluloid, covered on one side—or, less frequently, both sides—with a thin coating of emulsion. The base of the emulsion is gelatin, which holds in suspension fine crystals of silver chloride, bromide, or iodide. The silver salts, when exposed to light, as well as to certain wave lengths of invisible radiation, are affected in some way which enables certain reducing agents, known as developers, to reduce the light-exposed silver salts to metallic silver. The number of silver halide grains rendered developable—that is, subject to reduction by the developer—depends upon the intensity of the light which strikes the plate or film. Thus, if a photographic plate is exposed to the focused image of an object, in the brightest areas of the image registered on the plate, a larger number of crystals of silver halide will be rendered developable than in less intense areas. When the plate is developed, the brightest parts of the image will appear dense and dark because of the greater amount of reduced silver, whereas the thin and transparent parts of the plate will represent the dark areas of the image where little, if any, of the silver salts have been affected. The plate after development is called the negative, because it is complementary with respect to the object.

Photographic plates and films possess the following five characteristics, which are of importance from the standpoint of photomicrography: (1) color sensitivity, (2) emulsion speed, (3) latitude, (4) resolving power, and (5) available contrast. All these characteristics other than color sensitivity are so interrelated that it is impossible to obtain an extreme of one without accompanying changes in the others. Thus, plates possessing the desirable property of high resolving power will in general be accompanied by a slow emulsion speed and shorter latitude than plates manufactured specifically to produce these latter properties.

Color Sensitivity. With regard to color sensitivity, plates and films may be divided into three groups: (1) ordinary, which are sensitive to only the shorter wave lengths of radiation, such as ultraviolet and blue; (2) orthochromatic, which are sensitive to radiation from the blue through the green; and (3) panchromatic, which are sensitive to all colors of the visible spectrum and may in certain instances be sensitive to infrared. This difference in color sensitivity is introduced in the process of manufacturing and generally is accomplished by the addition of various organic dyes. The selection of the proper plate on the basis of color sensitivity depends upon the wave lengths of light transmitted by or reflected from the object.

Emulsion Speeds. Plates and films in the same group with regard to color sensitivity may be compared on the basis of their emulsion speeds, *i.e.*, the exposure time required to produce a satisfactory negative under identical conditions of subject illumination. Extreme rapidity generally involves the sacrifice of properties often more desirable in photomicrography, such as resolving power or fineness of grain, latitude, contrast, and resistance to fogging during development.

The speed of a photographic plate or film is not generally of great significance in photomicrography, because the brightness of the projected image is usually sufficiently great to permit appropriate photographic registration under moderately short times of exposure. However, in some exceptional cases involving low image brightness, the time of exposure may be so long that the use of slow-speed plates would be impractical. Faster speed plates may then be used, and some of the undesirable characteristics associated with such fast emulsions can be rendered less pronounced through appropriate choice of developer and careful control during development.

Latitude. The latitude of a photographic emulsion may be defined as its ability to render a satisfactory negative with regard to image contrast over a range of exposure times. Thus, emulsions that require photographic exposures within rather narrow limits of time are said to possess short, or narrow, latitude, whereas those which may be exposed over relatively wider time limits are said to possess long, or wide, latitude.

As illustrated in Fig. 74, the relatively straight portion of the time-density curve, defined by *B-C*, represents the latitude of the photographic emulsion, and hence, the limits of correct exposure times. In this region of the curve, the density (or negative darkness) of the silver deposits after development

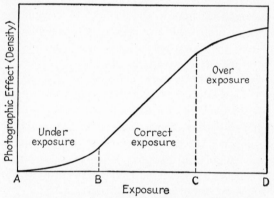

Fig. 74. Relationship between exposure time of a photographic emulsion and the resulting photographic effects.

will increase in proportion to the increase in exposure time. Underexposure, however, as represented by the portion of the curve *A-B*, will render a negative possessing less pronounced image contrast or differences in density so that, for example, details of low brightness in the image will either be completely absent or poorly represented in the negative. The portion of the curve *C-D*, representing overexposure, describes a negative wherein the bright details of the image will not be faithfully reproduced photographically, owing to the highlights being completely lost in, or merged into, the brighter areas of the image.

A plate possessing a wide latitude, therefore, will produce a more satisfactory negative than will a plate of narrow latitude, when the exact exposure time required is not known. Plates possessing wide latitude not only permit greater tolerance in exposure times, but are capable of reproducing more accurately greater differences in image brightness, *i.e.*, a greater range in subject contrast.

Resolving Power. It was shown in the preceding chapter that resolution or the separation of fine detail in the object, depended primarily upon the characteristics and, in particular, the numerical aperture of the microscope objective. However, to reproduce faithfully such fine details as separate entities in the photographic negative requires the use of emulsions possessing relatively high resolving power. This property is dependent upon the particle size of the silver salts in the emulsion and, in general, the smaller the particle size the greater will be the potential resolving power of the plate or film.

Available Contrast. One of the most desirable characteristics in a photographic negative is image contrast, *i.e.*, well-established differences in the densities of dark and light areas of the negative. A negative of low contrast is one exhibiting small differences in photographic densities; one of high contrast exhibits rather pronounced differences. In the discussion of negative contrast, a distinction must be made between contrast and over-all negative density. The over-all density of a negative may be either high (dense and dark) or low (thin and nearly completely transparent), but in either case, with regard to the densities of different areas, the contrast may or may not be of good quality.

The maximum contrast inherently available upon full photographic development depends upon the characteristics of the photographic emulsion; for example, plates and films of low to moderately high speeds are capable of rendering higher negative contrast than are faster speed emulsions. Rendering of maximum contrast also depends upon the type of developing solution used and the time of development. Thus, within limits, negative contrast in any given emulsion may be improved under certain circumstances by increasing the time of development.

In the selection of a photographic emulsion suitable for photomicrography, attention should be directed to plates or films that will render the degree of negative contrast desired, and from this group, a material selected that will best respond to the quality of illumination employed. Photographic plates and films suitable for most all types of photomicrography are shown in Table 15.

PRINCIPLES OF PLATE AND FILM DEVELOPMENT

Chemistry of Development. The purpose of development is to reveal and make visible the latent, or invisible, image implanted on the plate or film during photographic exposure. Essentially, developing solutions complete the reduction of the silver salts affected by light and are selective in their action so that only those grains of silver salt that were exposed to light are actually reduced.

All developing solutions contain three essential ingredients: (1) a reducing agent to reduce the exposed silver salts to metallic silver; (2) an alkali

TABLE 15. SOME RECOMMENDED PHOTOGRAPHIC PLATES AND FILMS FOR METALLOGRAPHY AND OTHER GENERAL PHOTOMICROGRAPHY

Characteristics of emulsion	Orthochromatic type	Panchromatic type
Very high available contrast.....	Kodak Metallographic plate* Kodak Contrast Process Ortho film Ansco Process film	Kodak "M" plate Kodak Contrast Process Panchromatic film
High available contrast.........	Kodak Super Ortho Press plate and film Ansco Commercial Ortho film	Kodak Panchromatic plate Kodak Process Panchromatic plate Defender Process Panchromatic film Hammer Panchromatic Dry plate
Medium available contrast......	Kodak Commercial Ortho plate and film Kodak 50 plate Defender Commercial film Defender Pentagon film Hammer Slow Ortho plate	Kodak Panatomic X plate and film Ansco Isopan film Hammer Commercial Panchromatic Dry plate
Rapid emulsion speed..........	Kodak Ortho X film Kodak Super Ortho Press plate and film Ansco SSS Ortho film Defender Ortho 7 film	Kodak Tri X Panchromatic plate and film Kodak Super Panchro Press plate and film Ansco SSS Pan film Defender Arrow Pan film

* Most suitable for general metallographic purposes. The emulsion is capable of high resolving power, and the contrast may be developed to a gamma of 3.0.

to act as an activating agent for the reducer; and (3) a preservative to prevent rapid oxidation of the developer by the air, by dissolved oxygen contained in the water solvent, or by use as a photographic developer.

Reducing agents, suitable for use in photographic development, must be inherently selective in their action and must be responsive to circumstances which tend to influence their rates of reaction. A number of reducers are available having these characteristics, of which metol, pyro, and hydroquinone are representative. These reducing agents, however, lack sufficient energy when in aqueous solutions to function effectively as a photographic developer unless activated by some alkali, such as sodium carbonate. By judiciously increasing the alkalinity of the developing solution, the rate of development, or the rate of reaction of the reducing agent, is increased with an attending gain in negative contrast for a given time of development. However, should the activity of the developer become too great, owing to the addition of too much sodium carbonate, the developing solution will reduce both the light-exposed and unexposed crystals of silver halide and, as a consequence, will cause a general chemical fogging to take place over the entire negative. In order, therefore, to restrain the action of the developer within limits, and to prevent attendant troubles, a soluble bromide compound, such as potassium bromide, is added to the developer in carefully controlled amounts.

Photographic reducing agents, like other chemical reducers, have a rather high affinity for oxygen. Unless a developer preservative, such as sodium sulphite, is added to the developing solution, the developer will become so highly oxidized upon storage or during use that it will soon be rendered useless for the purpose intended. Sodium sulphite functions as a preservative by combining with the developing agent to form a complex sulphonate, which is less readily oxidized than either the developing agent or sulphite alone. Sodium bisulphite or metabisulphite is sometimes used in place of sodium sulphite. These are both acid sulphites and are preferably used in those developers which are very easily oxidized.

Frequently, developing solutions contain two reducing agents, such as metol and hydroquinone. Metol, when present in the solution as the only reducer, produces detail of the image very rapidly, the density and contrast of the negative increasing as developing proceeds. The action of hydroquinone, however, differs from that of metol in that the highlights of the image appear first and the half tones and shadows follow more slowly. A combination of the two in the same developing solution will result in a developer having well-balanced propensities for producing detail and density in the negative.

The choice of a developing solution depends upon the type of photographic emulsion involved and the contrast desired in the finished negative. As was mentioned heretofore, the degree of contrast established in a given negative emulsion depends upon the type of developer used and the time of development. For any particular developing solution there is a limit to the degree of contrast obtainable in the negative upon prolonged development. If still greater contrast is desired in the negative, within limits of the available contrast of the plate or film, a high-contrast developer should be used.

Whenever possible the developer formula recommended by the manufacturer of a particular plate or film should be used exclusively. There are, however, four developer formulas that are suitable for development of the photographic plates and films generally used in photomicrography. Developer formula D-19, Table 16, page 436, is particularly suitable for rendering very high contrast in emulsions of high available contrast, such as Kodak "M" and Kodak Metallographic plates, and Ansco Process films. Formulas DK-50 and D-42, Tables 17 and 19, pages 437 and 438, respectively, are suitable for moderately high-contrast development; for low- and medium-contrast development of Kodak plates, formula D-41, Table 20, page 438 is ideally suited. A normal-contrast developer, formula D-61a, Table 21, page 439, is recommended for plates and films of moderate available contrast and for those of rapid photographic speeds.

Time of Development. In the process of development, the rate at which the developer reduces the exposed grains of silver halide is practically a constant, but as development proceeds the number of grains available for development becomes less and less, with the result that a curve representing the relation between the increase in density and the time of development is almost truly exponential. A time of development is then reached where, because of the very few exposed grains remaining, continued development is not worth while. When practically all the exposed grains have been developed, further development will reduce the unexposed grains, resulting in a general chemical fog and decreased contrast over the negative.

For properly exposed negatives, the correct developing time to produce a certain degree of contrast in the negative has been accurately determined for different developing solutions, and such data are generally included on instructional sheets enclosed within the box of plates or films. It is important that the correct developing time be noted and followed. Although it is possible, by careful selection of printing papers, to compensate for overdeveloped and underdeveloped negatives, nevertheless such a negative produces a print which is photographically deficient compared to one obtained from a correctly developed negative.

It is commonly believed that overexposed plates can be underdeveloped (developed for a shorter time than is recommended) to compensate for the exposure error. Since correct contrast, and not simply density, is what is most desired in a negative, underdeveloping cannot be expected to remedy this error. An overexposed plate will have too little contrast, even when developed for the normal length of time; and the contrast will be reduced still further if the plate is underdeveloped.

Likewise, an underexposed plate cannot be corrected by overdeveloping. The number of grains which are made developable in an underexposed plate is normally less than if the correct exposure is given, with the result that if developed even for the proper length of time, the negative will be lacking in density and in shadow detail. Overdeveloping will increase the contrast of the negative, but it will not add shadow detail; thus it is impossible to secure from such a negative a print with satisfactory representation of the shadow portions of the subject. It is advisable, therefore, that the plate be developed for the recommended length of time, regardless of the exposure. No amount of manipulation with the developer or developing time will render a negative so satisfactory as one produced by developing for the correct time and at the proper temperature.

Temperature of Developing Solutions. Perhaps one of the most important factors associated with the developing of photographic materials is the temperature at which the process is conducted. Usually some narrow range of temperatures is recommended by the manufacturer of the plate or film involved, and generally this temperature range is 20 to 25°C (65 to 75°F). Unless the temperature of solutions is rather carefully controlled during negative development, it will be difficult to reproduce results; the contrast established may vary from one negative to another owing to the dependence of correct developing time to solution temperature; and the developer may become highly oxidized at higher temperatures and cause considerable staining of the emulsion.

Developing at temperatures lower than those recommended will increase the development time considerably and, in the case of metol-hydroquinone developers, the reducing action of the hydroquinone will be retarded more than that of the metol. As a consequence, the developer will no longer behave properly but will act as a solution containing an excess of metol, rendering a negative of lower contrast and over-all density.

As one means of preventing physical damage to photographic emulsions, it is essential to maintain all solutions, including the wash water, at about the same temperature. The mechanical stresses introduced in the gelatin of the emulsion due to great differences in temperature between the various processing solutions, may be sufficiently great to cause reticulation, *i.e.*, local separation or rupture of the gelatin. Depending upon the precise circumstances causing separation of the gelatin, which may not always be due to temperature (see page 168), reticulation may exist in varying degrees from mild to severe cases, as illustrated in Fig. 75.

Chemistry of Fixation. The purpose in photographic fixing is to render the silver halide remaining in the film soluble in water, so that it can be removed by washing and thus leave an image of metallic silver implanted in the gelatin emulsion. Probably the best solvent for the purpose intended is an aqueous solution of sodium thiosulphate, better known as hypo. Inasmuch as a hypo solution during use becomes discolored and contaminated, owing to developer solution being inadvertently added when the negative is transferred from the developer, there is danger that the negative emulsion will become badly stained.   To prevent this, and to harden the gelatin emulsion and prevent its softening and attendant troubles, an acid hardening and fixing bath combination is usually employed, in preference to a plain aqueous solution of hypo.   An acid fixing and hardening bath ordinarily contains (1) a weak acid, such as acetic, which acts as an anti-staining agent by keeping the fixing bath acidic and neutralizing the developer brought over on the plates or films; (2) a preservative, which is usually sodium sulphite; and (3) a hardening reagent, such as potassium alum, the purpose of which is to harden the gelatin and prevent blisters and similar troubles.

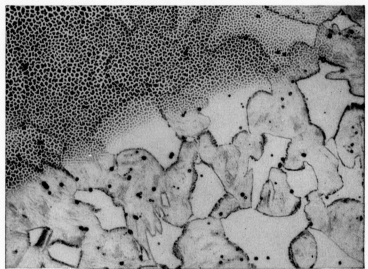

FIG. 75.   Illustrating partial reticulation of a photographic-plate emulsion.

Prepared fixing powders are available that contain all of the compounds mentioned.   They are merely dissolved in a given amount of water and the bath is then ready for use.   Tables 22 and 23, pages 439 and 440, give general formulas for preparing fixing baths in the laboratory.

PHOTOGRAPHIC PRINTING PAPERS

Print papers are sensitized sheets of paper which, when placed in contact with the negative and exposed to white light, will form an invisible positive image of the negative.   This latent image will be revealed as a positive when developed in a suitable developing solution.   Printing papers may be of either single- or double-weight paper stock.   The emulsion is similar to that on a plate or film, except that it is less sensitive to light, and it is made in this way for the primary purpose of obtaining better exposure control during printing.   Papers may be obtained in a variety of surface textures, such as matte, semimatte, velvet, glossy, etc., but for photomicrographs, glossy paper is always used because it reveals much finer details of structure in the finished print than do other types of surfaces.

The contrast in the finished print depends entirely upon the available contrast of the paper. Unlike plates and films, the contrast of the print cannot be controlled by varying the time of development.

Most glossy-surface papers are available in four to six grades of contrast, which are designated either by numbers such as 0, 1, 2, 3, 4, and 5, or as extra soft, medium, contrast, etc. The purpose of such a variety of contrasts is to provide a means of obtaining from photographically deficient negatives a print having the desired contrast. Table 26 gives the more common printing papers that are used extensively for photomicrographs.

TABLE 26. COMMERCIAL GRADES OF SMOOTH, GLOSSY PHOTOGRAPHIC PRINTING PAPERS SUITABLE FOR CONTACT-PRINTING PHOTOMICROGRAPHS

Manufacturer	Proprietary name	Available Weight	Contrast grades in order of increasing contrast
Eastman Kodak....	Azo F	Single and double	0, 1, 2, 3, 4, 5
	Velox F	Single	0, 1, 2, 3, 4, 5
	Illustrators' Azo F	Single and double	0, 1, 2, 3, 4
Defender..........	Apex R	Single	0, 1, 2, 3, 4, 5
	Apex T	Double	0, 1, 2, 3
Haloid.............	Halo grade F	Single	0, 1, 2, 3, 4, 5
	Halo grade FF	Double	0, 1, 2, 3, 4, 5
	Industro grade F	Single	1, 2, 3, 4
	Industro grade FF	Double	1, 2, 3, 4
Ansco.............	Convira	Single	0, 1, 2, 3, 4, 5
	Convira	Double	1, 2, 3, 4

TABLE 27. RECOMMENDED PAPERS FOR THE PRODUCTION OF SATISFACTORY PRINTS FROM FAULTY NEGATIVES

Negative fault	Characteristics of negative	Recommended‡ paper to be used
Underexposed*....	Lacks detail	High-contrast paper, such as Nos. 4 and 5
Overexposed*.....	Dense and lacks contrast	High-contrast paper
Underdeveloped†..	Lacks contrast	High-contrast paper, such as Nos. 3 and 4
Overdeveloped†....	Dense and high contrast	Low-contrast paper, such as Nos. 0, 1, and 2

* Normal development.

† Correct exposure.

‡ The contrast numbers of the paper are given merely as a guide. The actual choice of the paper depends entirely on the appearance of the negative and should be determined by trial.

It must be clearly understood that papers are to be selected on the basis of the contrast of the negative and not upon the over-all density.   The density of the negative affects only the exposure time in printing, being relatively short if the density is low and long if the density is high.

There are four common faults of negatives that may be improved somewhat by the proper selection of the paper.   Such defects, with suggestions as to the grade of contrast paper to be used, are given in Table 27.   The photographic effect of negatives which have been underexposed and overdeveloped, or overexposed and underdeveloped, cannot be corrected by the type of paper selected.   In these cases it is recommended that the picture be retaken with greater attention to correct exposure and development.

Figure 76 illustrates the effect of printing negatives of different contrasts on various contrast papers.   It will be noted that for each negative correctly exposed, but underdeveloped, correctly developed, and overdeveloped respectively, a different contrast paper is required to produce a satisfactory print.   It should not be construed that, because papers are available in a variety of contrast, the correct exposure and development of the negative may be neglected.   Only by careful control and correct procedures can a truly satisfactory print be obtained.

Processing of Printing Papers.   The principles associated with processing photographic printing papers are similar to those already described for photographic negative emulsions.   As more fully discussed on pages 168 to 174, the procedure consists of developing the exposed paper in an appropriate developer at a recommended temperature of 20 to 25°C (65 to 75°F), followed by rinsing, fixing, washing, and finally drying the finished print.   Development of a given grade of paper is best achieved by using the developer and the developing conditions recommended by the paper manufacturer.   Lacking such information, however, the developing solution recommended in Table 24, page 440, may be used satisfactorily for the papers given in Table 26.

DARKROOM TECHNIQUE

A darkroom, as the name implies, is a room which is completely lighttight to white light from the outside and is used solely for the purpose of processing photographic plates and films and making prints.   Such a room, because of the nature of the work carried on in it, must be illuminated by "safe" lights of such a color and intensity that they will not photographically affect the plates, films, or printing papers.   For the handling of orthochromatic plates and films, a deep-red light may be used; and for the more sensitive films, such as panchromatic, a special green light can be used, but preferably these films should be handled in total darkness.   Light filters of appropriate transmission characteristics may be obtained for any of these photographic materials. Printing papers, because they are much less sensitive to light than are negative materials, may be handled and processed under a yellow or orange-red light, depending upon their precise sensitiveness.

It cannot be emphasized too strongly that darkrooms should be kept clean and orderly and that all equipment, such as developing trays, etc., should be scrupulously clean.[1]

All trays, tanks, stirring rods, etc., that are used in the darkroom should be preferably of glass, bakelite, hard rubber, or of a material that will not be attacked chemically by the various solutions. Metal ware, with the exception of stainless steel, should never be used unless protected by enamel. Reaction between the solution and the container will not only spoil the solution, but the products of reaction may often stain the plates or prints, or produce a general fog on them.

All operations of developing and fixing for both plates and prints should be carried out under the recommended safelights and at no time during the process should the sensitive material be exposed to white light.

Development of Plates and Films. All developing processes consist of reducing to metallic silver those grains of silver halide on the plate or film that were exposed to light. The process may be carried out by either tray- or tank-development methods.

Plates and films may be developed conveniently in a tray containing sufficient developing solution to completely cover the plate. This method of developing has the advantage that the developing process can be watched under filtered lighting. The plate or film is placed in the tray with the emulsion side up and is developed for the proper length of time according to the contrast desired. If more than one plate or film are developed at the same time, care should be taken that they be separated from one another, thereby eliminating the danger of scratching the emulsion surfaces. During development, the developing solution is agitated frequently by rocking the tray, thus ensuring uniform development and the breaking up of all air bubbles that may be adhering to the surface of the plate. It is recommended that the tray be covered during negative development, so as to eliminate the possibility of fogging due to too high intensity of light from the safelight.

Tank development consists in placing the plate or film upon a suitable rack and immersing both in a tank of developer. This method of development has the advantage that the tank of developing solution may be left standing and be ready for instant use at any time. The solution should be kept covered, however, when not in use to prevent excessive oxidation. The rack containing the film or plate should be lifted out of the tank, drained, and returned several times during development, to ensure even development. Generally, tank development is about 20 per cent longer

---

[1] See Table 25, page 440, for tray-cleaning solution.

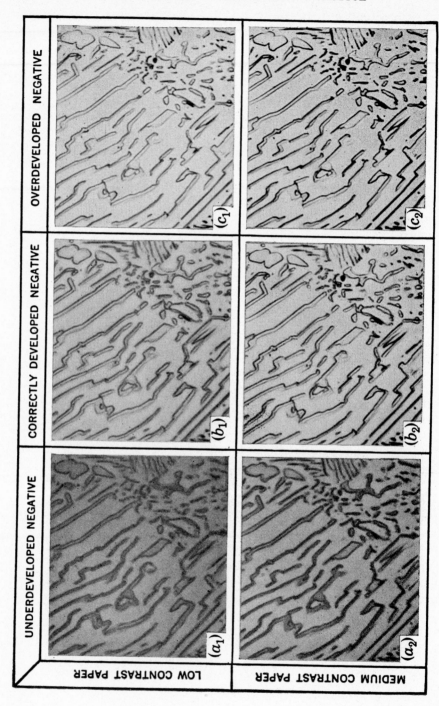

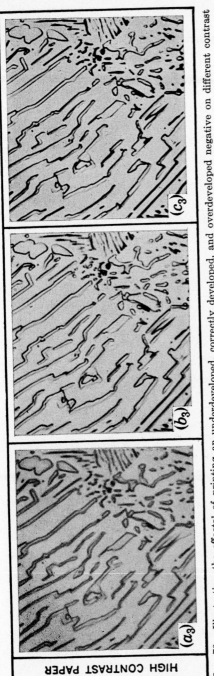

FIG. 76. Illustrating the effects[1] of printing an underdeveloped, correctly developed, and overdeveloped negative on different contrast papers. All negatives exposed for the correct length of time.

(a_1) For this type of negative originally lacking in contrast, a low-contrast paper produces a print that is lacking in both contrast and brilliancy.

(a_2) An improvement is noted over (a_1). Somewhat better contrast and definition.

(a_3) Better contrast and better quality of print than either (a_1) or (a_2).

(b_1) For this type of negative a low-contrast paper produces a print that is lacking in contrast.

(b_2) A properly balanced print. Contrast, brilliancy, and definition are good.

(b_3) Some fine detail is lost. Contrast is somewhat too high for a print of good quality.

(c_1) For this type of negative originally possessing high contrast, a low-contrast paper produces a print that is not entirely satisfactory, but is of better quality than either (c_2) or (c_3).

(c_2) Fine detail is lost. Contrast is somewhat better than in (c_1).

(c_3) More detail is lost than in (c_2). Contrast is too high and the print is of poor quality.

[1] Because some detail is always lost in reproductions, the comparison of the qualities of the above photomicrographs has been made on the original prints.

than development in a tray with the same developer to obtain the same degree of negative contrast.

When the plate has developed for the proper time, as indicated by a suitable darkroom clock, the plate is quickly removed from the developer and rinsed in clean water. In removing the plate or film from the developer and in handling it during the subsequent operations, it should be grasped at the edges to prevent scratching or otherwise damaging the emulsion. It should be remembered that the gelatin is soft when wet and in this condition is easily scratched and damaged if carelessly handled.

The plate, after being thoroughly rinsed, preferably in a weak acid bath (1 oz of 28 per cent acetic acid, 32 oz of water), is placed either in a hardening bath or directly in the fixing solution.

Chrome Alum Hardener. It is recommended, especially during warm weather, that the developed emulsion be hardened by a special hardening bath, in addition to the hardening produced by the fixing solution. This hardening treatment may be carried out by immersing the rinsed plate in a freshly made solution of the following composition:[1]

Water	1000 cc	32 oz
Potassium chrome alum	30 g	1 oz
Sodium sulphate (desiccated)	60 g	2 oz

The plate is allowed to remain in this solution for approximately 5 min. During the first few minutes of hardening, the plate is agitated to prevent streaks forming on the negative. The hardening properties of this solution decrease very rapidly upon standing and it is recommended that a fresh solution be made at least daily—more often if necessary.

When the hardening process is completed, the plate is again rinsed in clean water and placed directly in the hypo fixing bath.

Fixation. As was previously discussed, the purpose of fixation is to remove all unreduced silver salts from the plate by rendering them soluble in water. The plate is carefully placed in a tray containing sufficient acid fixing and hardening bath to completely cover the plate, and is allowed to remain there for a length of time approximately twice that required to completely clear the negative. During fixation the milkiness of the plate will disappear—an indication that fixation is proceeding. When the plate has remained in the hypo for the proper length of time, it may be removed and examined in daylight.

Fixing solutions should be discarded when they froth or when the time required to fix the plate completely is double that of a fresh hypo bath.

Washing the Negative. When the plate has been properly fixed, it is removed to a washing tray or tank. Washing is for the purpose of removing all chemicals from the emulsion. Preferably, running water should be

[1] Formula SB-4, Eastman Kodak Co.

used, at approximately the same temperature as the developing and fixing solutions.   When running water is available, washing should be continued for approximately 30 min.   Plates and films may be washed satisfactorily in still water by changing the water at least 10 times, allowing 5 min before each change.   By either method of washing, most of the hypo solution will be removed during the first few minutes, but washing must be continued in order to remove the last traces of hypo, which if not removed would cause eventual discoloration of the negative.

To determine whether or not the last traces of hypo solution have been washed from the plate, the water drippings from several plates may be collected and added to a dilute solution of the following:[1]

Potassium permanganate..........................	0.3 g
Sodium hydroxide................................	0.6 g
Distilled water to make..........................	250 cc

For use, make dilute solution in the following proportions:

Water..	250 cc
Above solution.................................	1 cc

If a small trace of hypo solution is present in the drippings, the test solution, originally purple in color, will turn orange in approximately 30 sec. Larger concentrations of hypo will produce a green-yellow color.   In any event, if hypo solution is indicated to be still present, the plates should be returned to the wash water until further tests show the hypo has been completely removed.

Drying the Negative.   When the negative has been completely and thoroughly washed until free from hypo, it is placed on a rack or suspended from clips and allowed to dry.   Drying should be carried out in a dust-free room so that dust particles will not adhere to the emulsion and result in a spotted negative.   Frequently, when negatives must be dried in a hurry, an electric fan may be employed to circulate the air.

Defects in the Negative.   Although care may have been taken during processing of a plate or film, it often happens that the finished negative exhibits one or more processing defects.   The appearance of the more common faults encountered and their possible causes are as follows:

1. Dense negatives may be caused by
   a. Overexposure—characterized by a dense negative lacking in contrast in the high lights.
   b. Overdevelopment—characterized by a dense negative possessing high contrast.
   c. Too strong a developing solution.
   d. Temperature of developing solution higher than recommended.

[1] Formula HT-la, Eastman Kodak Co.

2. Thin or weak negatives may be caused by
   a. Underexposure—characterized by a lack of detail in the shadows.
   b. Underdevelopment—characterized by a thin negative possessing shadow detail but lacking contrast.
   c. Temperature of developing solution lower than recommended.
   d. Error in compounding developer or in time of development.

3. Uneven development of the plate, characterized by areas of different densities may be due to
   a. Improper immersion of the plate in the developer.
   b. Two or more plates in contact with each other during development.
   c. Emulsion in contact with the tray bottom.
   d. Lack of agitation of the developer during development.

4. Yellow stains on the negative may be caused by
   a. Exhausted hypo solution.
   b. Contaminated developing solution.
   c. Incomplete fixation.

5. A white crystalline deposit adhering to the plate may be due to
   a. Insufficient washing of the negative.
   b. Precipitates from the washing water.
   c. Fixing bath incorrectly compounded. Contains too much acid.

6. The emulsion of the negative has suffered reticulation (see Fig. 75). This condition may arise from
   a. Too strong a developer, one used at too high a temperature, or one containing an excess of alkali.
   b. Difference in temperature between the developer, hardening solution, fixing bath, and wash water.

Printing the Negative. Photographic printing is a process whereby the image of the negative is latently formed on the sensitized emulsion of the printing paper, and by subsequent processing in a suitable developing solution this latent image is revealed as a positive reproduction. The latent image may be implanted on the emulsion surface of the paper by one of two conventional photographic methods—contact printing or projection printing. The latter method, however, is used almost exclusively to secure prints of larger or smaller image dimensions than those of the original negative. As the name implies, contact printing consists of placing the emulsion side of the printing paper in intimate contact with the emulsion side of the negative, and then exposing the paper, through the negative, to either artificial light or daylight for some appropriate length of time. The negative and photographic paper may be conveniently held together in a conventional contact printing frame or, preferably, exposure may be made directly in a contact printing box. A photographic printing box is merely a convenient arrangement whereby the negative and printing paper, after being appropriately placed together on a glass supporting plate, can be exposed to light from one or more electric lamps located within the box.

It is frequently desirable to edge the finished print with a white border. This is accomplished during printing by adjusting the masking strips provided for the purpose on most commercial printing boxes, or by using a printing mask of black paper with an opening cut out to conform to the size and shape of the finished print. Usually photomicrographs are finished in either rectangular, square, or circular shapes. To ensure sharp reproductions during printing, the printing mask should always be interposed between the light source and the negative (in close contact with the film or plate) and never between the emulsion sides of the negative and the printing paper.

It is a common occurrence that the negative to be printed is of varying density from one edge to the other—a circumstance usually associated with uneven illumination of the object during photographic exposure. From such a negative, satisfactory prints may be secured by application of dodging techniques during printing. In principle, dodging consists of limiting the exposure time during printing in that region of the negative which is thin and of low density, and allowing, as a consequence, relatively longer photographic exposure in that portion of the negative of higher density. In conformation to this principle, dodging may be carried out in a printing box in one of two ways. One method consists of placing below the thin parts of the negative torn pieces of tissue or onionskin paper. On most commercial printing boxes a sheet of clear glass is interposed between the negative (about 2 to 4 in. below the negative) and the light source to serve as a supporting medium to the dodging paper. By trial and error, and through the inspection of trial prints, it is relatively easy to determine if the correct amount of tissue paper has been used and whether or not the paper has been properly placed with respect to the areas of the negative requiring dodging. This method of dodging has the advantage over the one to be described in that more than one area of the negative can be dodged at the same time.

A second method of dodging, applicable only in cases where the printing box has a slot opening below the plane of the negative, consists of moving continuously an opaque shield beneath the thin part of the negative during exposure of the paper. After inspection of a few trial prints, the dodging manipulations can be corrected to yield a uniformly exposed print, as illustrated in Fig. 77.

Unevenly exposed negatives can be permanently dodged through the use of New Coccine. New Coccine is a water-soluble, red organic dye, which

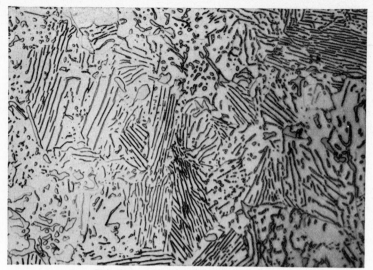

a. Printing paper exposed without dodging.

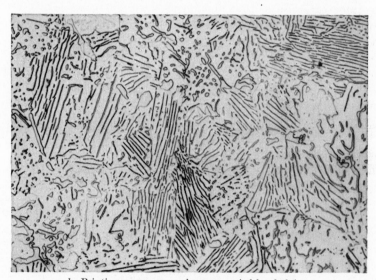

b. Printing paper exposed accompanied by dodging.

FIG. 77.   Illustrating the improvement secured in a finished photomicrograph by appropriately dodging during printing an unevenly illuminated negative.

when appropriately applied to the thin areas of the negative will serve to filter out some of the wave lengths of light to which paper emulsion is most sensitive. The dye solution can be applied to large areas of the negative by means of cotton, or to smaller areas by means of a spotting brush, and may be applied to either side of the negative. Whether or not the increase in negative density is sufficient to render an evenly exposed print can be ascertained by examination of trial prints. When necessary, additional dye solution may be applied, and if the density of the negative becomes too high through error, the color may be removed through washing the negative by conventional methods.

Determining Correct Exposure Time. Inasmuch as the latitude of exposure in photographic papers is rather narrow, it is essential to know with some degree of accuracy the correct exposure time for a particular grade of paper. A print that is overexposed during printing, for example, must be removed from the developing solution before development is complete in order to prevent the print from becoming too dark. In general such an underdeveloped print will be unsatisfactory owing to the poor rendition of photographic tone values in the image. An underexposed print, even when developed for times greater than normally recommended, will usually be deficient in print quality because of insufficient darkness of the image and absence of detail and contrast in those parts of the image corresponding to the dark portions of the negative.

The time of printing exposure varies inversely with the intensity of the printing light and directly with the over-all density of the negative and the contrast grade of photographic paper. Determination of correct exposure times under a given set of circumstances is best determined by trial, either by using small pieces of printing paper that are exposed for different lengths of time and subsequently developed, or by the strip method, which in principle is the same as that described for exposure determinations in photomicrography (see page 135). After some experience, correct exposure times can generally be estimated by mere examination of the negative.

Developing the Print. When the print has been exposed, it is removed from the printing box or frame and placed, emulsion side up, in a tray of developing solution. Generally the developing time is between 1 and $1\frac{1}{2}$ min. During development the tray is continually rocked back and forth to agitate the developer, thus ensuring that fresh developer is in contact with the paper emulsion and that all air bubbles adhering to the surface are broken. The development of the print is closely watched and when it becomes of the proper tone, depth, or darkness, it is removed from the developer and allowed to drain free of excess developing solution. It is presupposed that the exposure is such that the print is of the proper darkness after being developed for approximately the recommended length of time. The print is then placed in a shortstop solution for a few seconds before fixing. Such a shortstop bath may be prepared as follows:

Water	1000 cc	32 oz
Acetic acid, 28 per cent[1]	48 cc	1½ oz

A shortstop bath instantly stops the action of the developer and prevents stains from forming on the print. Its use also permits a larger number of prints to be fixed before the hypo solution need be discarded, because it retards neutralization of the acid in the fixing bath by the alkaline developer carried over on the prints after development. Shortstop solutions should be used fresh and, after once being used for a batch of prints, should be discarded.

Fixation. Prints, as in the case of plates and films, must be fixed in a hypo solution to remove all unreduced silver salts. The fixing bath is identical with the one used for some plates and films and may be made in the laboratory according to the formula given in Table 23, page 440. To avoid staining the prints they should not be fixed in a bath which has been used previously for films. The prints are allowed to remain in the fixing solution for approximately 15 min if the bath is fresh, and proportionally longer in solutions that have been used. During fixation, the prints should be frequently agitated, especially when a large number of prints are in the same solution, to ensure complete fixation of all parts of the prints.

Washing the Print. After fixation the prints are placed in a large tray and washed, preferably in rapidly running water. The stream of water should not be allowed to impinge on the prints directly because of the danger of bending and breaking the paper fibers. If running water is not available, the prints should be kept moving in the tray by hand agitation, and the water should be changed at least 10 times during the course of an hour. Prints are generally washed for a longer period of time than plates and films so as to ensure complete elimination of any hypo in the paper fibers. Usually the prints are sufficiently clean and free from hypo after washing for one hour. The hypo test on page 167 may be used, if necessary, to determine if hypo is completely eliminated.

Drying the Print. Glossy papers are generally dried on a ferrotype plate to obtain an additional gloss on the print. Ferrotype plates are essentially flat metal sheets that have been either chromium plated or black japanned to impart a high luster to the surface. The luster of the japanned plates is usually improved by polishing the surface with a paraffin-benzol stock solution made up in the following proportions:

Paraffin	1 g
Benzol	60 cc

[1] To make 28 per cent acetic acid from glacial acetic acid, dilute 3 parts of glacial acetic acid with 8 parts of water.

A thin coating of this wax solution is evenly applied over the ferrotype plate and, after the benzol has evaporated, the remaining layer of wax is polished with cotton or a soft cloth. Polishing should be continued until the entire surface is free from dull spots.

The prints are taken one at a time from the wash water and placed, emulsion side down, on the polished ferrotype plate. They are then rolled with a rubber roller to ensure thorough contact between plate and print and to remove excess water.

The prints are allowed to dry on the ferrotype plate. Drying generally takes about 20 min in a warm room, but may be hastened by the use of an electric fan to circulate the air. If the ferrotype plate has been properly polished, the prints will drop off as soon as they become completely dry.

Electrically heated print-drying units are available, consisting essentially of a chromium-plated ferrotype sheet and an appropriate heating unit, usually thermostatically controlled, that decrease the time of print drying to about 5 min. Because of the comparatively high temperatures reached during drying, such units are not suitable for use with japanned plates previously waxed.

Defects in the Print. Often, certain defects in the print can be traced to improper processing and unclean conditions in the darkroom. The more common defects and their possible causes are as follows:

1. A print that is too dark may be due to
   a. Overexposure of the paper.
   b. Overdevelopment of the print.
   c. Developer used at a temperature higher than recommended.
2. A print that is too light may be due to
   a. Underexposure of the paper.
   b. Underdevelopment of the print.
   c. Developer used at a temperature lower than recommended.
3. A print that is covered with brown stains may be caused by
   a. Old and oxidized developer.
   b. Old fixing bath that lacks sufficient acid.
4. Yellow stains on the print may be due to
   a. Developer too weak, old, or used at too high a temperature.
   b. Insufficient agitation of the prints when placed into the fixing solution.
   c. Insufficient fixing and subsequent washing (appears only after a lapse of some time).
   d. Underexposed print that was overdeveloped.
5. A general fogging of the print may be caused by
   a. Unsafe lighting conditions in the dark room.
   b. Old printing paper.

    c. Overdevelopment of the print.

    d. Incorrect composition of the developer, especially insufficient potassium bromide.

6. White deposit on the print is generally due to

    a. Insufficient washing.

    b. Incorrect composition of the fixing solution.

    c. Lime or similar substances in the wash water.

7. The print is too contrasty

    a. Wrong grade of paper. Use softer grade.

8. The print is lacking in contrast

    a. Wrong grade of paper. Use more contrasty grade.

References

"Elementary Photographic Chemistry," Eastman Kodak Co., Rochester, N.Y.

Henney, K. and B. Dudley: "Handbook of Photography," McGraw-Hill Book Co., Inc., New York, 1939.

Lester, H. M.: "Photo-Lab-Index," Morgan and Lester, New York, 1947.

Mees, C. E. K.: "The Fundamentals of Photography," Eastman Kodak Co., Rochester, N.Y., 1935.

Neblette, C. B.: "Photography, Principles and Practice," D. Van Nostrand Company, Inc., New York, 1938.

Neblette, C. B., F. W. Brehm, and E. L. Priest: "Elementary Photography," The Macmillan Company, New York, 1937.

"Photomicrography," 14th ed., Eastman Kodak Co., Rochester, N.Y., 1944.

Vilella, J. R.: "Metallographic Technique for Steel," ASM, Cleveland, Ohio, 1938.

CHAPTER 5

MACROSCOPIC EXAMINATION OF METALS

Metallurgical data obtained by a chemical and metallographic analysis of a metal or an alloy are usually not representative of the entire piece, owing to the heterogeneous nature of metals. Rather, such data represent the characteristics of the metal only at the particular location in the section from which the data were secured.

Without question, a single or multiple chemical and metallographic analysis of a service section is of considerable value, but the data are usually insufficient to permit an intelligent decision to be rendered concerning the serviceability of the piece in question. The data, for example, do not indicate with regard to the section of interest the general distribution and variation in size of nonmetallic inclusions; the uniformity of structure; the location and extent of segregation; the presence of fabricating defects, such as seams or hammer bursts; and residual ingot defects, such as pipe. These factors profoundly influence the over-all properties of a given section, and consequently are determining circumstances with respect to serviceability.

The nature of inhomogeneities in a metal, and the extent to which they exist therein, are best determined by macroetching a representative piece and subsequently examining the conditions thereby revealed with the unaided eye or with the aid of a low-powered microscope or magnifying glass. Such an examination is referred to as a macroscopic, or macro, examination. It is distinguished from a conventional metallographic examination by the low magnifications employed (usually not over $10\times$), and by the fact that large areas of the specimen are examined—the interest being in defects, etc., over the entire section or selected parts thereof—rather than microscopic areas at one or more locations. Macroetching and subsequent examination of the etched sections are widely used as control methods in the fabrication of steel and nonferrous metals and alloys. By macroetching appropriately selected sections at different stages in a forming process, for example, defective material can readily be detected early in the sequence of operations, and thereby the expense of continuing fabrication only to produce, finally, an inferior and low-quality product is avoided.

Although in principle macroscopic examinations and the etching techniques associated therewith are relatively simple to carry out, the various

175

conditions of the material as revealed by etching are sometimes very difficult to interpret correctly.   Competent evaluation of what is revealed requires considerable skill and experience on the part of the observer; in fact, the macroetch test is of no greater value than the accuracy with which the results are interpreted.

Macroscopic examination of metallic sections are not necessarily confined solely to the visual examination of etched specimens, but include also the examination of metal fractures.   The latter may reveal many characteristics of a metal with regard to strength, ductility, grain size, etc.   Macroetched sections, on the other hand, may reveal conditions in the metal that are related to one or more of the following heterogeneous circumstances:

1. *Crystalline heterogeneity*, the presence and extent of which depend upon the manner of solidification and the crystalline growth of the metal or alloy.

2. *Chemical heterogeneity*, owing to impurities in the metal or alloy and to localized segregation of certain chemical constituents.   Such segregation may be intentional (the introduction of carbon into the surface of steel during the process of case carburizing), or may be unintentional and undesirable, as for example, the segregation of sulphur or phosphorus that is so often found in cast steels.

3. *Mechanical heterogeneity*, arising from cold-working or any process that introduces permanent stresses into the metal.   Such heterogeneity seldom occurs in cast metals, but its presence is of importance in cold-rolled metals, forgings, etc.

The interpretation of macroetched sections, which reveals the above conditions of heterogeneity, will be discussed later.

PRINCIPLES AND PROCEDURE OF MACROETCHING

The entire process of macroetching a metal or alloy consists of three steps:

1. Obtaining a suitable sample of the metal appropriate for the purpose of the examination.   In the majority of cases where the finished section is small, permitting convenient handling, the entire piece may be used.

2. Carefully preparing the surface which is to be etched and subsequently examined (not always desirable or necessary, see page 177).

3. Appropriately macroetching the prepared or unprepared section of interest by procedures to be described.

The inhomogeneities in a metal or an alloy are revealed by macroetching only because they react at inherently different rates with the usual macroetching reagents.   Surface defects, inclusions, segregated areas, etc., are selectively dissolutioned by the etching reagent and, as a consequence, their precise locations are manifested finally by some visible etching characteristic, such as cavities, striations, local discoloration, etc.   Before etching, many of the inhomogeneities and defects associated with the section of interest may be extremely small in size—even entirely invisible. However, during macroetching the areas occupied originally by certain of these inhomogeneities will increase in size far beyond their original dimensions and eventually reach a size that will be visible to the unaided eye. Unless full cognizance is taken of this enlargement effect during macroetching, considerable confusion may arise, in some cases, when attempting to interpret the results. For example, the cavities occupied by manganese sulphide inclusions in steel may be enlarged during normal macroetching by a factor of two to three hundred times, with the attending possibility that the condition of the steel as revealed by etching may be incorrectly interpreted in terms of porosity and sponginess.

The successful application of macroetching depends upon four important factors, namely:

1. Condition of the metal surface that is to be etched, *i.e.*, whether or not it is rough, smooth, or polished.

2. Chemical composition of the etching reagent selected.

3. Temperature of the etching reagent.

4. The length of time the specimen is etched.

Selection of Sample. Surface Preparation. The selection of a representative sample for macroetching, to be removed from the whole, should be made with proper consideration for the aim of the test. Frequently, the piece may be of so convenient a size that it may itself, in its entirety, serve as the sample. Large sections should be avoided whenever possible, as they are difficult and cumbersome to handle. Likewise, long and narrow sections are unsatisfactory, as they require reagent containers of unusual shapes and designs, as well as an unnecessary excess of the etching reagent. Generally, thin slabs and disks are the most convenient to handle and manipulate during the etching process. They should not exceed, if possible, cross-sectional dimensions greater than 12×12 in.

Under circumstances where the condition of the specimen surface is of primary interest, even when the surface is covered with oxide scale, no preparation is required other than cutting the specimen to a convenient size. The action of the etching reagent will generally remove the scale and subsequently reveal the surface condition beneath. It is important, however, that adhering grease and oil first be removed from the scaled surface, in order to ensure rapid and efficient removal of the scale and subsequent etching of the specimen.

When a suitable specimen has been selected and removed from the whole by hacksawing or some similar method, the surface generally needs further preparation prior to etching. Often the presence of cutting marks so seriously efface details revealed by etching that it is difficult to interpret correctly the macro test results. The degree of surface smoothness depends upon the etching reagent selected and the condition that is most likely to be revealed by the etch.   Generally, the more drastic the action of the etching reagent, the coarser the surface finish may be.   When it is desirable to reveal fine and delicate details of the structure, the surface must, of necessity, be correspondingly smoother but it need not approach the degree of polish required for microscopic examination.

The kerf marks produced by the sawing operation may be removed from the surface by means of filing, machine grinding, or machining.   Finer surface finishes may be secured by grinding the specimen on No. 00 or No. 000 emery papers, but generally this procedure is unnecessary for iron and steel surfaces that are to be etched in the conventional hydrochloric acid solution.   Usually, round sections may be cut with a parting tool on a lathe, the surface resulting therefrom requiring no further treatment for routine examination.   It is obvious that the tool should always be sharp and properly set to avoid damaging the surface by dragging.

Whatever method is used in producing a smooth surface, it is important that during the operation the specimen be kept sufficiently cool to prevent heating the surface to an excessively high temperature.   Unless such precautions are taken, especially when grinding is carried out on high-speed grinders, the frictional heat evolved may cause structural changes to occur in the surface metal which may, after etching, lead to misinterpretation of the macrostructure.

Etching Reagents.   Any serviceable macroetching reagent, from the standpoint of practical use, should possess the following characteristics:

1. The reagent should be of such composition that it will give good all-around results and reveal the greatest number and variety of structural characteristics and irregularities present in the specimen.

2. The reagent should be simple in composition and stable, so that its concentration will not change appreciably upon standing or during use at moderately elevated temperatures.

3. The reagent should have constant characteristics at a particular temperature, so that the conditions of etching can be easily reproduced.

4. The reagent should be noninjurious and nontoxic to the person conducting the work.

Macroetching reagents for iron and steel are generally dilute aqueous solutions of one or more inorganic acids.   A solution consisting of equal volumes of hydrochloric acid (sp gr about 1.20) and water is a reagent quite suitable for general-purpose macroetching of iron and steel and has been accepted as a standard reagent for these alloys.   This reagent, when mixed in the proportions given, fulfills the above requirements for a serviceable and satisfactory etchant.   A solution consisting of 38 per cent hydrochloric acid, 12 per cent sulphuric acid, and 50 per cent water is also recommended for general use, and in some respects it is better than the first reagent mentioned because of the improved and more distinct etch produced by the presence of sulphuric acid. Although both of these reagents may be continually reused within reasonable time limits, it is recommended that fresh solutions be used for each specimen when the best possible results are desired.

In addition to the above general reagents, there are a great many solutions devised for the primary purpose of revealing specific defects and inhomogeneities in the metal. Such reagents for iron and steel are given in Table 28, page 442. For convenience, Table 29, page 443, lists alphabetically the conditions which the reagents are intended to reveal. Recommended macroetching reagents for the more common nonferrous metals and alloys are included in the tables of microscopic etching reagents given in Tables 3 to 11, pages 419 to 430.

It must be emphasized that for any particular reagent, the time of etching and the temperature of the etchant should be closely followed according to the recommendations accompanying these reagents.

Temperature of Etching. The rate at which the etching reagent attacks the specimen of interest depends mainly upon the temperature at which etching takes place. The precise influence of temperature, however, varies according to the composition, heat-treatment, and other conditions of the alloy.

The recommended temperature for the standard 50 per cent hydrochloric acid solution is 70°C (158°F). At this temperature the reagent is vigorous in its action without appreciable loss of solution or alteration in its concentration due to evaporation. The appearance of the macrostructure as influenced by the temperature at which etching takes place is illustrated in Fig. 78. It will be noted that it is difficult to make a clear distinction between the satisfactory and unsatisfactory steels etched at temperatures of 90°C (194°F) and 60°C (140°F) respectively. A distinction, however, can readily be made between the two when they are compared at the same etching temperature. It is therefore apparent that the temperature of the reagent, as one controlling factor, should be constant for all macroetching, especially if comparative results are desired.

Time of Etching. The time of etching is perhaps one of the most important factors contributing to successful macroetching and attendant appearance of the macrostructure. For example, if the time of etching is short as compared to that appropriate for a particular grade of material, the macroetched structure will not be completely developed nor will there be sufficient details revealed to permit accurate interpretation of the test. However, too long a time of etching is just as unsatisfactory as one too short, owing to details of the macrostructure being thereby obscured to

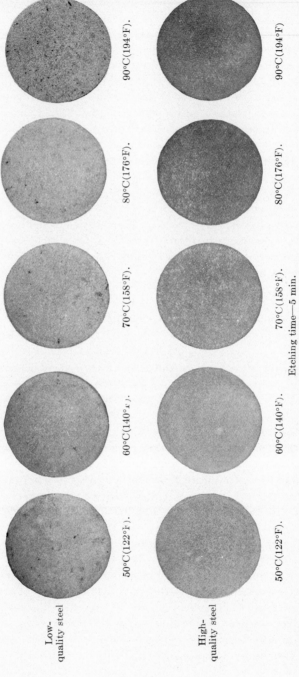

Low-quality steel: 50°C(122°F). 60°C(140°F). 70°C(158°F). 80°C(176°F). 90°C(194°F).

High-quality steel: 50°C(122°F). 60°C(140°F). 70°C(158°F). 80°C(176°F). 90°C(194°F).

Etching time—5 min.

Fig. 78.   The influence of temperature of macroetching on the appearance of the etched surface.   1×.   Macroetchant, 50 per cent aqueous solution of hydrochloric acid.

varying degrees and, frequently, some parts of the structure being completely obliterated.

Under certain circumstances, prolonged etching may produce an etched surface that will indicate the metal to be in much better condition than is really the case or than would have been indicated by shorter etching time. This anomaly is associated with the fact that a short time of etching attacks and selectively dissolves only the most readily dissoluble constituents, leaving relatively sharp edges and abrupt discontinuities over the etched surface. During prolonged etching, however, these abrupt surface irregularities will be evened out and, as a consequence, the surface will appear smoother and be misleading when evaluating the precise condition of the metal.

The time of etching depends upon the condition of the metal (normalized, hardened, etc.) and the temperature of the acid reagent. For SAE steels in a softened condition, Table 30, page 448, gives the recommended times of etching in a 50 per cent hydrochloric acid solution at a temperature of 70°C (158°F). Frequently cases may arise where the recommended etching times are inconveniently long. Approximately the same results may be obtained by etching for a shorter length of time at a proportionally higher temperature. The relation between etching time and temperature is best determined by trial, selecting as a comparative standard specimens of the same metal which have been etched at the recommended time and temperature.

The effect of time of etching at constant temperature on the appearance of the macroetched surface is illustrated in Fig. 79. It will be noted, particularly in the case of low-quality steel, that a prolonged etching treatment darkens the etched surface considerably and almost completely masks delicate details that are of evaluating importance.

Macroetching Technique. When the specimen surface has been carefully prepared, as described before, all adhering dirt and grease should be removed by a suitable solvent, such as benzol or gasoline. The cleansed specimen is then placed in a suitable receptacle containing water heated to a temperature approximately the same as that of the etching reagent to be used. Depending upon the size of the specimen, the time that the specimen is allowed to remain in the heating bath will vary. In any event, the time should be sufficiently long so that the specimen will be nearly at the same temperature throughout. The purpose of preheating the specimen is to obtain better control with regard to time of etching and to permit etching conditions to be more easily reproduced.

The preheated specimen is transferred to a vessel containing the etching reagent, already heated to the proper temperature, and is allowed to remain submerged in the reagent for a length of time corresponding to that recom-

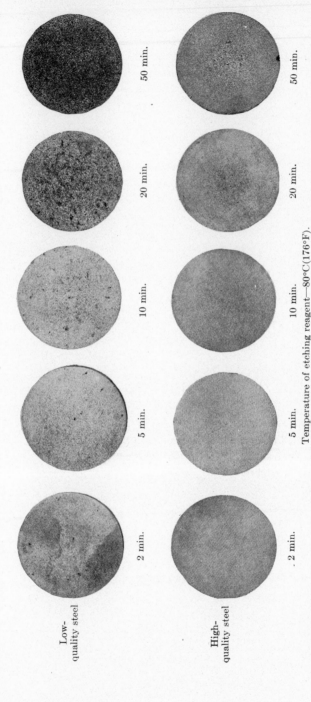

Fig. 79. The influence of time of macroetching on the appearance of the etched surface. 1×. Macroetchant, 50 per cent aqueous solution of hydrochloric acid. Temperature of etching reagent—80°C(176°F).

mended for the material of interest. The acid container used should be of sufficient capacity so that the specimen is completely immersed in the reagent and it should be constructed of a corrosion-resisting material, such as pyrex glass, porcelain, or duriron.

The acid reagent may be conveniently heated by placing the reagent dish on a gas or electric hot plate, or by the use of an electric immersion heater adequately protected by an acid-proof protection tube. An ordinary glass thermometer serves well for measuring the temperature. It is desirable, although not absolutely essential, to agitate the reagent by means of a stirring arrangement, so as to prevent smut and scum from adhering to the specimen surface during etching. Such a procedure tends to promote a more even etch.

When etching is completed, the specimen is removed from the hot acid reagent by tongs or sling wires and is then thoroughly rinsed in a stream of running water. During rinsing, the surface is carefully scrubbed with a soft-bristle brush to remove dirt and other deposits formed on the specimen during etching. After repeated washings, the specimen is rinsed in alcohol to remove the last traces of water that may be entrapped in cavities, cracks, etc., and then finally dried in a blast of hot air from an ordinary hair drier. Drying may also be quickly and effectively accomplished by placing the specimen directly on a clean hot plate.

An alternative method for cleansing the specimen after etching is to subject it to the erosive action of live steam. Scum and smut are effectively removed by this treatment, and the rise in surface temperature of the specimen due to the relatively high temperature of the steam promotes rapid drying without appreciable rusting.

Macroetched surfaces may be preserved over comparatively long periods of time by coating the surface with a thin layer of clear lacquer. For temporary preservation, the surface may be coated with either a light machine oil, such as 3-in-1, or glycerin. Very often, however, specimens will rust appreciably upon standing, even though precautions have been taken to prevent this from occurring. Rusted surfaces may be cleaned through the use of fine emery papers. In cases where this is not feasible, the rust may be removed by soaking the specimen in either a hot or a cold 5 to 20 per cent aqueous solution of sodium or ammonium citrate, a 50 per cent aqueous solution of phosphoric acid to which a little sugar has been added, or a 5 per cent alcoholic solution of phosphoric acid. Following this rust-removal treatment the specimen should be thoroughly washed in warm running water, rinsed in alcohol, and finally dried in a blast of warm air.

Methods of Recording Macrostructures. *Photographic.* Conventional photographic methods are probably the best available for permanently recording details of macroetched specimens or the fracture characteristics of metal failures. In some cases, particularly when the specimen of interest is relatively small and the requisite mode of illumination is simple and direct, macroscopic photography can be carried out on a metallograph. However, to do so requires the use of a metallograph that can be arranged for the purpose and the use of special auxiliary equipment. Because of the limitations associated with the best of such arrangements, modern metallographic

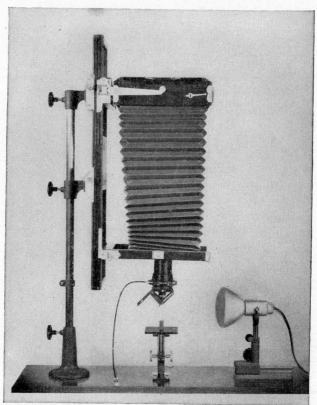

FIG. 80. Vertically mounted studio-type camera and auxiliary equipment arranged for macroscopic photography. (*Courtesy of Bausch and Lomb Optical Co.*)

design no longer includes macroscopic features, other than the feature of securing photomicrographs of flat surfaces at low magnifications.

Perhaps the most versatile type of camera for photographing either large or small macro specimens is a studio camera mounted horizontally on an adjustable tripod or optical table; of less versatility is a vertical arrangement, illustrated in Fig. 80, which is commercially available especially for photomacrography purposes. A camera designed to accommodate a plate or film of 5 × 7 in. maximum is one convenient to manipulate, and the negative size is sufficiently large for most purposes. If larger finished prints are desired, however, they may readily be secured from a 5 × 7-in. negative by projection-printing methods.

Almost any photographic lens of high quality and appropriate focal length, such as a Zeiss Tessar, Leitz Summar, or Kodak Ektar, is ideally suited for photomacrography. These lenses represent a group which are well corrected for aberrational errors and which are capable of excellent resolving power and the formation of an extremely sharp image. Furthermore, these lenses have an adjustable iris diaphragm—a requisite of the lens that is essential for photomacrography purposes—which may be "stopped down" to improve the depth of focus and image sharpness.

The use of a studio-type camera usually requires artificial lighting facilities to illuminate properly the specimen of interest. In some cases, however, particularly if the specimen is flat and exhibits many specular or high-reflectivity areas, ordinary daylight illumination may be more appropriate than artificial lighting. Because of the many different directions in which daylight will illuminate such specular areas, they will, as a consequence, be rendered more natural in appearance in the finished print, *i.e.*, they will be of light tone values. This circumstance is readily demonstrated by comparing the reproductions of the high-reflectivity areas in Figs. 101 and 104*b*, which were photographed with daylight illumination, with such areas in Fig. 87, which was photographed with artificial lights appropriately arranged.

To secure high-quality photographs of irregularly shaped specimens or uneven surfaces, such as fractures, requires the use of artificial illumination provided by two or more photoflood lamps or their equivalent, and possibly by one or more photographic spot lights, preferably equipped with adjustable Fresnel lenses. With such an array of lighting equipment it is possible to secure an image that appears natural and realistic, and through appropriate arrangement of the lights it is possible to emphasize certain details of interest in the specimen. For any given size and shape of specimen, the proper arrangement of lights is best determined by trial. The conventional method of approach consists of observing the illuminated image on the focusing screen of the camera, and then adjusting and moving the lights until the image appears to be adequately shadowed and highlighted and is rendered realistic in appearance with respect to the object. To ultimately achieve such an illuminating arrangement may require the use of more or less lighting units than were originally contemplated, or the introduction of one or more appropriately placed spot lights.

A medium-speed, medium-contrast panchromatic type of film is usually preferred for most macroscopic photography, owing to the rather wide latitude and excellent tone gradation characteristic of this type of photographic emulsion. However, objects that are lacking in color and particularly those that reflect predominately the shorter wave lengths of visible radiation (steel specimens as compared to some high-colored nonferrous specimens) may be photographed successfully on orthochromatic emulsions. To ensure appropriate rendition of colors in terms of black and white tonal values, orthochromatic films should be used in conjunction with an appropriate lens filter, preferably one that is light, medium, or deep yellow in color.[1]

Contact Printing. An alternative method for recording the details of a macroetched specimen is by contact printing. It is to be recommended when photographic equipment is not available. This method, however, requires a certain amount of skill in order to produce satisfactory results, especially when the etched surface is not perfectly flat.

A thin coating of black printer's ink is evenly distributed over the etched surface by means of a rubber squeegee roller. To ensure an even distribution of ink on the roller, it is passed back and forth over a sheet of glass on which a small amount of ink has been previously spread with a spatula.

The inked surface of the specimen is placed in contact with a sheet of heavy white paper and the two are firmly pressed together. The paper should be backed with a suitable stiff material, such as heavy pasteboard, to ensure an even distribution of pressure over the paper. When the specimen is removed, the details of the etched surface will be imprinted on the surface of the paper. Reasonable care should be exercised in removing the specimen from the paper so that the two surfaces are not rubbed together, causing the reproduction to be blurred.

The correct amount of ink to be applied to the surface of the specimen is best determined by trial. Too little ink will result in a light print, with loss of detail due to lack of registration; whereas, if too much ink is used, a smeary print will be produced, with loss in detail due to blurring.

Sulphur Printing. Although sulphur printing cannot be used as a general method for recording macroetched structures, it nevertheless affords a convenient and specialized means for detecting and permanently recording the distribution of sulphur in steel. Sulphur may exist chemically in steel in one of two forms, either as manganese sulphide or as iron sulphide, the amount of the latter depending upon the amount of manganese originally present.

The surface of interest to be tested for distribution of sulphur should be reasonably smooth and free from foreign matter such as dirt and grease. Grinding the surface on No. 00 or No. 000 emery paper and subsequent

---

[1] For a detailed account of light filters and their use, see Eastman Kodak Co.: "Kodak Data Book on Filters," Rochester, N.Y., 1944.

thorough washing will generally produce a surface satisfactory for the purpose.

Photographic bromide paper is soaked in a 2 per cent aqueous solution of sulphuric acid for approximately 3 or 4 min. This operation, as well as the entire process of sulphur printing, may be carried out in daylight; contrary to the usual requirements for the handling of photographic paper. Although in principle any surface finish on the paper will produce the desired results, it is recommended that a semimatte paper be used, so as to minimize the danger of slippage when the paper is placed in contact with the specimen surface. The action of the dilute acid on the gelatin of the paper tends to make it soft and slimy, this condition being decidedly worse on glossy-finished paper than on semimatte.

The paper is removed from the acid solution and allowed to drain free from excess solution. The emulsion side of the paper is then placed in direct contact with the prepared specimen surface and allowed to remain in contact under moderately applied pressure for 1 or 2 min. Care must be taken that all entrapped air bubbles between the paper and the specimen surface are eliminated. This may be assured by carefully pressing with a squeegee print roller.

The reaction of the sulphuric acid with the sulphide regions of the steel produces hydrogen sulphide gas, which reacts with the silver bromide in the paper emulsion, forming a characteristic brown to gray-black deposit of silver sulphide. These reactions may be expressed as follows:

$$FeS + H_2SO_4 \leftrightarrows FeSO_4 + H_2S \tag{1}$$

or

$$MnS + H_2SO_4 \leftrightarrows MnSO_4 + H_2S \tag{2}$$

$$H_2S + 2AgBr \leftrightarrows Ag_2S + 2HBr \tag{3}$$

When the reaction has proceeded for approximately the recommended length of time, the photographic paper is removed from the surface of the specimen, rinsed in clear running water, and then fixed permanently by placing it in a photographic fixing solution for about 15 min.[1] When fixation is completed, the print is again washed in running water for approximately 30 min and subsequently dried in the usual manner.

The examination of a properly prepared sulphur print will disclose quite clearly, because of the presence of darkly colored areas of silver sulphide, the precise location of sulphur inclusions on the prepared surface of the metal. A grouping or agglomeration of such silver sulphide areas indicates the presence of sulphur segregation, whereas a random dispersion of the

[1] For further detailed information on fixation and print washing, see p. 172.

spots denotes a more uniform, and perhaps less harmful, distribution of the sulphur inclusions.   A typical sulphur print obtained from a low-carbon steel plate is shown in Fig. 81.

Phosphorus Printing.   In a manner similar to the procedure described for sulphur printing, contact prints may be obtained which will indicate the distribution of phosphorus on the surface of steel.   The surface of the specimen to be so tested is usually prepared by successively grinding through No. 000 emery paper; as in the case of sulphur printing, the surface must be cleaned after grinding by washing in warm running water and carefully drying.

Photographic or filter paper is first soaked in a solution consisting of ammonium molybdate (5 g per 100 ml of water) to which 35 ml of nitric acid (density 1.2) has been added.   The paper is then removed from the reagent,

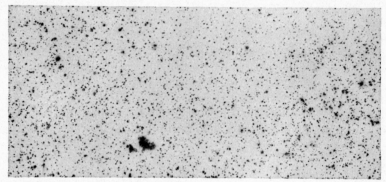

FIG. 81.   Typical appearance of a sulphur print secured from a steel of relatively high sulphur content.   1×.

drained nearly free of excess solution, and applied directly to the prepared surface of the metal, care being exercised to effect intimate contact between the paper and the surface.   The paper is allowed to remain in contact with the steel for approximately 5 min.

After being removed from the surface of the specimen, the paper is developed for approximately 4 min in an aqueous solution of 35 per cent hydrochloric acid, to which a little alum and 5 ml of a saturated stannous chloride solution have been added.   During development the complex phosphomolybdate areas formed on the paper during the first stage of the process are partly reduced.   This reduction effects a change in the color of these molybdate areas from a yellow-brown to a characteristic blue.   Thus blue areas on the print indicate the location of phosphorus-rich areas on the surface of the metal, and by noting the relative intensity of the blue color the amount of phosphorus present may be estimated.   Generally the intensity of the color increases with increased amounts of phosphorus.

Oxide Printing. A satisfactory contact method for revealing oxide inclusions in steel has been developed in which a technique is used similar to that described for sulphur printing. This method[1] is a modification of that first proposed by Niesner, and is intended to reveal most types of oxide inclusion except those that are very small in size and alumina inclusions that are low in ferrous oxide content.

The method consists essentially in placing a sheet of photographic paper, previously soaked in a 1:20 aqueous hydrochloric acid solution, in intimate contact with a well-prepared surface of the specimen under examination. The paper is allowed to remain in contact with the surface for approximately 1 to 2 min, after which it is removed and developed in a solution containing 20 g of potassium ferricyanide per liter of solution. Development is allowed to proceed for about 10 min, or until a desired bluish tint is obtained. When development is completed, the print is washed in running water for approximately 10 min, fixed in a photographic fixing solution, rewashed for approximately 30 min, and subsequently dried in the usual manner.

Examination of the finished print will disclose the relative size, location, and distribution of the oxide inclusions on the surface of the specimen.

INTERPRETATION OF CONDITIONS IN STEEL AS REVEALED BY FRACTURES AND MACROETCHING

It is beyond the scope of this book to discuss and to illustrate all the conditions that may be revealed by macroetching the many nonferrous and ferrous metals and their alloys. Inasmuch as macroetching finds wide application in the steel industry, the discussion will be confined to the interpretation of the more common types of fractures and macroetched structures characteristic of steel.

Examination of Fractures. The examination of fractures of metallic specimens that either have failed in service or have been tested to destruction in testing machines often reveals certain characteristics pertaining to the properties of the metal and may indicate to some extent the nature of the failure.

Fractures of comparatively large sizes are most conveniently examined under good illumination with the unaided eye. Smaller size fractures or local areas of interest on larger ones may be critically studied with the aid of a hand magnifier or a low-powered binocular-type microscope. In many cases a binocular microscope is preferred as it will impart depth and a natural perspective to the observed image, provided the depth of focus of the objective system is sufficiently great.

---

[1] Newirth, F., R. Mitsche, and H. Dienbauer: Anwendbarkeit des Oxydabdruckverfahrens nach M. Niesner, *Arch. Eisenhüttenw.*, Vol. 13, 1940.

To describe fractures adequately, even qualitatively, is sometimes difficult because no standards are available upon which to base such a description.   It is, therefore, common practice to describe fractures in terms of
their appearance, such as woody, fibrous, silky, fine, coarse, tough, cupped,
etc., or in other terms that seem appropriate and provide little chance
for misunderstanding.

It is a common occurrence for metal drive shafts, axles, machine mem-
bers, etc., when in service and subjected to vibratory or alternating stresses,
to fail by a process of progressive failure known as fatigue.   When a metal
is subjected to alternating stresses of sufficient magnitude to cause eventual
failure, even though the applied stress would, in general, be statically ineffective, small microscopic cracks are formed, usually originating on the

Fɪɢ. 82.   Fatigue fracture of a 6-in. drive shaft, illustrating the typical coarse- and fine-
appearing areas of the fracture surface.   ⅓×.

surface of the member.   Surface imperfections such as minute pits or
scratches, tool marks, sharp-cornered fillets, and inclusions often serve as
nucleuses for these cracks.   With repeated alternate stressing of the metal,
these small incipient cracks increase both in size and number and gradually
they are propagated to within the interior of the piece.   Eventually the
cross section of the member is so weakened by lack of supporting metal
that it fails abruptly and usually with no warning whatsoever of impending
failure.

A typical fatigue fracture of a drive shaft is shown in Fig. 82.   As characteristic of many fatigue failures, it will be noted that the approximate
upper half of the fracture appears relatively smooth and silky whereas the
lower half appears coarse and irregular.   The smoothness of the upper part
of the fracture may be attributed to the fact that it was through this particular region of the cross section that the fatigue cracks were propagated during progressive failure of the member.   This circumstance permitted the

two cosurfaces to rub together during repeated stressing of the shaft and thus become smooth.

When failure had progressed inwardly to such an extent that insufficient solid metal was present to withstand the applied stress, the shaft suddenly failed, producing the coarse-appearing portion of the fracture. In this region of the cross section no time was allowed for the cosurfaces to be in rubbing contact with one another and thus become smooth. Because of this coarse-appearing part of the fatigue fracture and because the mechanism of fatigue failure was not clearly understood, it was common in the past to attribute the cause of fatigue failure erroneously to "crystallization" of the metal.

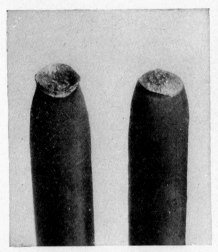 

a. Fracture indicative of relatively good ductility. Note well-defined cup-cone fracture.

b. Fracture indicative of relatively poor ductility. Note absence of cup-cone fracture.

Fig. 83.   Typical appearance of tensile fractures of relatively ductile and brittle steels.   1×.

The appearance of a fracture resulting from such common modes of failure as tension, impact, torsion, etc., depends to a large extent upon the chemical composition, grain size, and structural condition of the metal. These variables in many cases, with the exception of the first, may be directly associated with the prior heat-treatment given the metal. The appearance of such fractures, as influenced either singly or by a combination of these variables, may range from one that is coarse and "woody" to one that is relatively straight and silky. There are intermediate fractures of varying degrees of coarseness and fineness.

The examination of a fracture resulting from a tensile test will generally indicate whether or not the metal is uniform and ductile. A well-formed cup-cone type fracture, as shown in Fig. 83*a*, indicates that the metal possesses relatively high tensile ductility, whereas, a metal of comparatively low ductility reveals by the same test, as shown in Fig. 83b, a more abrupt and straight break with little if any "necking down" at the point of failure.

The fracture of a soft tough metal, i.e., one that possesses a relatively high resistance to rupture after permanent deformation has occurred, is usually of the woody, or fibrous-appearing, type, as typified by the fracture of wrought iron shown in Fig. 84. The appearance of the fracture in this case may be attributed to the comparatively soft and ductile ferritic matrix and the discontinuities within the structure due to the dispersion of elongated slag inclusions.

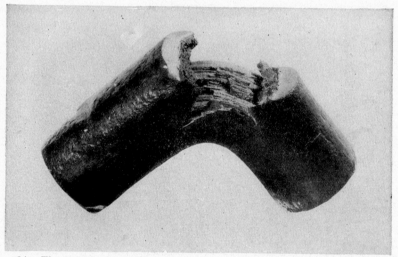

Fig. 84. The appearance of a fibrous type of fracture, typical of wrought iron. 1×.

The examination of a fracture resulting from an impact failure may often indicate, among other things, the relative grain size of the failed metal which in turn may permit an estimation of the relative toughness and ductility of the piece. In the case of a normalized steel, a straight, relatively smooth and silky fracture, as shown in Fig. 85a, generally indicates a comparatively fine-grained metal having greater impact strength and ductility than that possessed by a coarse-grained steel of identical composition and exhibiting a coarse, bright, and crystalline-appearing fracture, as shown in Fig. 85b.

The relative grain size of fully hardened steel may be estimated in like manner; in general, the finer the grain size present, assuming constant grain size throughout the piece, the finer and more silky the fracture of the hardened zone will appear. However, in contrast to the fracture appearance of normalized or similarly heat-treated steels, the corresponding fractures of fully hardened steels will usually be finer and more silky in appearance and will never appear "woody" or "teary." Because of the relationship existing between martensitic grain size, austenitic grain size, and the appearance of the fracture of fully hardened steel, a fracture method has been established for determining the size of the prior austenite grains existing in the steel at the quenching temperature. Although details of this method are discussed fully on page 278, the procedure is essentially one of visually comparing and matching the fracture of fully hardened steel with one out of a series of ten standard grain-size fractures, the grain size of each standard fracture being known and correlated with the comparative standards used in metallographic methods.

a. Indicative of relatively good toughness and fine grain. 2×.

b. Indicative of relatively poor toughness and coarse grain. 2×.
Fig. 85.   The appearance of some typical impact fractures of steel.

It is evident from the few illustrations cited that the fracture appearance of any given steel is distinctly characteristic, depending among other things upon the structural condition of the metal. Since this condition is usually directly related to prior heat-treatment, it is therefore relatively easy to note by examination of an appropriate fracture whether or not the steel section has been surface-hardened, and, if so, to determine with some degree of accuracy the depth of the hardened case. As shown in Fig. 86, the fracture of the hardened surface appears relatively smooth and silky, whereas the fracture of the softer and more ductile core appears woody and more or less coarse. Further evidence that the piece has been surface-hardened is shown in the macroetched section of Fig. 87. Although the hardened case etches more lightly than the softer core, it appears dark in Fig. 87 because of scattered light reflections present when the section was photographed (see page 185).

Examples of Macroetched Structures of Steel.[1]   Before the various conditions in steel as revealed by macroetching can be discussed intelligently, certain fundamental concepts of the mechanism of solidification must be fully understood.

When molten steel is poured into an ingot mold, the liquid metal in immediate contact with the walls of the mold is the first to solidify, forming a shell of solid metal of more or less uniform thickness.   This preferential solidification occurs mainly because the relatively cold walls of the mold

FIG. 86.   Longitudinal fracture of a surface-hardened, hollow-type air-hammer piston.   1×.
Note smooth appearing type of fracture adjacent to outer and inner surfaces.

FIG. 87.   The fracture surface of Fig. 86 prepared and macroetched to reveal the hardened case.   1×.   Macroetchant, 5 per cent nital.

impart a local chilling effect to the liquid metal, causing the temperature of the metal to drop rapidly in these areas of contact with attending spontaneous nucleation of solid metal at many independent centers.

The thin shell of solid metal first formed during solidification is generally composed of small heterogeneous equiaxed grains, *i.e.*, grains of approximately equal "diameters" in all directions.   The actual size of these equiaxed grains depends to a great extent upon the relative cooling rate; in

¹ Many of the photographic illustrations that follow have been furnished by O. V. Greene of The Carpenter Steel Company, Reading, Pa.

general, the faster the rate of cooling, the smaller will be the size of the grains.

As solidification proceeds with fall in temperature, the liquid metal solidifies inwardly from this outer shell of equiaxed grains and, if the temperature of the remaining liquid metal drops sufficiently fast and uniformly, small equiaxed grains will continue to form throughout the remainder of the ingot. In a less ideal case, however, such as encountered in commercial steelmaking practice, solidification will usually continue by the growth of long columnar grains, growing inwardly from the shell of solid metal first formed. Owing to the low rate at which heat is conducted away from the central region of the liquid metal, and the attending circumstance of a plentiful supply of liquid metal, the growth of grains is more or less restricted to a direction normal to the bottom and side walls of the ingot mold. Growth in other directions will be limited owing to the presence of similar adjacent growing grains.

In many cases the columnar grains will continue to grow in their preferred direction until the ingot is completely solid, giving rise to a situation where the columnar grains from opposite sides of the ingot mold actually meet at the center of the ingot. In other cases, however, where the conditions of cooling are somewhat different, the temperature of the liquid metal may drop below its freezing point before the columnar grains actually meet and, as a consequence, equiaxed grains may be formed in this central region.

The mechanism by which metallic grains grow from the liquid, whether their ultimate external shape is equiaxed or columnar, is by a process of dendritic growth. The dendritic mechanism is one associated with different rates of growth in different crystallographic directions. In a rather gross analysis, the solid phase is initiated by the formation of randomly oriented nucleuses, and from each nucleus growth takes place at different rates in different directions, being manifested in certain stages of the process by the existence of a branched skeleton crystal resembling very much the physical shape of a fir tree. As solidification proceeds, the interstices between the dendritic branches are filled with solid metal, eventually forming, when solidification is complete, a metallic grain. The grain so formed, in the case of such a complex alloy as steel, is not chemically homogeneous; and the external shape or symmetry of the grain is influenced principally through contact during solidification with neighboring dendritic growths.

The liquid metal that is the last to freeze, *i.e.*, the liquid metal between the dendritic branches and the various growing dendrites, is usually richer in carbon, phosphorus, sulphur, and both metallic and nonmetallic impurities than are the first branches of the dendrites to form. The degree to which such chemical heterogeneity exists is dependent primarily upon the conditions of solidification and rate of cooling.   Fortunately carbon diffuses at a rather rapid rate at elevated temperatures, and hence is dispersed uniformly throughout the grain.   Sulphur and particularly phosphorus, however, diffuse much more slowly, leaving undissipated concentration gradients with regard to these elements in any one grain.   Although the dendrites themselves lose their identity as such, dendritic segregation

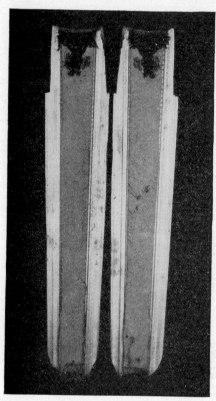

remains—as evident in an appropriately macroetched section—unless such segregation is eliminated by diffusion of the segregating constituents.

The transition of liquid to solid steel as described is always accompanied by an evolution of gases and a decrease in total volume, *i.e.*, shrinkage.   Owing to these circumstances, an ingot mold completely filled with liquid metal will exhibit after solidification a hollow top cavity or pipe, as illustrated in Fig. 88.   Since the metal immediately adjacent to the pipe is the last portion of the ingot to undergo solidification, there is found therein rather gross segregation of impurities, dirt, nonmetallic inclusions, etc.   In order, therefore, to secure wrought products of high quality, it is essential, in conformation to good steelmaking practice, that this portion of the ingot is cropped and discarded before the ingot is mechanically worked into final shape and form.

FIG. 88.   An ingot of steel split longitudinally to show pipe formation.

Dendritic Segregation.   Dendrites, as previously discussed, are the result of the crystallization characteristics of the ingot.   The presence of a dendritic structure generally indicates inhomogeneities in the steel in the form of either segregated metallic or nonmetallic constituents.   In the normal solidification and subsequent cooling of an ingot, part of the nonmetallic constituents are generally entrapped in the interdendritic fillings and remain segregated in these parts because of their inability to diffuse throughout the grains.   A dendritic pattern formed in this manner may reduce the strength and ductility of the steel and may (although this is not definitely known) have some influence on premature service failures. If the dendritic pattern is formed by metallic elements that are capable of diffusion, the effect of the pattern on the mechanical properties is considered negligible. Once formed, however, and particularly so if due mainly to phosphorus, the structure persists and is difficult to eliminate by either heat-treatment or mechanical working. A typical dendritic pattern as present in one end of a 1.0 per cent carbon-steel ingot is shown in Fig. 89.

Columnar Structure. Figure 90 shows large columnar grains formed in an ingot during solidification. Such a structure is usually found only in

Fig. 89. A mild dendritic pattern in a 1.0 per cent carbon-tool-steel ingot. Note horizontal grinding marks not completely eliminated during preparation of the surface. Macroetchant, 50 per cent aqueous solution of hydrochloric acid.

ingots that have been slowly cooled, *i.e.*, where the ingot has remained liquid for a comparatively long time. During the process of solidification the columnar grains projecting from each side grow at approximately the same rate until further growth is obstructed by contacting similarly growing grains from adjacent and opposite sides. It will be noted that such grain growth is terminated along diagonal planes, which extend the full length of the ingot. Such dividing planes, known as planes of weakness, are sometimes more pronounced than those shown in Fig. 90, and are potential sources of weakness when the ingot is mechanically worked. Planes of weakness can generally be eliminated in the ingot by casting into ingot molds provided with rounded corners and edges.

Pipe. If the ingot pipe is not completely removed and discarded before rolling, forging, or other working operations on the ingot are complete, it may be present to some extent in the finished piece. The presence of pipe will lower considerably the mechanical properties of the steel, not only because of the gross segregation of dirt, impurities, etc., associated with the pipe, but also because of discontinuities within the metal. These discontinuities are principally the result of the pipe cavity not being completely welded shut during the initial break-down of the ingot—a circumstance

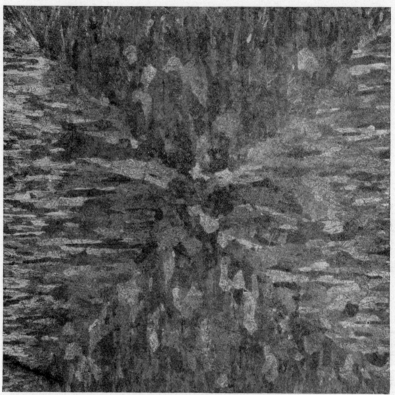

Fig. 90.   Transverse section of a slowly cooled ingot of tool steel, showing a typical columnar structure.   Macroetchant, 50 per cent aqueous solution of hydrochloric acid.

associated with the oxidized condition of the pipe cavity which was initially present or brought about during the hot-working operation.

A typical example of remanent piping in a steel billet, as revealed by appropriate macroetching, is shown in Fig. 91a.

Very often piping in defective steel will exist on such a gross scale that it may be detected without macroetching, as illustrated in Fig. 91b. In such cases, however, the appearance of the pipe may closely resemble that of hammer bursts (internal rupturing of the metal due to faulty forging tech-

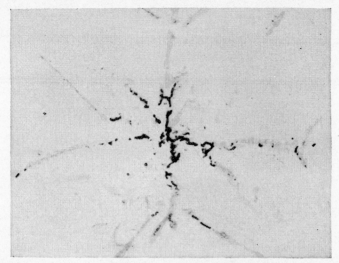

a. Macroetching required to reveal the presence of pipe. Macroetchant, 50 per cent aqueous solution of hydrochloric acid.

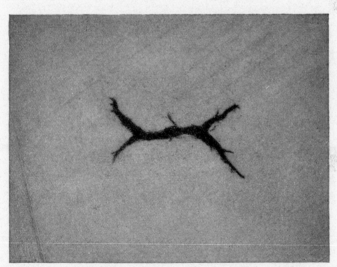

b. Macroetching not required to reveal the presence of pipe. Specimen unetched.

FIG. 91. Typical appearance of remanent ingot pipe under two different circumstances.

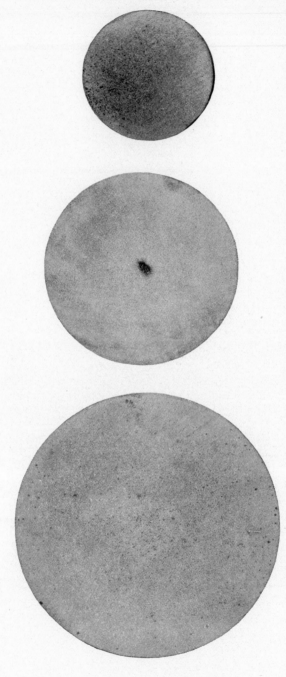

FIG. 92. Typical examples of excessive and unsatisfactory distribution of nonmetallic inclusions in steel. 1×. Macroetchant, 50 per cent aqueous solution of hydrochloric acid.

Fig. 93. Typical examples of small and satisfactory distribution of nonmetallic inclusions in steel. 1×. Macroetchant, 50 per cent aqueous solution of hydrochloric acid.

niques), but the two defects can usually be distinguished by etching. Material that is piped will generally exhibit spongy and porous areas surrounding the defect, whereas these conditions will not be present after etching if the defect is a hammer burst.

Segregation. The macroetched appearance of low-quality steel, owing to conditions of segregation, and large amounts of nonmetallic inclusions and unfavorable sizes, is illustrated in Fig. 92.

Segregation in steel as revealed by macroetching techniques is not always an unequivocal indication of defective material. Metallographic examination of such segregated areas is recommended when in doubt, in order to

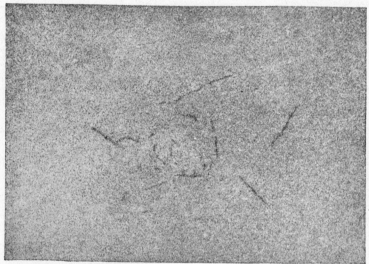

FIG. 94.   Internal cooling cracks in a steel plate revealed by macroetching.   1×.   Macro-etchant, 50 per cent aqueous solution of hydrochloric acid.

confirm whether or not the areas comprise metallic segregation, concentration of impurities, or a manifestation of crystalline orientation.

Examples of satisfactory distribution of included material in steel and of particle sizes not considered objectionable are illustrated in Fig. 93. As will be noted, the inclusions are small in size and they are dispersed rather uniformly over the entire cross section of each bar. These steels, from considerations of nonmetallic inclusions and their distribution, are of much better quality than those represented in Fig. 92.

Internal Cracks. Internal cracks, sometimes called flakes, hairline cracks, or cooling cracks, are shown in Fig. 94 and are the result, in this particular case, of unfavorable cooling conditions after hot-working. These cracks are actual discontinuities in the metal and are potential sources of failure when the piece is subjected to further mechanical working or heat-treatment.

The disastrous results of heating and quenching this particular piece (Fig. 94) for purposes of hardening are shown in Fig. 95. It will be noted that the piece has been completely split apart, and although not actually discernible in the illustration, the path of fracture was initiated by, and followed through, some of the original internal cracks.

Porosity. Porosity in a steel ingot and the eventual presence of this condition in fabricated material, can usually be related to insufficient feeding of molten metal during solidification of the ingot and to evolution of gases. As in the case of pipe remanents, porosity may be associated with actual discontinuities within the metal that are readily detectable without

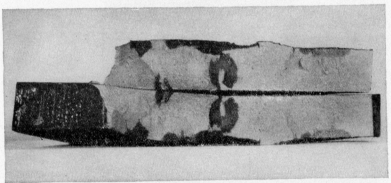

FIG. 95.   The full section of Fig. 94 showing the fracture resulting from water quenching during hardening.

etching, as illustrated in Fig. 96b. More often, however, macroetching is required to reveal porous areas through circumstances of selective dissolution of segregated inclusions and reasonable enlargement of small defects, particularly an agglomeration of small blowholes.

The typical appearance of porosity in a rolled bar, as carried over from the initial ingot, is illustrated in Fig. 96a.

Surface Seams. Because of adverse effects on steel quality, the presence of surface seams in fabricated or semifinished material is considered objectionable. This is particularly true if the piece or member is to be subjected to vibratory stresses, since surface seams and other surface defects readily serve as initiating sources for fatigue failure.

Surface seams can originate from a number of attending circumstances, most notably from faulty rolling or drawing operations; by forcing oxide scale into the metal during mechanical working operations; or from original imperfections on the surface of the ingot and elongated blow-hole defects that became oxidized during the initial stages of hot-working and are thus rendered unweldable.

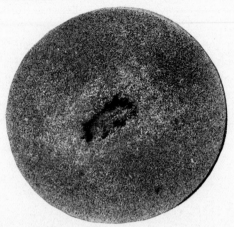

a. Center porosity in a rolled bar revealed by macroetching. 1×.

b. Center porosity in a tool-steel billet.   Macroetching not required in this case to reveal this defect, although, as noted, specimen is macroetched.   1×.

FIG. 96.   Typical appearance of porosity in steel.   Macroetchant, 50 per cent aqueous solution of hydrochloric acid.

Generally, macroetching is not necessary to reveal surface seams if the examination is conducted on an appropriate transverse section.   However, when the defect is to be detected by merely examining the worked surface, such as illustrated in Fig. 97*b*, macroetching is appropriate and essential, particularly if the surface of interest is covered with oxide scale.

Flow Lines. It is sometimes desirable to determine whether or not a finished piece has been forged, cut, or cast into shape and to note, in the

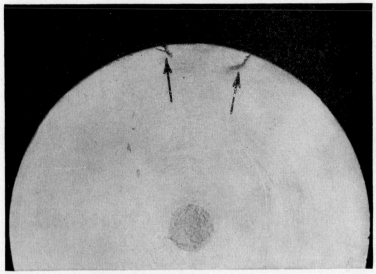

a. Surface seams (indicated by arrows) visible on the transverse section of a rolled billet without macroetching. Approximately $\frac{1}{2}\times$.

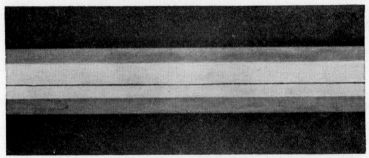

b. Surface seams on the surface of a rolled bar revealed by macroetching. $\frac{1}{2}\times$. Macro-etchant, 50 per cent aqueous solution of hydrochloric acid.

FIG. 97. Typical appearance of surface seams in rolled steel.

event that it was forged, the direction of metal flow.   Flow lines as revealed by macroetching in forgings are not indicative of defective material since these etching characteristics are a natural consequence of applied mechanical working.   If, however, the flow pattern shows highly selective etching characteristics, it is likely that the material may be defective and may contain an excessive amount of inclusions and segregated areas. Whether or not this condition is prevalent can be determined readily through microscopic examination of specimens secured from the regions of interest.

A macroetched forging, shown in Fig. 98, discloses a directional flow pattern consisting of streaks and striations. The orientation of this pattern with respect to the plane of the prepared surface indicates the direction of metal flow during deformation. The flow lines are made visible because the elongated inclusions and other elongated heterogeneous areas are selectively attacked by the etching reagent. Another contributing factor in revealing the flow pattern is the differential attack by the etching reagent of areas that have been stressed unequally during deformation. The reagent

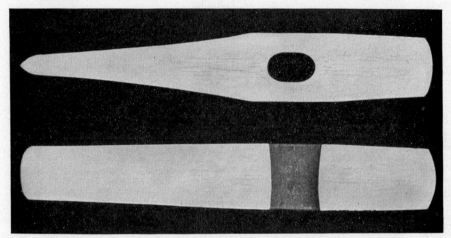

Fig. 98.   Appearance of flow lines in a typical steel forging. Upper illustration is a top view of the section; lower illustration is a longitudinal cross section of the forging.   Macroetchant, 50 per cent aqueous solution of hydrochloric acid.

is usually more reactive in those areas that have been highly stressed than in those regions which have been only slightly or moderately deformed.

Grinding Cracks.   During machine grinding of fully hardened steel it is not uncommon for surface cracks to form that may be immediately discernible to the unaided eye or that may be of such fineness that macroetching is necessary to make them visible. Grinding cracks, an example of which is illustrated in Fig. 99, arise from the introduction, or unequal relief, of surface stresses due to local heating of the surface during the grinding operation; and possibly to stresses introduced by localized transformation of retained austenite attending this frictional heating effect. However, even with the utmost care exercised during grinding, cracks may appear because of prior improper hardening of the piece of interest, or because of embrittling effects imparted to the steel through overheating.

Quenching Cracks. It is not uncommon for certain classes of steel, particularly those of high carbon and alloy content, to suffer cracking when quenched from some appropriate temperature for purposes of rendering a fully hardened structure. The internal stresses introduced into the piece, owing to the inevitable expansion associated with the transformation of

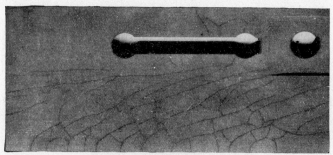

FIG. 99. Grinding cracks on the surface of a hardened steel die. ½×. Macroetchant, 50 per cent aqueous solution of hydrochloric acid.

austenite to one or more transformation products, may be sufficiently great and sufficiently unbalanced to cause rupture. Depending upon the conditions of the heat-treating procedure, as well as the size and shape of the steel piece (which would affect the distribution of the stresses), the cracks

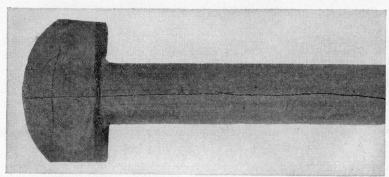

FIG. 100. Cracks produced in an upset bar of steel when water-quenched during heat-treatment. 1×. Macroetchant, 50 per cent aqueous solution of hydrochloric acid.

formed may be relatively large or they may be small and difficult to detect with the unaided eye. In the latter case, macroetching is usually necessary to reveal the cracks, as shown in Fig. 100.

It is well known that in pickling or macroetching fully hardened or otherwise highly stressed steel, the action of the usual etching reagent may cause surface cracks in perfectly sound material, owing to the unequal relief of surface stresses. If this situation occurs, it may cause considerable confusion when one attempts to evaluate the results of the test and correlate the appearance of the etched surface with the preceding treatment the metal has received.   It is therefore recommended that, whenever possible, fully hardened steel be tempered prior to macroetching.   In cases where this procedure is not feasible, but where it is still desirable to reveal surface cracks, the section of interest may be coated with either finely divided whiting or zinc oxide suspended in kerosene or penetrating oil. The suspended whiting will penetrate the surface cracks and, after the mixture has dried and the surface of the specimen has been carefully wiped free of excess whiting, the surface cracks will appear as fine white lines.[1]

Depth of Hardness.   Because there are very pronounced differences in the etching characteristics of hardened and unhardened steel, macroetching may be conveniently used to determine the depth of hardness, or hardness

FIG. 101.   Transverse section of a water-quenched steel bar, macroetched to reveal the depth of hardening.   1×.   Macroetchant, 2 per cent nital.

penetration, in a steel appropriately heat-treated.   The depth of hardness is best revealed by etching a prepared transverse section of the piece with an etching reagent somewhat milder in its action than the usual hydrochloric acid solution.   Nital,[2] in concentrations of either 2 or 3 per cent, is an ideal reagent for this purpose, developing good contrast between the hardened and the unhardened zones.   As noted in Fig. 101, the hardened case produced by brine quenching from an appropriate temperature etches more lightly, and thus is readily distinguishable from the more severely etched core of somewhat lower hardness.

Welded Sections.   It is often desirable, and frequently necessary, to determine the soundness of a welded joint and to observe macroscopically over a cross section of the weldment the various zones wherein structural changes have occurred.   This may be conveniently accomplished in part by macroetching the prepared surface with the usual hydrochloric acid solution, or in some cases with a milder reagent such as 2 or 3 per cent nital.

[1] Other methods of crack detection are discussed on p. 318.

[2] See reagents Nos. 1 and 8, Table 2, p. 412.

Excellent macroetching techniques applicable to ferritic and austenitic welded sections have been developed by Miller and Houston.[1]   In the case of ferritic welds, the specimen of interest is prepared in a manner described for metallographic specimens (see Chap. 1) and finally alternately polished and etched in saturated picral to remove disturbed metal.   The prepared surface is then etched for 10 to 20 sec in 5 per cent nital, after which the surface is thoroughly washed and lightly rubbed on a metallographic polishing cloth until the columnar grains in the weld metal show distinctly. This procedure is repeated several times to lessen the light reflectivity

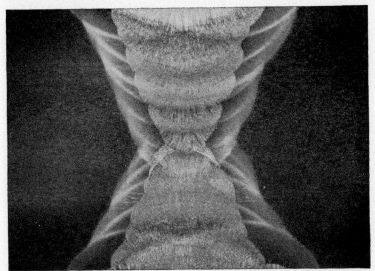

Fig. 102.   Transverse section of a ferritic weld in a base metal of low-alloy steel.   2×. Macroetchant, nital and picral reagents as described.   (*Courtesy of U.S. Steel Corp. Research Laboratory.*)

characteristics of the surface and to produce some relief of the macrostructure.   The specimen of interest is then etched by immersion for about 2 min in saturated picral, followed by thorough washing in cold running water and swabbing with a tuft of cotton to remove the loosely adhering reaction products formed on the surface.   The specimen is finally rinsed in alcohol and dried in a stream of warm air.   The excellent rendition of the macrostructure by this technique of macroetching is illustrated in Fig. 102.

Because of the passive characteristics of stainless steels of high chromium content to the commonly used metallographic etching reagents, the above macroetching procedure is not suitable for revealing the macrostructure of austenitic stainless-steel weldments, unless the primary interest is in the

[1] Miller, O. O., and E. G. Houston: Macro-etching and Photomacrography of Ferritic and Austenitic Welded Joints in Low-alloy Steel, *Supplement, J.Am. Welding Soc.,* Vol. 12, No. 10, 1947.

plain carbon or low-alloy-base steel.   Austenitic stainless-steel filler metal can be appropriately macroetched, as illustrated in Fig. 103, by electrolytic etching techniques, using chromic acid as the electrolyte.

The specimen of interest is polished metallographically, as in the case of ferritic weldments, and finally alternately polished and etched electrolytically in chromic acid to remove all traces of disturbed metal.   Macroetching of the section is then carried out in a 10 per cent aqueous solution of chromic acid, maintained at a temperature of 15 to 40°C (60 to 150°F). The specimen of interest serves as the anode, and a sheet of stainless steel or platinum as the cathode; the distance between electrodes is from 1 to 3 in.

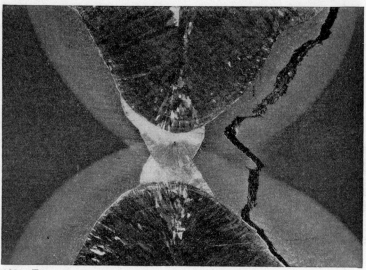

FIG. 103.   Transverse section of an austenitic weld in a base metal of low-alloy steel.   2×. Note crack in heat-affected zone of parent metal.   Macroetched electrolytically in chromic acid as described.   (*Courtesy of U.S. Steel Corp. Research Laboratory.*)

The current density is 1 amp per sq in., and the time of etching is from 1 to 7 min.   The criteria of a suitable etch are complete revelation of and appropriate establishment of contrast in the macrostructure of both the heat-affected zones of the base steel and the austenitic cast or filler metal.

Carburized and Decarburized Surfaces.   Macroetching is a convenient method of determining with some accuracy the transverse depth of a carburized case in annealed and normalized steels.   The carburized case, being of higher carbon content than the underlying base metal, will etch more rapidly and will after etching appear darker than the core.   As shown in Fig. 104a, a distinction is readily noted between the two regions.

In like manner, decarburized surfaces may be detected by macroetching; and, because these areas are very low in carbon content, and hence, etch relatively light, they may be readily distinguished, as shown in Fig. 104*b*, from the darker-etching unaffected zone.

In cases similar to those cited, it is recommended that a mild etching reagent, such as nital, be used in the etching procedure, as the standard macroetching reagent, when used according to the recommendations previously discussed, may cause the zones of comparatively low carbon content to etch out in relief.

a. Carburized case.	*b*. Decarburized case.

Fig. 104. Transverse sections of heated-treated steel bars macroetched to reveal, respectively, the extent of carburization and decarburization. 1×. Macroetchant, 2 per cent nital.

References

ASM: "Metals Handbook," Cleveland, Ohio, 1948.

Enos, G. M.: "Visual Examination of Steel," ASM, Cleveland, Ohio, 1940.

Gill, J. P., and H. G. Johnstin: An Interpretation of the Deep Acid Etch Test as Applied to Tool Steels, *Trans. ASST*, Vol. 21, 1933.

Miller, O. O., and E. G. Houston: Macro-Etching and Photomacrography of Ferritic and Austenitic Welded Joints in Low-Alloy Steel, *Supplement, J. Am. Welding Soc.*, Vol. 12, No. 10, 1947.

Newirth, F., R. Mitsche, and H. Dienbauer: Anwendbarkeit des Oxydabdruckverfahrens nach M. Niesner, *Arch. Eisenhuttenw.*, Vol. 13, 1940.

Palmer, F. R.: "Tool Steel Simplified," The Carpenter Steel Co., Reading, Pa., 1937.

Pulsifer, H. B.: "Inspection of Metals," ASM, Cleveland, Ohio, 1941.

Sauveur, A.: "The Metallography and Heat-treatment of Iron and Steel," University Press, Cambridge, Mass., 1926.

Stoughton, B., and A. Butts: "Engineering Metallurgy," 3d ed., McGraw-Hill Book Company, Inc., New York, 1938.

Williams, R. S., and V. O. Homerberg: "Principles of Metallography," 5th ed., McGraw-Hill Book Company, Inc., New York, 1948.

Woldman, N. E.: "Physical Metallurgy," John Wiley & Sons, Inc., New York, 1930.

Yatsevitch, M. G.: Essential Factors in Conducting the Macroetching Test under Usual Practical Conditions of Production Work, *Trans. ASST*, Vol. 21, 1933.

CHAPTER 6

HARDNESS TESTING

Although the term "hardness" with regard to metals and alloys is a comparative consideration of great engineering importance, it is not considered in the light of present knowledge to be a fundamental property of matter. The index of hardness is a manifestation of several related properties of the metal, which may well include a combined effect of yield point, tensile strength, ductility, work-hardening characteristics, resistance to abrasion, etc. Depending upon the manner by which the hardness index of a material is secured, different combinations of these properties are effective in influencing the hardness. For example, there are good reasons to believe that the hardness designation obtained from a Brinell hardness test employing a 3000-kg load is the manifestation of a different group of properties than that associated with the same test using a 500-kg load; and the combination of properties inherently influencing the Rockwell hardness index are different both in kind and degree from those related to the Brinell or other hardness tests. Because of these circumstances, uncertainties exist when attempting to correlate hardness designations secured by different methods of measuring the hardness. Although hardness conversion data are used in engineering practice (see page 466), they must be used and interpreted with caution and full cognizance of the attendant limitations. Hardness conversion data secured experimentally for any one class of metals or alloys cannot be extended with assurance to other classes of metals whose work-hardening characteristics are different; and even under the most favorable circumstances, such conversion data are to be considered only approximate.

In view of the past discussion, it is apparent that the kind or type of external stress imposed upon a metal will result in different manifestations of the hardness characteristic. For example, glass will usually scratch hardened steel, but will fail completely in a comparative test with steel as a machining tool. Many comparatively soft metals are hard in the sense that they are capable of resisting severe abrasion stresses, whereas harder materials, from the standpoint of indentation hardness, may fail completely under identical circumstances. The rebound hardness of hard rubber, as determined by the scleroscope method, is approximately the same as that of mild steel. If, however, an indentation method of testing is employed, such as the Brinell test, the hard rubber will behave very much like a soft

212

material and during the test it will flow to such an extent that it will be impossible to obtain a worthwhile hardness measurement.

It is evident from these few illustrations that the term "hardness" is ambiguous, and that hardness data secured on a given material must be interpreted in relation to the type of hardness test used.

The hardness of a metal or an alloy may be manifested in one or more of the following forms:

1. Cutting hardness—the resistance of metals to various cutting operations.

2. Abrasive hardness—the resistance of metals to wear when subjected to rotative or sliding motion.

3. Tensile hardness—the strength of metals in terms of elastic limit and ultimate strength.

4. Rebound hardness—the resistance offered by metals to strike and rebound, which is principally a measure of its elastic properties.

5. Indentation hardness—the resistance offered by metals to indentation, which is mainly a measure of its plasticity and density.

6. Deformation hardness—a measure of the distortion properties of metals, which is of primary importance in the case of sheet metals.

Hardness testing plays an important role in determining the characteristics of a metal and alloy and often the results of such measurements are a determining factor in the acceptance or rejection of semifinished or finished products. This is especially true in any heat-treating or hardening process. It is important, therefore, that the principles of the various hardness testers used in both the fields of research and routine shop or laboratory inspection be well understood, in order that the data obtained may be correctly evaluated.

MOHS SCALE OF HARDNESS

One of the first systematic hardness scales ever devised was proposed by Friedrich Mohs. It provides a means whereby the scratch hardness of a material may be quickly and effectively determined. Although it is rarely, if ever, used in the testing of metals and alloys it nevertheless finds a wide application in the field of mineralogy. The scale consists of ten standard minerals arranged in order of increasing hardness, each mineral being numbered according to its position in the series. The standard Mohs scale of hardness is as follows:

1. Talc	6. Orthoclase
2. Gypsum	7. Quartz
3. Calcite	8. Topaz
4. Fluorite	9. Corundum
5. Apatite	10. Diamond

Thus, for example, if a material is noticeably scatched by the mineral apatite (No. 5) and not by fluorite (No. 4), the assigned hardness value will be between 4 and 5.   Similarly, the hardness of a material scratched by corundum and not by topaz will be between 8 and 9.

THE FILE HARDNESS TEST

One of the most simple methods for determining the abrasion hardness of a metal is by subjecting the test piece to the cutting action of a file and noting whether or not a visible cut is produced.   Although this method of hardness testing requires skill on the part of the technician, the technique of the test can nevertheless be quickly learned so that the factors which influence the accuracy of the test may be readily accounted for.   Extensive use of the test is limited by the fact that the results are generally not reproducible enough so that physical standards may be established between different manufacturers.   This may be due to the flexibility of the terms used to define and interpret the test results and the inevitable variations in characteristics of ordinary files.   The test as used is a comparative measure of the hardness between the material in question and a predesigned standard. Such a standard is often a sample of the material of the correct hardness, as determined by more precise hardness testing methods.   Because of the rapidity of the test, however, file testing affords a simple and routine method for control checking of hard surfaces in production.   To illustrate, the hardness of each tooth of a gear may be determined within a few seconds by file testing, whereas by other testing methods the time element would be considerably longer.   The very nature of the test permits the hardness of a material to be determined in inaccessible places where hardness measurements would be normally impossible to secure with a conventional testing machine.

The material to be tested is either held in one hand or securely clamped in a vise.   A standard file is slowly but firmly drawn over the surface and, as soon as it is apparent whether or not the file will cut, it is removed.   The test piece need not be prepared in any way, as the file will generally remove enough adhering surface scale so that the accuracy of the test will not be affected.

Comparative tests with a file depend upon three important factors, namely (1) size, shape, and hardness of the files, (2) speed, pressure, and angle of filing during the test, and (3) the composition and heat-treatment of the metal under test.

Types of Files.   The size, shape, and hardness of the files can generally be standardized by using files made especially for this purpose.   A typical set of test files for the testing of steel are as follows:

1. Six-inch Pillar Testing File No. 0 and No. 1—to be used on flat, ob-

long, or square test pieces. File No. 0 is best suited for fully hardened steel, whereas the finer No. 1 file is best suited for testing quenched and tempered steel.

2. Eight-inch Pillar Narrow Testing File—to be used in testing hardened objects of the shapes given in 1.

3. Six-inch Three-square Testing File No. 1—to be used in testing objects of irregular shapes in which there are crevices and grooves.

New files are generally standardized by noting whether or not they will cut pieces of steel or samples of the product that are of the proper hardness as determined by other hardness-test methods. Such a procedure of standardization will also indicate whether or not a used file is worn beyond its usefulness and will serve further as a check when the hardness of the object of interest is doubtful. Files should be discarded when they no longer produce satisfactory cuts on the test standards or when they show any non-uniformities, such as worn or chipped teeth.

Speed, Pressure, and Angle of Filing. It is recommended that in the file test the file be drawn slowly over the test surface, since the hardness indications are then usually more precise and valid than when the cut is made too rapidly. Fast cutting will remove metal, not only from the test surface, but also from the file, and thereby indicate a material softer than is actually the case.

No general rule can be formulated with regard to the amount of pressure that is to be applied to the file. It is recommended, however, that in testing many samples of the same material the pressure be the same for each test. The angle of contact between the surface of the test piece and the file should be standardized and such conditions maintained throughout subsequent tests. This is of importance because it is generally much easier to produce a cut on projections and sharp edges than on flat surfaces.

Influence of Composition and Heat-treatment of the Test Section. The limitation of the file test is partially governed by the relative composition and heat-treatment of the test piece. The test is limited to certain very high hardness values, with a minimum approximately Rockwell C 58 to C 60. For use on quenched and tempered pieces, the test may be limited to a still higher range. This is illustrated by the fact that a quenched carbon tool steel indicating a Rockwell hardness of C 63 to C 65 may not be "touched" by a file, but after it is tempered at 190°C (375°F) it will be readily cut, even though the Rockwell hardness still measures C 62 to C 64.

THE BRINELL HARDNESS TEST
Standard Testing Units

In 1900 Dr. J. A. Brinell, of Sweden, published his fundamental researches describing a method whereby the hardness of a metal could be determined by measurement of the impression made by a steel ball forced into the metal under definite static loads.   The decided advantages of this method of hardness testing were that without regard to the relative hardness of the metal tested, only one theoretical linear scale of hardness was necessary, and that the hardness of the metal on this scale showed a close relation to its maximum tensile strength.

One of several models of the standard Brinell hardness tester is shown in Fig. 105.   It is essentially a machine, operating under hydraulic pressure, that forces a standard-size steel-ball penetrator into the metal under test by a predetermined static load.   The magnitude of the applied load is determined in part by the size of the penetrator and the relative hardness of the material under test.   To ensure against the application of a greater load than is desired, the maximum pressure is controlled by means of a weighted yoke.   Thus, if a 3000-kg load is to be applied in the test, weights proportional to this load are distributed equally on the two pans of the yoke. When the maximum load of 3000 kg has been attained by means of the hydraulic pump, a piston raises and "floats" the yoke and weights. As long as the piston is maintained in a raised position by occasional pumping, the pressure will remain constant.   This arrangement consists of a small hydraulic accumulator and by the proper

Fig. 105.   The standard Brinell hardness tester.   (*Courtesy of H. A. Holz.*)

selection of pan weights any desired load may be constantly applied.   The Bourdon spring gauge, mounted on the top of the tester, is for the purpose of indicating to the operator the approach of the maximum load to which the yoke has been weighted.   When the test is completed, the pressure is released by actuating a release valve located on the hydraulic unit.

Penetrators and Loads.   The penetrators used in the standard Brinell hardness test are spherical balls with diameters of either 5 or 10 mm. Indenters having other diameters, such as 1.25, 2.50, and 7.00 mm are available for special testing purposes, although the use of these odd-size penetrators is not considered to be a standard Brinell test. The standard penetrators are available in three grades, (1) standard balls made from ordinary high carbon steel, (2) high-carbon steel balls that are heat-treated to render them hard (Hultgren balls), and (3) balls made from tungsten carbide. The appropriate use of any one of these penetrators depends upon the hardness of the material being tested.

When the indenter is forced into the specimen during test, the ball penetrator inevitably suffers deformation, the magnitude of which depends upon the relative hardness of the material being tested, the hardness of the indenter, and the applied load. When the hardness of the material exceeds a Brinell number of about 525, the deformation of the ordinary steel-ball penetrator is sufficiently great to cause some unreliability in the hardness measurements. This upper limit of 525 Brinell may be extended to about 600 Brinell by the use of a Hultgren ball, and to as high as approximately 725 Brinell by use of a tungsten carbide ball. However, at hardnesses beyond about 600 Brinell, regardless of the penetrator used, the hardness measurements must be interpreted with caution.

In full awareness of the deformation that accompanies the penetrator ball during test, and the error subsequently introduced in calculation of the final Brinell hardness number, the penetrators described are recognized as standard when used under standard conditions of testing to be discussed. To minimize as far as possible the ultimate adverse effects of ball deformation on hardness numbers, and thus tend toward a theoretical hardness scale at high hardness values that is based on proportionality and independence of applied load, the load should be so adjusted that the diameter of the impression is between 0.25 and 0.50 of the diameter of the ball. To maintain this ratio range of diameters, however, would require in many instances departure from standard testing specifications. Because of this, and unfortunately so, little attempt is made in practice to conform to this relationship.

In securing Brinell hardness measurements on steel and alloys having a hardness similar to steel, the British Standards Institution specifies either a load of 3000 kg on a 10-mm-diameter ball or a load of 750 kg on a 5-mm ball; for nonferrous metals and alloys, such as copper and brass, a 10-mm-diameter ball is specified under loads of either 1000 or 500 kg.

In the United States, the Brinell hardness test is usually conducted in conformation to test specifications established by the American Society for Testing Materials. These standards specify the application of a 3000-kg load for at least 10 sec on a 10-mm ball when testing iron and steel; for nonferrous metals and alloys, a load of 500 kg on a 10-mm ball for at least 30 sec. Frequently, because of conditions relating to the specimen of interest, such as size or thickness, the above procedure of standard testing cannot be adhered to. Under such circumstances, as will be discussed later, the material may be tested with either smaller loads or with indenters of smaller diameter, since the same Brinell hardness number should be obtained, at least theoretically, if the load is varied in direct proportion to the square of the ball diameter. Such departures, however, from standard test specifications should not be considered standard Brinell hardness tests.

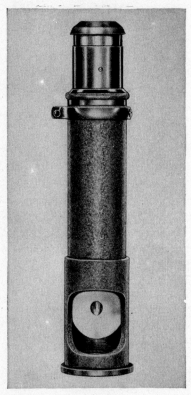

FIG. 106. Portable microscope designed for measuring the diameter of the impression formed by the Brinell indenter. (*Courtesy of Bausch and Lomb Optical Co.*)

Methods of Measuring Ball Impressions. In order to determine the Brinell hardness number of a material after the test has been made, it is necessary to measure the diameter of the impression (the most common method) or the depth of impression made on the test piece by the ball indenter.

The measurement of the diameter of the impression is readily accomplished by the use of a special measuring-type microscope, one design of which is illustrated in Fig. 106. The eyepiece tube of the microscope contains a micrometer-scale reticle graduated in tenths of a millimeter, from 0 to 7 mm, which permits measurements to be estimated to 0.05 mm. For more precise measurements, as is often required when the impression is very small in diameter, portable precision microscopes are available, which permit measurements to be made to 0.001 mm. With either type of microscope, a number of measurements must be made across different diameters and the average calculated to represent the diameter of the impression.

The measurement of the depth of the impression is made by a special device attached to the hardness tester. It has the one advantage that without interrupting the test, the simultaneous values of the load and depth of the impression can be determined.

The Brinell hardness number calculated by the measured diameter of the impression differs from that when the calculation involves the depth of the impression. This difference may be attributed to errors associated with each method of measuring, arising from the raised ridge of metal around the impression in certain soft to medium-hard metals, and from the depressed surface metal around the impression so often observed in brass, bronze, and cast iron. As a consequence, the diameter of the impression will usually appear to be too large, whereas the depth of the impression will appear to be too small. In general, however, the calculated spherical surface area of the impression is more exact if the diameter of the impression is involved instead of the depth; this circumstance is responsible for the widespread use of diameter measurements in determining the Brinell hardness number. This greater exactness in area is related to the greater accuracy afforded in measuring diameters than depths, owing to the larger dimensions associated therewith.

Principles. The principle of the Brinell tester is based upon the fact that with ball penetrators of different sizes, similar geometrical impressions having proportional diameters are produced which give the same Brinell hardness number, provided that the loads applied are proportional to the square of the ball diameters.

It is well established that any deviation from the standard test conditions with regard to ball diameter and applied load should be made according to the following load–ball diameter ratio:

$$P = 30D^2 \text{—for iron and steel} \tag{1}$$
$$P = 5D^2 \text{—for brass, bronze, and other nonferrous metals} \tag{2}$$

where P = load, kg
D = ball diameter, mm

This means, considering the ratio $P = 30D^2$, that with a 7-mm ball and a load of 1470 kg, or a 5-mm ball and a load of 750 kg, Brinell hardness numbers will be obtained that are on the same standard hardness scale as tests performed with a 10-mm ball under a load of 3000 kg. This is shown to be quite conclusive at least for low hardness values, by the consistency of the following data:

Diameter of ball, mm	Load, kg	Diameter of impression, mm	Brinell hardness number
10.00	3,000.0	6.300	85
7.00	1,470.0	4.400	85
5.00	750.0	3.130	87
1.19	42.5	0.748	86

As these test data indicate, the Brinell hardness number is independent of the ball diameter and test loads used, provided the two are used in the proper ratio.

The numerical value of the Brinell hardness number, expressed in units of kilograms per square millimeter, is equal to the quotient obtained by dividing the load in kilograms by the area in square millimeters of the spherical impression formed by the ball. That is,

$$\text{Brinell hardness number} = \frac{P}{A} \qquad (3)$$

where P = applied load, kg

A = area of spherical impression in the metal under test, sq mm

The area of the spherical surface made by the penetrator ball can readily be calculated in terms of the ball diameter D and the diameter of the ball impression D_1, as determined by the impression-measuring microscope. For purposes of derivation, Fig. 107 illustrates the penetrator ball seated in the metal directly after the test has been made.

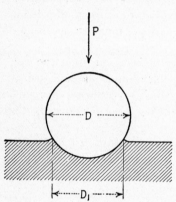

Fig. 107. Illustrating schematically the Brinell indenter seated in the impression formed on the test surface.

It can readily be shown by trigonometric relationships that the zone area of the ball between the plane of the great circle of diameter D and the plane of the small circle of diameter D_1 is equal to $\frac{\pi D}{2}\sqrt{D^2 - D_1{}^2}$. The area of one-half the ball penetrator is equal to $\pi D^2/2$. Therefore, the surface area of the impression is equal to the difference between one-half of the total ball area and the zone area of the ball between the great and small circles of diameters D and D_1 respectively:

$$\text{Area of impression} = \frac{\pi D^2}{2} - \frac{\pi D}{2}\sqrt{D^2 - D_1{}^2} \qquad (4)$$

Rearranging and substituting this term for the area A in Eq. (3), a workable and convenient equation is obtained for determining the Brinell hardness number.

$$\text{Brinell hardness number} = \frac{P}{\dfrac{\pi D}{2}(D - \sqrt{D^2 - D_1{}^2})} \qquad (5)$$

where P = applied load, kg

D = diameter of ball, mm

D_1 = diameter of ball impression, mm

If the steel penetrator ball was in no way deformed during test, then the above equation would be applicable without error to any combination of applied load and ball diameter. As mentioned heretofore, however, there is always some deformation of the ball, and attending elastic recovery of the impression when the load is released, so that the impression resulting therefrom is not perfectly spherical. This, of course, introduces errors in the hardness values as computed by Eq. (5), since the equation derivation has assumed a perfectly spherical impression. These circumstances not-withstanding, hardness numbers so computed are nevertheless accepted as valid for all practical purposes, provided standard conditions of testing are maintained. Table 31, page 449, gives Brinell hardness numbers cor-responding to different diameters of impression resulting from the standard 10-mm ball under the two standard loads of 500 kg and 3000 kg, respect-ively. The values so given are merely solutions of Eq. (5).

Condition of Test Surface. The surface of the specimen onto which the impression is to be made should be reasonably smooth and free from surface defects. Generally, a satisfactory surface may be obtained by grinding or filing with a relatively fine file. Polishing of the surface is not necessary, but care should be taken that no oxide scale or grit particles are adhering to the surface at the point of ball penetration. In the testing of round sections or curved surfaces, unless merely comparative data are desired, a spot must first be made flat by filing or grinding.

Minimum Thickness of Test Section. Considering that maximum loads of 500 kg and 3000 kg are used in the Brinell hardness test, the specimen must be of such thickness that the anvil support will not influence the pene-tration of the ball. Any anvil effect will, of course, produce erratic hard-ness measurements. In general, the thickness of the test piece should be at least 10 times the depth of the impression, and the distance from the edge of the test piece to the center of the ball impression should not be less than 2.5 times the diameter of the impression. The minimum thickness of test sections, recommended in terms of different applied loads and appro-priate ball diameters, may be summarized as follows:

Minimum thickness of test section		Diameter of ball, D, mm	Test load to be used, kg	
Mm	In.		$P=30D^2$ kg Steel, not hardened	$P=5D^2$ kg Nonferrous metals
6	¼	10	3000	500
3	⅛	5	750	125
1.2	0.05	2.5	187.5	31.25
0.6	0.024	1.25	46.875	7.812
0.4	0.016	0.625	11.72	1.953

Brinell Hardness Number vs. Tensile Strength. The Brinell method of hardness testing is distinct in that the hardness numbers may be used to determine the approximate ultimate tensile strength of metals which, because of the size or nature of the piece, cannot be conveniently determined by destructive testing methods. It is evident that the method to be described is not applicable in cases where such data are to be known to a high degree of accuracy.

There is a close relationship between the ultimate tensile strength of a metal and its Brinell hardness number. This relation is a direct one, but the constants used in the correlation are different for different metals and alloys and depend to a great extent upon the structural condition of the metal. It is beyond the scope of this book to discuss all the metals and alloys in this respect, but plain carbon steel and a few alloyed steels will be considered. For plain carbon steels up to a maximum of 1.00 per cent carbon, rolled and annealed, the constants have been accurately determined and are as follows:

Hardness numbers	Constant	Position of test
Below 175 Brinell.....................	515	Transversed to rolling direction
	504	In direction of rolling
Above 175 Brinell.....................	489	Transversed to rolling direction
	461	In direction of rolling

Thus, for a plain carbon steel having a Brinell hardness number of 310, as determined transverse to the rolling direction, the approximate ultimate tensile strength will be

$$489 \times 310 = 151,590 \text{ lb per sq in.} \qquad (6)$$

For certain alloyed steels[1] that have been heat-treated to bring out the most in their physical properties, the following equations are applicable in computing the approximate ultimate strengths. In these equations, $U =$ ultimate tensile strength in pounds per square inch, and $B =$ Brinell hardness number.

For nickel steels of different carbon contents:

$$U = 710B - 32,000 \qquad (7)$$

For chromium-nickel steels of different carbon contents, nickel 3.5 per cent, chromium 1 per cent:

$$U = 710B - 33,000 \qquad (8)$$

[1] Clark, J. J.: Hardness of Metals, *Can. Min. Jour.*, May 22, 1925.

For chromium-nickel steels of different carbon contents, nickel 1.5 per cent, chromium 0.5 per cent:

$$U = 680B - 22,000 \qquad (9)$$

For chromium-vanadium steels of different carbon contents:

$$U = 710B - 29,000 \qquad (10)$$

To illustrate, the ultimate tensile strength of a nickel-carbon steel having a Brinell hardness number of 300 will be

$$U = (710 \times 300) - 32,000 = 181,000 \text{ lb per sq in.} \qquad (11)$$

Fig. 108. The King portable Brinell hardness tester. (*Courtesy of Andrew King.*)

Portable Testing Units

The King portable hardness tester, illustrated in Fig. 108, is one of several available portable instruments designed to operate on the same basic principles as the standard Brinell hardness tester. In comparison to the standard Brinell instrument (see Fig. 105), the King portable tester is of lighter construction and, consequently, more readily moved about; test specimens that are too large or cumbersome to be handled in the standard tester can readily be tested with the portable unit; the King tester operates in any position, even upside down, provided the unit can be appropriately attached to the specimen of interest; and because the test head is removable from the regular base as illustrated, the test head may be attached to special holding devices devised to meet certain requirements of testing.

The King Brinell hardness tester is available in two designs of nearly identical appearance. One of the units is specially designed for measuring the hardness of iron and steel and is capable of applying calibrated loads of 500, 700, 1000, 1500, and 3000 kg. The second instrument is designed for testing nonferrous metals and alloys, and is capable of producing calibrated loads of 125, 250, 500, 750, and 1000 kg. Both units are equipped with an adjustable limiting valve that prevents the applied load from exceeding that for which the valve is set. In conformation to the ASTM specification for the Brinell hardness test, the King instrument is equipped with interchangeable ball penetrators of 5- and 10-mm diameters, and for special testing purposes, the conventional Vickers pyramid diamond indenter (see page 225) is available.

THE ARMSTRONG-VICKERS HARDNESS TEST

The Armstrong-Vickers hardness tester, the original of the so-called Vickers instruments, is illustrated in Fig. 109. This instrument, owing to the shape of the diamond penetrator normally employed and the wide latitude available in adjusting the applied loads, is applicable to measuring the hardness of thin sheets and superficial hardened surfaces, as well as massive sections of the softest and hardest metals and alloys. The Vickers hardness tester operates on the same basic principles as the Brinell instrument, and the hardness scale is identical with that of the Brinell, *i.e.*, the designations are expressed in terms of load and surface area of the impression.

The Vickers instrument is definitely a precision unit that normally requires care in operation, rather than skill on the part of the operator. The hardness tester is semiautomatic in operation, in that after the specimen surface is brought into contact with the indenter, the preset load is applied for some definite length of time, after which the load is then automatically removed. The penetrator is appropriately loaded by means of a lever arm (ratio—1 to 20) and weights attached thereon. By means of a unique cam arrangement and a permanent auxiliary weight, the preset load is applied slowly and at a diminishing rate for the last part of the load, thereby eliminating practically all errors that might arise from inertia effects and sudden application of the entire load. By means of an oil-dashpot arrangement, the time of load application and load duration can be adjusted, the usual times varying between 10 and 30 sec.

After an impression has been appropriately made on the test surface of interest, the dimensions of the impression may be measured directly without removing the test piece from the platen of the instrument. This is accomplished by first lowering the specimen and supporting platen by means of a hand wheel, and then bringing into place above the impression a special measuring microscope attached to the side of the instrument, as shown in Fig. 109.

Fig. 109. The Armstrong-Vickers hardness tester. (*Courtesy of American Machine and Metals Inc., Riehle Division.*)

Penetrators and Loads. The standard Vickers indenter consists of a diamond in the form of a square-based pyramid with an included angle of 136 deg between opposite faces. This design of penetrator conforms to the specifications established by the British Standards Institution, and because of the precise shape of the indenter, the Vicker hardness numbers, more commonly referred to as diamond pyramid hardness numbers, are on the same theoretical hardness scale as the Brinell hardness numbers. As mentioned heretofore, a more nearly proportional hardness scale would be obtained in the Brinell test, particularly at relatively high hardness values, if the load and ball diameter were always adjusted so that the diameter of the impression would be 0.25 to 0.50 (average 0.375) that of the ball diameter.   If tangents are drawn to the edge of an impression whose diameter is 0.375 that of the ball, the angle will be 136 deg, equivalent to the included angle between opposite faces purposely specified in the Vickers indenter.

As illustrated in Fig. 110, the Vickers and Brinell hardness designations are practically identical up to about 300 Brinell.   Beyond this hardness value, however, the two curves illustrated tend to diverge, and to rather serious proportions at about 600 Brinell.   The departure of the Brinell curve from proportionality is related to deformation of the penetrator ball

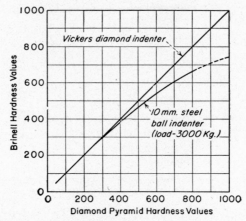

FIG. 110.   Illustrating the approximate divergence of Brinell from Vickers hardness numbers at high hardness values.

at high hardness values, whereas the Vickers curve is practically straight owing to the insignificant amount of distortion of the diamond indenter at these high hardness levels.   For all practical purposes, the Vickers hardness measurements should be considered the only reliable evaluation of hardness at hardness values above about 600 Brinell.

In addition to the standard diamond indenter, steel balls of 1-mm and 2-mm diameters are available for special testing purposes, and when appropriately used, constitute an extremely accurate form of light-load Brinell test.

By appropriate selection of weights, the Vickers instrument is capable of applying calibrated loads to the indenter of 1, 2.5, 5, 10, 20, 30, 50, 100, and 120 kg.   The load selected for a given test should be as large as possible, with due consideration for the dimensions of the test piece (particularly the thickness), the relative hardness of the material to be tested, and the purpose of the test.   For most routine hardness testing, a load of 30 or 50 kg is appropriate, although when measuring the hardness of cast iron and certain nonferrous alloys, higher loads are desirable if an over-all average hardness of a heterogeneous structure is desired.

Method of Measuring Impression. In order to determine after test the diamond-pyramid hardness number, it is necessary to measure accurately the length of the two diagonals of the square surface impression made by the standard indenter; then compute the arithmetic average of the two; and finally substitute this average measurement in the hardness-number equation to be discussed, or use the average value as a basis of reference to appropriate hardness-number tables. Precise measurement of the diagonals is accomplished by means of a metal-lurgical-type microscope mounted on the side of the instrument, which can be swung into position above the surface impression whenever necessary. The microscope is equipped with a Filar-type measuring eyepiece (see page 103) which can be rotated through 90 deg or more about its principal axis, thereby enabling measurements to be conveniently made of both diagonals. The measuring eyepiece of the Armstrong-Vickers instrument consists of movable knife-edges, instead of the conventional hairlines, and is capable of length measurements to 0.001 mm. To facilitate determining the actual movement of the knife-edges, a digit counter, coupled to the actuating vernier screw, is mounted on the eyepiece housing.

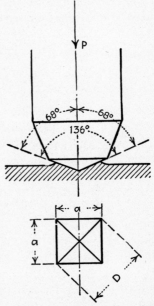

Fig. 111. Illustrating schematically the Vickers indenter and the impression formed on the test surface.

In some models of the Vickers hardness tester (not the Armstrong-Vickers), a magnified image of the impression can be projected by a reflex arrangement upon a focusing screen mounted above the instrument. This feature facilitates securing diagonal measurements, and is ideally suited to save time in routine production hardness testing.

Hardness Scale. The numerical value of the diamond-pyramid hardness number is equal to the applied load divided by the surface area of the pyramidal impression. Thus,

$$\text{Diamond-pyramid hardness number} = \frac{P}{A} \qquad (12)$$

where P = applied load, kg

A = lateral area of pyramidal impression, sq mm

The lateral area A of the pyramidal impression is equal to the perimeter of the square surface impression $4a$ multiplied by one-half the slant height S (see Fig. 111).

$$A = 2aS \tag{13}$$

From Fig. 111 it is seen that the relation between the diagonal length and a side of the square impression is

$$D^2 = 2a^2 \qquad \text{or} \qquad a = \frac{D}{\sqrt{2}} \tag{14}$$

Therefore, from Eqs. (13) and (14),

$$A = \frac{2D}{\sqrt{2}}S \tag{15}$$

The included angle between opposite faces of the indenter and, therefore, the corresponding included angle of the impression, is 136 deg. The slant height S may then be expressed as

$$S = \frac{a/2}{\sin 68°} = \frac{D/2\sqrt{2}}{\sin 68°} \tag{16}$$

Substituting this value of S in Eq. (15) the total lateral area of the impression is as follows:

$$A = \frac{D^2}{2 \sin 68°} \tag{17}$$

If the value of A in Eq. (17) is substituted in Eq. (12) and $2 \sin 68$ deg is evaluated, a working equation for determining the diamond-pyramid hardness number is obtained and may be expressed as

$$\text{DPH} = 1.854 \frac{P}{D^2} \tag{18}$$

where DPH = diamond-pyramid hardness number

P = applied load, kg

D = diagonal length of square impression, mm

Equation (18) is theoretically valid only when the impression has been made by a perfect indenter—a circumstance that is never fully realized in actual practice. However, the error attending the use of Eq. (18) is insignificant, provided the dimensions of the indenter are within the tolerances specified for such penetrators by the British Standards Institution.

In the practical performance of the Vickers test, hardness numbers corresponding to different average diagonal lengths may be determined directly from computed tables, thus eliminating computations with Eq. (18) after each set of measurements. Table 32, page 454, gives such a compilation based on an applied load of 10 kg. For loads other than 10 kg, the diamond pyramid number may be found by first multiplying the measured average diagonal length in mm by 10; then determining the diamond-pyramid number in Table 32 corresponding to the adjusted diagonal length; and finally correcting the diamond-pyramid number by multiplying it by the actual applied load in grams divided by 100.

Condition of Test Surface. The condition of the test surface, because of the very small impression normally produced, must of necessity be smooth and free from surface imperfections, oxide scale, etc. A polish, approaching that on a metallographic specimen, is to be recommended whenever possible. Decarburized or recarburized areas should first be removed by grinding, unless they themselves are of primary interest, as hardness measurements on such surfaces will not be representative of the entire piece. In doubtful cases, it is recommended that two or more tests be made under different loading conditions until the results check one another. Round or curved surfaces may be readily tested, but unless only comparative results are desired, the surface should be made flat by either filing or grinding.

Minimum Thickness of Test Section. The testing of thin sections requires that both sides of the piece be reasonably smooth and that full contact be made with the supporting anvil. In general, the necessary minimum thickness of the test piece will vary depending upon the relative hardness of the metal and the magnitude of the applied load. It is recommended that the thickness of the section be at least 1.5 times the diagonal length of the impression. On this basis, and in terms of applied load and hardness of the material under test, minimum thicknesses of test sections are shown in Fig. 112.

THE ROCKWELL HARDNESS TEST

In 1908 Professor Ludwig of Vienna described in his book, "Die Kegel-probe," a method whereby the hardness of a metal could be determined by a "differential-depth" measurement test. This method of hardness testing consisted of measuring the increment of depth of a diamond-cone penetrator that was forced into the metal by a primary and a secondary load. The "differential-depth" measurement method was of significance because errors due to mechanical defects of the system, such as back lash, were eliminated, as well as errors resulting from slight surface imperfections and varying qualities of contact between the penetrator and surface of the specimen under test.

The modern conception of Ludwig's principles is embodied in the Rockwell direct-reading hardness tester, one model of which is illustrated in Fig. 113.   This machine is a precision-built apparatus of weights and levers that automatically measures the increment of linear depth by one of several indenters to which is applied an initial minor load of 10 kg and a major load of either 60, 100, or 150 kg, depending upon the type of penetrator used.

Penetrators. Of the several indenters used in the Rockwell hardness test, there are two standard penetrators whose combined use under appro-

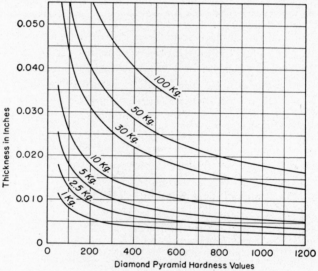

FIG. 112.   The minimum thickness of test section, as related to applied load and hardness, recommended in the Vickers hardness test.

priate major loads serves to measure the hardness of nearly all the common metals and alloys, irrespective of the hardness induced by heat treatment. These penetrators are a $\frac{1}{16}$-in.-diameter hardened-steel ball, designated as the B-scale indenter, and a spheroconical diamond penetrator, designated as the C-scale indenter, or the Brale (trade-mark).

With regard to the arbitrary Rockwell hardness scales, which will be discussed more fully later, the working range of the $\frac{1}{16}$-in.-diameter penetrator is from Rockwell B 0 to B 100. Materials under test having a hardness greater than Rockwell B 100 will tend to distort the indenter during test and cause an appreciable error in the hardness measurement. Notwithstanding deformation of the indenter, the $\frac{1}{16}$-in.-diameter ball would still be inappropriate for high hardness measurements owing to the lack of sensitivity of the spherical-shaped indenter to small differences in hardness under circumstances where the depth of impression is very shallow. It is because of this that balls of larger diameters are not used to minimize the adverse effects of indenter distortion at high hardness values—in fact, the loss in sensitivity at high hardnesses with balls of diameters greater than $\frac{1}{16}$ in. is sufficiently great to invalidate the hardness measurements.

The lower limit of hardness—Rockwell B 0—for which the $\frac{1}{16}$-in.-diameter-ball indenter is designed, should not be exceeded. Materials

Fig. 113. The standard Rockwell hardness tester. Model 3-JR. (*Courtesy of Wilson Mechanical Instrument Co., Inc.*)

having a hardness lower than Rockwell B 0 will introduce errors in the hardness measurements secured on the B scale, owing to the ball chuck possibly contacting the specimen surface, and because the $\frac{1}{16}$-in.-diameter ball becomes supersensitive at low hardness values and unduly magnifies structural inhomogeneities existing in the test specimen. For materials softer than Rockwell B 0, a Rockwell hardness scale other than the B scale should be appropriately selected to meet the specific requirements of the test, as indicated on page 233.

The Brale indenter consists of a spheroconical diamond point manufactured to very close dimensional tolerances. The included angle of the cone is 120 deg, and the tip of the indenter is spherical, being lapped tangent to the cone with a radius of 0.200 mm. Normal care is required in using the diamond penetrator, because an indenter that is chipped or cracked will introduce errors in the hardness measurements which may not always be detected from the data secured.

The useful range of the Brale indenter on the Rockwell C hardness scale is from Rockwell C 20 (equivalent to Rockwell B 97) to about Rockwell C 70 or slightly higher. The Brale penetrator should not be used for measuring hardnesses below Rockwell C 20, owing to certain errors of measurement being introduced at relatively low hardness values. Because any inaccuracies associated with shaping the diamond indenter during manufacture become more pronounced at the base of the cone than at the apex, the manufacturer does not bother to check the accuracy of the indenter at penetration depths greater than that equivalent to a hardness value of Rockwell C 20. Furthermore, when testing relatively soft materials, the spherical apex of the indenter will be forced into the test specimen a considerable distance, and unless the speed of contact between the indenter and specimen surface and the time interval between application of the minor and major loads are standardized, the hardness measurements will be quite erratic and unreproducible. These circumstances, however, are of little consequence when testing only relatively hard materials with the Brale penetrator or when testing softer materials with a ball indenter.

In addition to the two standard indenters, the Brale and the $\frac{1}{16}$-in.-diameter ball, there are available steel balls of larger diameters ($\frac{1}{2}$ in. diameter maximum) that are used on materials softer than can accurately be tested with one of the standard penetrators. These larger size indenters are particularly well suited for determining the hardness of plastic materials, cast iron with large graphite flakes, soft bearing alloys, solders, etc. The use of these penetrators, however, introduces a new arbitrary Rockwell hardness scale, which when used with a recommended major load will be included in one of the standard hardness scales designated in Table 33.

Rockwell Hardness Scales. At the present time, there are 15 standard Rockwell hardness scales, shown in Table 33, that will appropriately define the hardness of practically any engineering material. These hardness scales have been arbitrarily established, and are dependent upon the penetrator used and the major load applied. The minor load of 10 kg remains constant for all scales.

In reporting Rockwell hardness measurements, it is essential to define the conditions of the test by appropriate notation, referring to the Rockwell hardness scale involved. For example, a hardness measurement reported merely as a number as read from the instrument dial, say 60, has no meaning whatsoever. In other words, the hardness scale is not defined.

TABLE 33. STANDARD ROCKWELL HARDNESS SCALES
10-kg minor load

Scale or prefix	Penetrator	Major load, kg	Dial indications	Remarks
B	$\frac{1}{16}$-in.-diameter ball	100	Red numbers	Useful range from B 0 to B 100. At higher hardnesses, ball deforms and becomes less sensitive to small changes in hardness. For measuring hardness of such materials as rolled sheet stock, brass, annealed low- and medium-carbon-content steels, etc.
C	Brale	150	Black numbers	Useful range from C 20 to about C 70. For measuring hardness of materials harder than B 100. This includes such materials as fully hardened steel, quenched and tempered steel, etc.
A	Brale	60	Black numbers	Useful for measuring hardness of very hard materials, such as tungsten carbide; hard thin materials; case-carburized surfaces; etc. Also used under circumstances where the diamond may chip or be otherwise damaged if used under 150 kg major load.
D	Brale	100	Black numbers	Useful for applications where a major load is desired intermediate between those required for A and C scales. For example, certain case-carburized surfaces.
E	$\frac{1}{8}$-in.-diameter ball	100	Red numbers	Useful for measuring hardness of very soft materials (lower than B 0), such as bearing alloys.
F	$\frac{1}{16}$-in.-diameter ball	60	Red numbers	Useful for about the same purposes as the E scale.
G	$\frac{1}{16}$-in.-diameter ball	150	Red numbers	Useful for some materials slightly harder than B 100.
H	$\frac{1}{8}$-in.-diameter ball	60	Red numbers	
K	$\frac{1}{8}$-in.-diameter ball	150	Red numbers	
L	$\frac{1}{4}$-in.-diameter ball	60	Red numbers	
M	$\frac{1}{4}$-in.-diameter ball	100	Red numbers	Useful for special-purpose testing of very soft materials.
P	$\frac{1}{4}$-in.-diameter ball	150	Red numbers	
R	$\frac{1}{2}$-in.-diameter ball	60	Red numbers	
S	$\frac{1}{2}$-in.-diameter ball	100	Red numbers	
V	$\frac{1}{2}$-in.-diameter ball	150	Red numbers	

However, prefixing this number 60 with the letter C—which is in accord with the standard method of notation—so that the measurement now reads Rockwell C 60, immediately standardizes the test, defines the hardness scale, and informs the reader that the test was conducted with a Brale indenter and a major load of 150 kg. Accordingly, the material is relatively hard and corresponds to a Brinell hardness number of approximately 622.

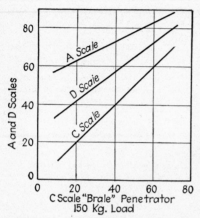

FIG. 114. Approximate relationship between the Rockwell C scale and the Rockwell A and D scales. (*Courtesy of the American Society for Metals.*)

Figures 114 and 115 illustrate the approximate relationship between some of the standard Rockwell hardness scales. Inasmuch as these curves are only a first approximation, they should be interpreted with caution and not used where great accuracy of conversion is required.

The Direct-reading Hardness Dial.

Before discussing the principles involved in the operation of the Rockwell hardness tester, it is essential to fully understand the characteristics of the direct-reading hardness dial, illustrated in Fig. 116.

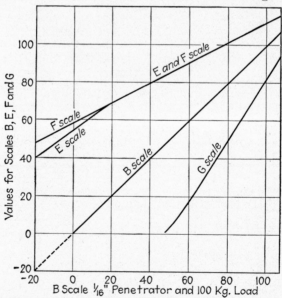

FIG. 115. Approximate relationship between the Rockwell B scale and the Rockwell E, F, and G scales. (*Courtesy of the American Society for Metals.*)

The hardness indicating dial, as illustrated in Fig. 113, is permanently located on the testing instrument directly above the shaft to which the indenter is attached. By means of a unique linkage system, any vertical motion of the penetrator is translated in terms of rotational motion to the large pointer of the dial.

The dial face is inscribed with 100 equal division marks, each division representing one point on the Rockwell hardness scales and an equivalent vertical motion of the indenter of 0.002 mm.   Two sets of numerals are in-

Fig. 116.   The   Rockwell   direct-indicating   hardness-tester   dial.   (*Courtesy   of   Wilson Mechanical Instrument Co., Inc.*)

dicated on the dial, one set printed in red, the other in black.   The red numerals are shown to be used for B-scale readings, although with the introduction of other Rockwell scales the red numerals are also used for such scales that involve the use of ball penetrators of a size other than $\frac{1}{16}$ in. diameter.   The black numerals, designated on the dial face as Rockwell C-scale numbers, are also used in reading the hardness number on other Rockwell scales that employ the Brale penetrator.   Obviously, the letters B or C should not prefix the hardness number solely because the reading was taken from either set of numerals, unless the major load and type of penetrator used were in exact accord with the specifications defining the B or C hardness scale.

The B-scale dial divisions differ from those of the C scale not only by letter designation and color but also in the precise location of specific numbers. The B scale is shifted 30 divisions in a counterclockwise direction, so that B 30 corresponds to the same position on the dial face as C 0. This was purposely done to avoid negative hardness readings on soft materials, such as brass, when tested on the Rockwell B scale; and to establish Rockwell B 100 as the upper limit of hardness that could be tested with a 100-kg major load and a $\frac{1}{16}$-in.-diameter ball without seriously deforming the ball penetrator.

The scale of the dial, with respect to motion of the pointer, is reversed so that a shallow impression when formed by the indenter will indicate a high Rockwell hardness number, and a deeper impression will indicate a correspondingly lower hardness number. The dial-face scale is free to move about its normal axis by either rotating the bezel of the dial directly or by rotating a hand wheel located just beneath the capstan wheel of the elevating screw in late models of the instrument. Movement of the dial scale is essential with regard to the principles of operation of the Rockwell tester, and specifically, this latitude of movement permits the set point of the scale to be made coincident with the large dial pointer after the minor load of 10 kg is applied. This arrangement in effect permits the hardness number to be read directly from the dial after the test is made, since the Rockwell hardness number is an expression of the increment of penetration due to increment in load between the minor and major loads applied.

The small pointer, located in the upper left-hand quadrant of the dial face, is for the purpose of indicating full application of the minor load. The minor load is applied by means of a calibrated spring, and becomes effective when the specimen surface is brought into contact with the indenter. As the specimen is further elevated by means of the elevating screw to be described, the indenter is forced into the specimen surface, and the total load of 10 kg is applied when the small pointer becomes coincident with an index mark.

Principle of Operation. The principles of operation concerning the Rockwell hardness tester are best understood by reference to Fig. 117. For purposes of illustration, the standard $\frac{1}{16}$-in.-diameter ball penetrator has been chosen and has been grossly exaggerated in size to illustrate better the penetration distances associated with the minor and major loads.

In Fig. 117 (1), the specimen of interest is placed on a suitable anvil at the upper end of the elevating screw. The dial pointers are idle, and neither the minor nor the major load has yet been applied.

In (2), the elevating screw is rotated by means of the capstan wheel so as to bring the specimen surface into contact with the ball penetrator. By further elevating the specimen, the minor load of 10 kg is slowly applied,

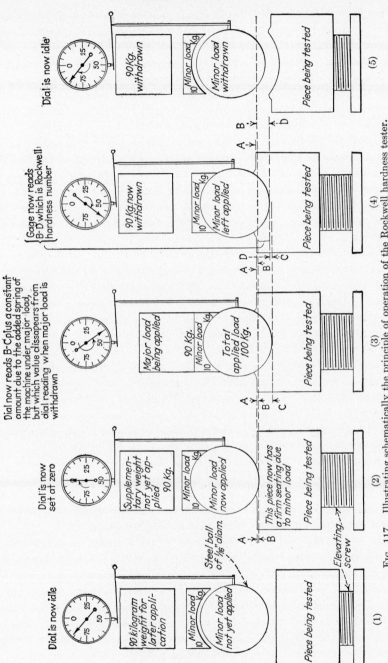

Fɪɢ. 117. Illustrating schematically the principle of operation of the Rockwell hardness tester.

and is fully effective when the specimen has been raised sufficiently high so that the small pointer is coincident with its index mark. During application of the minor load, and owing to the manner by which it is applied, the large pointer is rotated in a clockwise direction. When the minor load is fully applied, the large pointer will be in a near vertical position. The bezel of the dial is then rotated until the set-point mark is coincident with the large pointer. The minor load of 10 kg has forced the indenter into the specimen to a depth corresponding to $A - B$.

In (3), the major load of 100 kg (appropriate for this diameter penetrator ball) is applied by means of a release handle mounted on the side of the instrument. Through an oil-dashpot arrangement, the major load is applied at a definite rate—the rate being established by adjusting the flow valve of the dashpot so that the release handle completes its travel in 4 to 5 sec with no specimen beneath the indenter and with an applied major load of 100 kg.

The major load consists of the original 10-kg minor load plus an additional load of 90 kg. The application of the major load has forced the ball penetrator into the specimen to an additional depth corresponding to $B - C$. During application of the major load, the large pointer is rotated in a counterclockwise direction from set point to 40, which gives, as a matter of interest, a measure of the precise depth of additional penetration, *i.e.*, $(100 - 40) \times 0.002$, or 0.120 mm.

Without removing the minor load of 10 kg, the major load (90 of the total 100 kg) is withdrawn, allowing the impression to recover elastically an amount equivalent to $D - C$, as shown Fig. 117 (4). When the major load is removed, the large pointer rotates in a clockwise direction from numbers 40 to 60 on the dial. This difference of 20 divisions is actually a measure of the elastic recovery of the metal, corresponding to $D - C$.

With removal of the major load, the hardness test is complete. The hardness number is read directly from the dial as 60, and with regard to the test conditions, the hardness is appropriately designated as Rockwell B 60. It is of interest to note that this hardness number, B 60, is actually a reversed measurement of the increment of linear penetration $B - D$. That is, the depth $B - D$ is not equivalent to 60×0.002, or 0.120 mm, but corresponds to $(100 - 60) \times 0.002$, or 0.08 mm. This circumstance, of course, is associated with the direct-reading dial which by design indicates increment of penetration for increment of load between the minor and major loads, and does not indicate total depth of penetration of the indenter.

Both the setting of the dial at set point and observation of the final hardness number are conducted under identical stress conditions in the frame of the instrument, *i.e.*, stresses associated with the 10-kg minor load. This circumstance eliminates errors in final hardness numbers that would otherwise be present if the final stress conditions varied from one test to another.

Condition of Test Surface. The surface on which the hardness test is to be made should be as free as possible from surface imperfections. Generally, ordinary machined surfaces are satisfactory, but where a high degree of accuracy is required, the surface should be polished. Surface irregularities of any kind will offer unequal support to the penetrator, and as a consequence hardness data obtained under such conditions will be inaccurate.

Thickness of Test Section. The hardness readings are influenced not only by the hardness of the material at the point of penetration but also by the hardness of the material under the impression. In the testing of hardened-steel specimens, for example, the actual penetration of the Brale penetrator is about 0.0027 in., but the actual hardness reading is influenced by the hardness of the material at least ten times this distance below the point of impression. Consequently, if a soft layer is located within this depth, the impression will be greater, and the result a lower hardness reading. Because of this added influence on the hardness measurements, it is obvious that in testing thin sheet metal, the hardness indications may be higher, owing to the influence of the hard supporting anvil, than when the same material is tested in thicker form. No hardness measurements on thin sections or soft materials can be considered reliable if the compression or indenter mark appears on the reverse side of the material being tested. Soft, thin sheets may be tested on an anvil with a polished-diamond center, which gives a standardized anvil-surface condition. It is necessary in these cases to specify the thickness of the material, as well as the hardness number.

Generally, round pieces ⅜ in. in diameter or larger can successfully be tested without the use of a correction factor. Hardness numbers obtained on smaller diameter rounds must be corrected to convert the actual readings to the equivalent in hardness on the true Rockwell scale.[1] In any event, the diameter of the piece tested should be stated, together with the hardness number. The influence of the rounded surface may be eliminated by filing or grinding a flat spot on the curved surface.

Standardization. Standardization of Rockwell machines is secured by the use of standard test blocks. A series of such test blocks, calibrated initially by the manufacturer on a master Rockwell hardness tester, are available for all of the Rockwell hardness scales. Three or four test blocks are usually sufficient to cover the range of any one hardness scale; for example, the Rockwell C scale may be appropriately surveyed with three test blocks having respectively such measured hardnesses as

[1] See "Hardness Correction for Rounds," in the form of nomographs, *Metal Progress*, Vol. 52, 1947.

C 28.5 ± 1.0, C 41.0 ± 1.0, and C 62.6 ± 0.5.   To be aware of any departure of the hardness tester from standardization, the test blocks should be used frequently, particularly when a new or different indenter is used or when the major load is changed to define a different hardness scale.

Brinell Hardness-number Conversion.   The Rockwell hardness numbers on the two standard hardness scales, B and C, can be converted into Brinell hardness numbers by the following equations, developed by Petranko at the U. S. Bureau of Standards.   These equations are semiempirical and are accurate to within ± 10 per cent.

For Rockwell numbers B 35 to B 100:

$$\text{Brinell hardness number} = \frac{7300}{130 - \text{Rockwell B number}} \tag{19}$$

For Rockwell numbers C 20 to C 40:

$$\text{Brinell hardness number} = \frac{1,420,000}{(100 - \text{Rockwell C number})^2} \tag{20}$$

For Rockwell numbers greater than C 40:

$$\text{Brinell hardness number} = \frac{25,000}{100 - \text{Rockwell C number}} \tag{21}$$

THE ROCKWELL SUPERFICIAL-HARDNESS TEST

The Rockwell superficial-hardness tester, shown in Fig. 118, is a special-purpose tester, designed to be used on test sections where only a very shallow impression is permissible, where the determination of surface hardness is of primary importance, and for any other reasons where light testing loads might be desired.   The depth of penetration in any case is always less than 0.005 in.   For this reason, the superficial tester is particularly adapted for testing thin sheets, such as razor blades, and for determining the hardness of lightly carburized or decarburized steel surfaces, nitrided surfaces, etc.

Principle of Operation.   The Rockwell superficial-hardness tester operates on exactly the same principles as the regular Rockwell tester, so they need not be repeated here.   The same standard $\frac{1}{16}$-in.-diameter ball penetrator as used in the regular Rockwell test is employed in the superficial test for hardness measurements on the softer metals, such as brasses, bronzes, and unhardened steel.   For harder materials, such as hardened steel surfaces, a Brale penetrator is used, very similar to that used in the regular Rockwell, except that the spherical end of the indenter is shaped to within closer tolerance limits.   This particular Brale as used in the superficial tester is designated as the "N Brale."

The superficial direct-reading dial is similar to that used in the regular Rockwell, in that there are 100 equal divisions, each division representing one point on the hardness scale. The dial differs, however, in that there is only one set of numerals, from 0 to 100, and each division represents 0.001 mm vertical motion of the penetrator. The dial is, in general, more sensitive to small movements of the penetrator than is the one employed in the regular Rockwell, and is necessarily so because of the more shallow impressions made by the superficial tester.

Fig. 118. The Rockwell superficial-hardness tester. Model IJS. (*Courtesy of Wilson Mechanical Instrument Co., Inc.*)

The standard minor load employed in the superficial tester is 3 kg. The customary major loads, depending upon the work under test, are 15, 30, or 45 kg, as compared with the standard major loads of 60, 100, or 150 kg employed in the regular Rockwell. The rate at which the major loads are applied are standardized by adjusting the flow valve of the oil dashpot so that the release handle completes its travel in 6 to 7 sec with no specimen beneath the indenter and with an applied major load of 30 kg.

Inasmuch as the minor and major loads of the Rockwell superficial tester differ from those of the regular instrument, the standard hardness scales applicable to the latter cannot be used. As a consequence, arbitrary superficial-hardness scales have been established, and like the regular Rockwell scales, they are defined by the major load and the type of indenter used. The standard superficial-hardness scales are as follows:

Scale or prefix	Type of penetrator	Major load, kg
15N	N Brale	15
30N	N Brale	30
45N	N Brale	45
15T	$\frac{1}{16}$-in.-dia. ball	15
30T	$\frac{1}{16}$-in.-dia. ball	30
45T	$\frac{1}{16}$-in.-dia. ball	45

Minor load = 3 kg

When reporting superficial hardness numbers, it is essential to prefix each hardness number with the appropriate hardness-scale designation as given above, *e.g.*, 15N-35, 45T-20, etc. Just as in the case of reporting regular Rockwell hardness numbers, the hardness number alone has no significance.

Figure 119 shows the approximate relationship between the B and C hardness scales of the regular Rockwell and the 30T and 30N hardness scales of the superficial instrument. These relationships should be interpreted with caution, since they are only a first approximation for unhardened steel and nonferrous metals.

Condition of Test Surface. Inasmuch as the impression formed on the test surface by the $\frac{1}{16}$-in.-diameter ball or the Brale is extremely shallow, the condition of the test surface must, of necessity, be smoother and less irregular than that required in the regular Rockwell test. Polished surfaces are to be desired for accurate hardness measurements. Hard dirt particles or scale on the under surface of the test piece or on the anvil will cause inaccurate hardness measurements, due to sinking of the test section during test, which would falsely add to the depth of penetration as indicated by the direct-reading dial.

Standardization. Standardization of the superficial tester is accomplished by means of special test blocks, similar to those used in the regular Rockwell, which are appropriate to the N and T scales.

THE TUKON HARDNESS TEST

The Tukon hardness-testing instrument, illustrated in Fig. 120, is a special-purpose testing unit particularly useful for measuring the indentation hardness of very small objects and very thin sheet material, including carburized, decarburized, and nitrided surfaces. It is also applicable for determining the hardness of thin electroplated metals, structural phases of microscopic dimensions in duplex-type alloys, microscopic areas of segregation, and very hard and brittle materials, such as glass, porcelain, metallic carbides, etc. By appropriate selection of loads, with respect to the hardness of the material under test, the depth of the impression need not exceed about 1 micron.

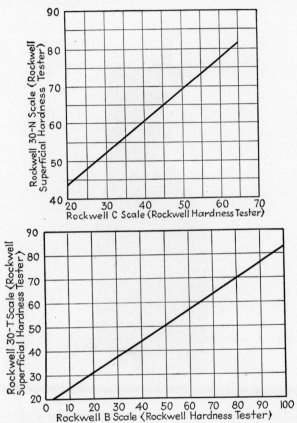

FIG. 119. Approximate relationship between the standard Rockwell hardness scales, B and C, and the Rockwell superficial-hardness scales, 30T and 30N. (*Courtesy of Wilson Mechanical Instrument Co., Inc.*)

The Tukon hardness tester is a precision-built apparatus of weights and levers whose cycle of operation, except for the manipulations involved in measuring the dimensions of the impression made by the indenter, is completely automatic and controlled by electronic devices. The specimen of interest is placed upon a special mechanical stage, illustrated in Fig. 121, and the area or location on the specimen surface where the impression is to be made is then located precisely with the permanently attached metallurgical-type microscope shown in Fig. 120. The upper platen of the stage is then moved backward by hand, carrying the specimen and the area of interest to a position directly below the indenter (within limits of about 0.0005 in.). By actuating the electronic circuit, the stage and specimen are automatically moved upward by a reversible-type motor coupled to the elevating screw until the specimen surface makes contact with the indenter.

FIG. 120.   The Tukon hardness tester.   Model LR.   (*Courtesy of Wilson Mechanical Instrument Co., Inc.*)

When this occurs, the beam on to which the indenter is attached is raised very slightly, which, in turn, breaks the electric-motor circuit and stops further elevation of the specimen. The preset load is then automatically applied to the indenter and by means of a timing unit the load is allowed to be effective for about 20 sec. Owing to the kinetic load of contact between the indenter and the specimen, and attendant formation of an enlarged impression at loads under about 200 g particularly, the total predetermined load is uniquely applied in increments to the indenter over a period of a few seconds. At the end of the approximate 20-sec time cycle, a portion of the total load is removed from the indenter; the penetrator is automatically removed from the impression by an upward motion of the indenter-supporting beam; and the stage and specimen are then automatically lowered so that the specimen stage can be again moved to relocate the area of interest and the impression beneath the microscope objective. This completes the cycle of operation, except for measuring the impression in a manner to be described.

Penetrators and Loads. There are two standard penetrators available for use with the Tukon hardness tester—the standard Vickers 136-deg diamond pyramid described on page 225, and a diamond indenter developed at the National Bureau of Standards,[1] known as the Knoop indenter.

Fig. 121. The mechanical stage of the Tukon hardness tester. (*Courtesy of Wilson Mechanical Instrument Co., Inc.*)

The Knoop indenter, a sketch of which is shown in Fig. 122a, consists of a pyramidal-shaped diamond cut to an included transverse angle of 130 deg 0 min, and an included longitudinal angle of 172 deg 30 min. Each Knoop indenter is certified by the National Bureau of Standards to be of specified quality and to be within certain dimensional and angular limits, expressed as the percentage of departure from a theoretical indenter constant C_p, derived on page 247. Acceptable Knoop indenters have an indenter constant within 1 per cent of the theoretical value.

The impression formed by the Knoop indenter, when viewed normal to the specimen surface as illustrated in Fig. 122b, is rhombic in shape with one diagonal perpendicular to and about seven times the length of the other.

[1] Knoop, F., C. G. Peters, and W. B. Emerson: A Sensitive Pyramidal-Diamond Tool for Indentation Measurements, *J. Research Nat. Bur. Standards*, Vol. 23, 1939.

The depth of the impression at the indenter apex is about one-thirtieth the length of the longer diagonal.

The Tukon tester is capable of applying loads to the indenter of from 25 to 10,000 g.[1]   The selection of an appropriate load depends upon the precise hardness and dimensions of the specimen area of interest, and is best determined by trial.   In general the load should be of such value that the impression formed by the penetrator will be as large as possible commensurate with the requirements to be met in the test.

Method of Measuring Impression. As mentioned heretofore, in all penetration hardness tests there is some elastic recovery of the impression formed by the indenter when the applied load is removed.   The amount of such elastic recovery and the final distorted shape of the impression depends

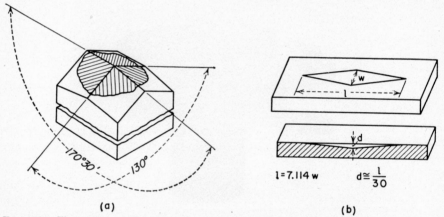

FIG. 122.   Illustrating schematically the shape of the Knoop indenter and associated shape of the surface impression.   (*From J. Research Nat. Bur. Standards, Vol.* 23, 1939.)

upon the size and precise shape of the indenter.   Owing to the shape of the Knoop penetrator, elastic recovery of the projected impression occurs principally in a transverse direction, *i.e.*, along the shorter diagonal, rather than in a longitudinal direction.   As a consequence, from the measured length of the longer diagonal and the constants of the indenter, dimensions of an impression closely related to the unrecovered length are secured.   Precise measurement of the longer, unrecovered diagonal of the impression is, therefore, of major interest since, as will be shown later, the Knoop hardness number is related to the unrecovered projected area of the impression. However, for special studies, particularly those related to elastic recovery of materials, recovered projected areas of the impression may be obtained by measuring both the transverse and longitudinal diagonals.

[1] In very recent models the loads may be varied from 25 to 50,000 g to extend the usefulness of the Vickers 136-deg diamond-pyramid indenter (see Fig. 120).

The longer diagonal of the impression is measured by means of the microscope attached to the front of the instrument. The microscope is usually equipped with a revolving nosepiece containing three objectives of different initial magnifications; and a Filar eyepiece of conventional design, as described on page 103. The Filar eyepiece must be appropriately calibrated in terms of millimeters of length, which is best accomplished by the use of a conventional stage micrometer. To secure accurate measurements of the longer diagonal, an objective should be selected that produces an image of the impression whose longer dimension is between 200 and 700 Filar units.

In the event that a Vickers impression is to be measured, the procedure and the calculation of the diamond-pyramid hardness number is precisely the same as described on pages 227 to 228.

Hardness Scale. The numerical value of the Knoop hardness number is equal to the applied load divided by the unrecovered projected area of the impression. It should be noted that the area referred to is the projected area and not the surface area of the impression as in the Vickers and Brinell equations. Thus,

$$\text{Knoop hardness number} = \frac{P}{A_p} \tag{22}$$

where P = applied load, kg

A_p = area of unrecovered projected area of impression, sq mm

The unrecovered projected area of an impression, formed by an indenter of theoretical dimensions and having perfect included transverse and longitudinal angles, is equal to

$$A_p = \frac{wl}{2} \tag{23}$$

where w = length of transverse (shorter) diagonal, mm

l = length of longitudinal (longer) diagonal, mm

Inasmuch as the unrecovered projected area A_p is more precisely related to the measured length of the unrecovered diagonal length l than to the recovered diagonal w, it is necessary to express this area in terms of the longer diagonal only. From considerations of the transverse included angle of 130 deg 0 min and the longitudinal included angle of 172 deg 30 min, the following relationship may be established between the short and long diagonals:

$$\frac{\tan 65°}{w} = \frac{\tan 86°15'}{l} \tag{24}$$

Substituting appropriate numerical values in Eq. (24),

$$l = 7.114w \tag{25}$$

or

$$w = 0.14056l \tag{26}$$

The value of w in terms of l [Eq. (26)] may be substituted in Eq. (23), and the area of the unrecovered projected area of the impression may be expressed as

$$A_p = \frac{0.14056l^2}{2} = 0.07028l^2 \tag{27}$$

where $0.07028 = C_p =$ indenter constant relating the longer diagonal l to the unrecovered projected area

If the value of C_p is now substituted into Eq. (22), the following working equation is obtained that is applicable for an impression made by a theoretically perfect indenter, or for one within limits of error defined by specification for a certified penetrator:

$$\mathrm{KHN} = \frac{P}{C_p l^2} = \frac{P}{0.07028l^2} \tag{28}$$

where KHN = Knoop hardness number

P = applied load, kg

C_p = Knoop indenter constant

l = length of longitudinal diagonal, mm

The Knoop hardness number corresponding to a measured length of the long diagonal may be determined from Table 34, page 460. Although this table is computed for a load of 1 kg, hardness numbers may be calculated for other loads by merely multiplying the hardness number in Table 34 that corresponds to the actual measured length of the long diagonal by the actual applied load in kilograms.

Table 34 is computed for an indenter having transverse and longitudinal included angles as described and an associated constant for the projected area of 0.07028. The data in this table is sufficiently accurate for most determinations with certified indenters, but under circumstances of testing where the precise value of the indenter constant should be considered, Eq. (28) should be used. Accordingly, the appropriate value of C_p for the indenter used, which is furnished with each Knoop indenter, should be substituted therein and each hardness number computed.

In reporting Knoop hardness numbers it is essential to report also the load applied to the indenter. Knoop hardness numbers are not independent of the applied load and, consequently, there is no assurance that, even on homogeneous material, values secured with one load will be the same as those secured with another load.

Condition of Test Surface. Inasmuch as the impression made by the Knoop indenter, particularly under very light loads, is extremely small in size, as illustrated in Fig. 123, the specimen of interest must be prepared metallographically to render a surface free from defects and scratches. Hardness measurements can be secured on specimens that are etched, as well as unetched, and when mounted for convenience in conventional metallographic plastics. It is essential, however, that the surface of the specimen, when placed upon the supporting stage, is normal to the vertical axis of the indenter. Otherwise the impression will be lopsided and introduce errors in the final hardness number. To achieve appropriate orientation of the specimen, it may require careful grinding of the back side of the specimen or specimen mount and accurately checking for parallelism with the prepared surface by means of a measuring micrometer.

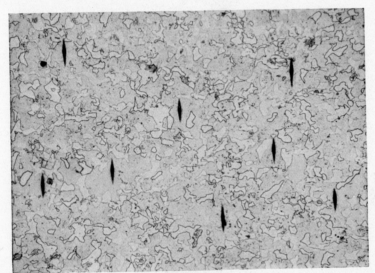

FIG. 123. Knoop indenter impressions on martensitic areas in a martensite-ferrite aggregate. 500×. Applied load during test, 50 g.

THE SHORE SCLEROSCOPE HARDNESS TEST

VERTICAL-SCALE TYPE, MODEL C-2

The Model C-2 scleroscope hardness tester, shown in Fig. 124, is a device for measuring the elastic or rebound hardness of a metal by noting the height to which a diamond-tipped hammer rebounds from the surface of the test piece after falling by its own weight from a definite height. The instrument consists essentially of a glass tube accurately aligned in a vertical position through which the hammer is guided during its downward fall. The hammer is raised to the top of the tube by an air-suction device and held in top position by means of a mechanical catch. When the hardness test is to be made, an air valve, actuated by means of a rubber air bulb that is squeezed with the hand, releases the mechanical catch and allows the hammer to fall.

Within the glass tube is inscribed the scleroscope hardness scale, against which the first rebound of the hammer is noted.   The top of the hammer serves as the indicator and, to facilitate accurate readings, either artificial light or daylight should fall downward on the instrument so that the glistening hammer top may be more readily observed.

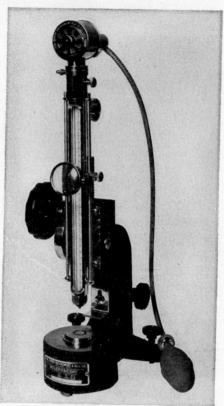

FIG. 124.   The Shore scleroscope hardness tester.   Model C-2, vertical-scale type.   (*Courtesy of the Shore Instrument and Manufacturing Co.*)

Mounted in front of the tube are a magnifier lens and needle pointer, which together may be moved to any part of the scale.   The purpose of the magnifier is to permit greater accuracy in determining the height of the hammer rebound, especially in testing soft metals, where the rebound is limited and the difference in hardness is significant but difficult to detect visually.

The scleroscope hammer, tapered at one end, is essentially a cylindrical metal plug weighing approximately 2.6 g.   The end of the plug is tapered and is fitted with a ground and polished diamond, which serves as the actual indenter and prevents distortion of the hammer end when striking the metal under test. The regular hammer provided with the scleroscope has a comparatively limited rebound when testing soft metals. This hammer may be replaced by another having a larger point area, which enables it to rebound higher, thus magnifying small but significant differences in the hardness measurements.

The glass tube and its housing are moved as a unit by a rack and pinion, which allows the test specimen to be securely clamped between the tube end and the supporting anvil. Plumb rods are provided on the side of the tester to ensure accurate vertical alignment of the instrument. Any contact of the falling hammer with the side walls of the tube will retard both its downward fall and its rebound from the test surface, with accompanying inaccuracies in the hardness measurement.

At the outset some difficulty may be experienced during the test in judging the exact height to which the hammer might rebound. It is recommended that a few trials be made on the test piece, in order to determine that particular part of the scale to which the rebound will be confined. This estimation may be facilitated by reference to Table 35, page 464. The magnifier lens and pointer needle may then be moved to this location, and slight variations in the rebound of the hammer easily noted. In making either trial or official tests, it is important that the hammer be allowed to strike only once on the same spot. Repeated tests without moving the test section will introduce errors in the hardness measurements and may cause the diamond point to crack or chip.

The scleroscope unit can readily be detached from its base and used as a portable hardness-testing instrument. When used in this manner, care must be exercised in securing appropriate contact between the end of the tube and the test surface and in aligning and maintaining alignment of the instrument during test with the specimen surface. Whether or not the axis of the instrument is perpendicular to the test section can be ascertained by noting the position of the plumb rod provided for just this purpose.

Principles. The scleroscope hardness scale, *i.e.*, the linear scale against which the rebound of the hammer is measured, consists of 100 equal divisions of the average rebound height from quenched high-carbon steel. On the scleroscope scale the rebound of the hammer from this particular grade of steel is between 95 and 105 divisions. For purposes of versatility the scale is extended to 140 divisions to enable hardness measurements to be secured on superhard and brittle materials, such as mineralogical crystals, glass, etc.

When the hammer strikes the surface of the test piece, after falling from the requisite height, there is exerted over a very small area on the surface a momentary stress exceeding 470 tons per sq in. This stress, of course, is sufficiently great to render a permanent impression on the test surface of a size and shape that corresponds to the diamond tip.   The impression so formed, however, is very small and its marring effects, even on the most finished surface, may usually be disregarded.

The kinetic energy associated with the falling hammer at the instant contact is established with the test surface is expended mainly in the formation of the impression and in the rebound of the hammer.   In general, the softer the material under test, the larger will be the impression formed by the falling hammer and, as a consequence, the greater will be the amount of energy expended in forming the impression with an attendant lesser amount available for the hammer rebound.   For example, of the total energy available from the falling hammer, as much as 98 per cent is consumed in forming the impression in lead, as compared to only 20 per cent in the case of fully hardened high-carbon steel.

Inasmuch as the registration of the scleroscope number is dependent fundamentally upon the amount of energy available for rebound of the hammer, the mass of the test section must be considered.   It can be demonstrated that a block of hardened steel in the form of a cube and weighing 1 lb. is of sufficient mass to overcome inertia effects or shock of the striking hammer.   As the mass decreases, however, there is an attending decrease in the hammer rebound owing to inertia effects becoming less pronounced. The divergence in scleroscope hardness numbers resulting therefrom are more pronounced in soft materials than in hard materials.   For example, if the mass of hardened steel referred to above is reduced to 0.5 lb or to 1 oz, the height of the hammer rebound will be reduced by about 1 and 20 per cent respectively; and approximately 2 and 40 per cent respectively for a metal about one-fifth as hard.

The mass effect of the test section, therefore, is perhaps the one factor influencing most reproducibility of the scleroscope hardness scale.   Although all test sections may not in themselves be of sufficient mass to render accurate hardness measurements, their lack of mass can readily be compensated for by appropriately clamping the test piece between the anvil of the instrument and the drop tube.   During test when the hammer strikes the test section (even when in the form of sheet material), an audible dull thud will be heard, provided the specimen is appropriately clamped to the anvil.   This sound will be similar to that heard when the hammer strikes the anvil directly.   Specimens that are not properly clamped and that do not lie flat and firmly against the anvil will produce a higher pitched sound when struck by the falling hammer—a sound similar in pitch to that associated with an insufficient mass and an attending shorter rebound of the hammer.   It is essential, therefore, to note carefully during the test the sound produced when the hammer strikes the test surface.   If a shrill note is rendered, the test should be repeated, with attention directed to the adequacy of clamping, until the characteristic dull thud is audible. This procedure will then ensure that the hardness measurements secured are reliable.

Condition of Test Surface. Usually the specimen surface must be smoother for scleroscope hardness measurements than is required for either Rockwell or Brinell hardness determinations. A rough or poorly prepared surface will introduce variations in the rebound of the hammer and will usually indicate a hardness lower than the true value.

Metals that are comparatively soft can be appropriately prepared by dressing the surface with a No. 2 or No. 3 flat file. Alloys in the hardened condition, such as fully quenched or quenched and tempered steel, are best prepared by grinding on the flat side of a medium-grit emery wheel. Adequate precautions must be taken to prevent the section from becoming tempered during the grinding operation.

Decarburized surfaces should always be removed by grinding in order to secure representative hardness measurements of the test section, unless the primary interest lies in the hardness of the decarburized layer. Metals and alloys containing large structural aggregates, such as become associated in cast iron and certain nonferrous alloys, should be tested at a number of places over the surface, with an average of the measurements taken to represent the hardness of the piece.

Minimum Thickness of Test Section. In the testing of thin sheets and flat material, it is important that the test piece be securely held in close contact with the anvil and that no foreign material be allowed to intervene between the anvil support and the test piece.

Hardened-steel sheets, such as razor blades, should have a minimum thickness between 0.005 and 0.006 in., whereas annealed sheets should be thicker than 0.010 in. Cold-rolled, unannealed brass and steel should be at least 0.015 in. thick. Thinner sections than those given above may be made to the required thickness by piling up two or more sheets. Whenever this procedure is followed, it is important that the material lie flat and be clamped tightly together, but not tightly enough to crush or distort the pack.

In the testing of round or irregularly shaped surfaces, generally no error is involved in the determination unless the irregularities are such that the hammer, on striking the test piece, rebounds to one side. This condition can be eliminated, however, by grinding a flat spot on the surface. Hollow shapes, such as cylinders, may give too low a hardness reading, due to insufficient mass and improper support. Errors arising from these sources may be eliminated by the use of a mandrel and appropriately clamping the test piece.

Standardization. One of the factors responsible for inaccurate hardness measurements is a chipped or cracked diamond tip.   Arising of this adverse condition can be prevented usually at the outset by proper handling of the instrument and by never allowing the hammer to strike the same spot on the test surface more than once.

The instrument may be calibrated by using a standard hardened-steel test block, supplied by the Shore Instrument and Manufacturing Co., that

FIG. 125.   The Shore scleroscope hardness tester.   Model D.   (*Courtesy of the Shore Instrument and Manufacturing Co.*)

has a hardness equivalent to 100 on the scleroscope scale.   The rebound for any given steel cannot be corrected by a proportionate decrease or increase if the hammer gives a rebound either less or greater than 100 on the hardened block of steel.

DIAL-RECORDING TYPE, MODEL D

The principles of the Model D scleroscope instrument, illustrated in Fig. 125, are precisely the same as those already discussed with respect to the vertical-scale-type tester.   The Model D unit, however, differs somewhat in appearance and design from the Model C-2, most notably in the manner by which the hardness numbers are indicated.

By an ingenious mechanical arrangement, the rebound of the hammer in the Model D instrument is automatically recorded on an instrument dial, calibrated directly in scleroscope hardness numbers. Owing to the space required for the necessary recording mechanism, the distance through which the hammer falls is only about 0.75 in. The energy relationships, however, are the same as in the Model C-2 instrument because the weight

FIG. 126. The Monotron hardness tester. (*Courtesy of the Shore Instrument and Manufacturing Co.*)

of the hammer is proportionately greater (2.6 as compared to about 37 g). This is further demonstrated by the similarity in hardness measurements on the same test specimen as secured with either instrument.

THE MONOTRON HARDNESS TEST

The Monotron hardness tester, shown in Fig. 126, is a direct-reading pressure instrument which registers on the upper dial the load required to force a standard penetrator into the metal under test to a predetermined depth. The hardness number is read from the dial in terms of kilograms per unit area while the load is applied, thus serving to eliminate errors arising from elastic recovery of the metal.

To operate the tester, the specimen is placed on the anvil and brought into contact with the penetrator. By means of a side arm, pressure is applied to the penetrator which forces it into the metal to a depth of 0.0018 in. The depth of penetration is indicated by the lower dial. Readings on the dial are facilitated by twin pointers, thereby making it unnecessary to reset the pointer to zero at the beginning of each test. The dial is inscribed with two similar scales, each scale divided into 100 equal numbered divisions. Each division represents 0.0002 in. (0.005 mm) depth of penetration of the penetrator. The inner set of divisions is numbered in a clockwise direction for standard constant-depth hardness testing, whereas the outer set of divisions is numbered counterclockwise for use in special tests, such as measuring the hardness in terms of flow and ductility under selected constant loads. Although the hardness number is read directly with the load still applied to the penetrator, the load may be released with an accompanying movement of the depth indicator toward zero. The displacement of the indicator, similar to that of the Rockwell tester, is a measure of the elastic recovery of the metal under test and the number of divisions above zero indicates the degree of permanent deformation. When the recommended depth of penetration is exceeded, the ductility and plastic flow of the softer metals may be determined and the effect of time may be observed under overloaded conditions.

In the standard test the load required to force the standard penetrator to a depth of 0.0018 in. is indicated on the upper dial. This dial consists of an inner set of 160 equal divisions, numbered in a clockwise direction, each division representing a load of 1 kg per unit area. The outer scale is calibrated directly in Brinell numbers, which may be read directly when standard tests conditions are used.

Penetrators and Hardness Scales. The standard Monotron penetrator is a 0.75-mm hemispherical diamond that is forced into the metal under test to a standard depth of 0.0018 in. This depth of penetration produces an impression 0.36 mm in diameter, which is almost entirely invisible to the unaided eye. For most finished surfaces this impression is not objectionable, but in cases where a smaller impression is to be desired, the method of partial penetration to be described later may be used.

The term "Monotron hardness" indicates that the 0.75-mm diamond point is used and, when the depth of penetration is standard, the values are referred to the M-1 scale, which reads directly in kilograms and Brinell units. For the testing of soft materials, where the comparative values are small as determined with the 0.75-mm diamond, larger penetrators of tungsten carbide are available, in diameters of either 1.53 or 2.5 mm. The measurements obtained with these penetrators are referred to as M-3 and M-4 scales, respectively. The M-2 scale requires a 1-mm-diameter diamond point, but because the M-1 scale generally serves the condition of testing required for the M-2 scale, the latter is seldom used.

Condition of Test Surface. In the use of the three standard penetrators, the surface may be tested in either a prepared or an unprepared condition. With prepared surfaces, such as a ground or polished finish, the pressure indicator is normally set at zero, which is actually 0.625 kg above the no-load point, so as to compensate for the weight of the indenter bar.

In the case of testing the underlying metal through a scaly, rough, or decarburized surface, the pressure indicator is set at a value of 10 or 20 kg or more below the zero mark, thus in effect applying a prepressure to penetrate through the unprepared surface. When the indicator is brought to the zero point by this method, the depth indicator hand is set at zero, and the test is made in the normal way.

Minimum Thickness of Test Section. Generally the minimum thickness of test sections that may be tested without anvil effect is 0.020 in. Sheets thinner than this require that an allowance be made for the anvil effect or—which is more desirable—that this condition be prevented by piling up sheets to the required thickness.

In the testing of thin carburized or nitrided surfaces or in cases where the size of the impression is objectionable on the test section, partial penetration is to be recommended. The penetrator is allowed to penetrate to only a fraction of the standard depth—say, one-third—in which case the pressure readings are multiplied by 3 to obtain the standard hardness value. Too deep a penetration in thin cases will produce inaccurate hardness values, due to the influence of the softer base metal beneath.

THE HERBERT PENDULUM HARDNESS TEST

The Herbert pendulum hardness tester, shown in Fig. 127, is primarily a special research instrument that is capable, through appropriate test procedures, of distinguishing between the hardness of a test specimen regarded as resistance to indentation and the hardness as manifested by resistance to work hardening.

The pendulum hardness test can be applied readily to measuring the hardness of extremely thin test sections, wires, carburized cases, etc., with a reasonably high degree of accuracy. It may be successfully employed in measuring the hardness of very hard materials that, under ordinary circumstances, would be inappropriate for determination by the Brinell test owing to attendant deformation of the ball indenter. The test is applicable to fragile articles or finished surfaces, without apparent injury to them, as the defacing mark produced during test is generally microscopic in size.

The instrument consists essentially of a weighted rocking device known as the *pendulum*, which, in the standard instrument, is an arched casting 12 in. in length and capable of spanning flat surfaces 6 in. wide or cylindrical objects 8 in. in diameter. Under the center of the arch a 1-mm-diameter hardened-steel ball is mounted in a suitable chuck for the purpose of forming a contact point with the test surface. The standard pendulum weighs 4 kg, with the center of gravity normally adjusted so that, in effect, a com-

Fig. 127. The Herbert pendulum hardness tester. (*Courtesy of Tinius Olsen Testing Machine Co.*)

pound pendulum is formed, 0.1 mm in length. The center of gravity may be shifted in two directions, horizontally and vertically, by means of three balancing screws. For special testing, pendulums may be obtained that weigh either 2 kg or 24 kg and that are equipped with 1-mm-diameter hardened-steel, ruby, or diamond balls.

The test surface should be reasonably smooth and free from surface imperfections so that the ball is afforded free motion during the swing. The pendulum is placed upon the surface of the test piece and, by noting the oscillations of the pendulum when swung, the following five distinct types of tests may be made:

Time Test. The "time hardness number" is determined by noting the time in seconds for 10 complete swings (5 over and 5 back) of the pendulum through a small arc. It is a measure of a form of indentation hardness of the metal and corresponds to Brinell hardness numbers, into which it is convertible. In the case of hardened steels, the Brinell number is determined directly by multiplying by 100 the number of seconds required for one complete swing.

Scale Test. If the pendulum is tilted through a definite angle and then released, the angle reached at the end of the ensuing half swing is recorded as the "scale hardness number." Inasmuch as the magnitude of the angle of the completed swing is dependent upon the energy absorbed at the point of contact between the ball and specimen surface, the scale hardness number is a measure of the specimen's resistance to deformation by rolling, drawing, or stamping. The flow hardness of a material is expressed by the scale-time ratio, which is obtained by dividing the scale hardness number by the time hardness number.

Time Work-hardening Test. The time work-hardening test is a measure of the induced hardness produced in the test specimen by rolling it with the pendulum ball and comparing the original time hardness number (the number of seconds for 10 complete swings) with that recorded after each swing.

Scale Work-hardening Test. This test measures the increase in work-hardening of the test specimen by repeated rolling of the ball on the same spot. Five successive scale tests are made by tilting the pendulum alternately in opposite directions. The work-hardening capacity of the specimen is then expressed by the difference between the original scale hardness number and the average of the other four determinations.

Damping Test. Although the damping test is not a measure of hardness, it is an indication of the rate at which energy is absorbed by the specimen when it is subjected to alternate stressing below the elastic limit. The numerical value of this property is obtained by noting the rate of decrease of the amplitude of the pendulum when swinging through a small arc.

THE MICROCHARACTER HARDNESS TEST

The Microcharacter, shown in Fig. 128, is primarily a research instrument for determining the scratch hardness of microscopic areas, constituents of alloys, thin layers of metals, and hardness gradients over small areas. The instrument is designed to be attached directly onto the stage of a table-type metallurgical microscope, and operated thereon. The Microcharacter consists of a well-balanced arm to which is attached on the underside a sharp-pointed diamond tool. It is so mounted that, when the arm is lowered and properly weighted, the diamond point rests directly on the surface of the specimen to be tested. The specimen is mounted beneath the diamond tool on a movable stage, and by actuating a micrometer screw the specimen is slowly moved under the diamond point. This operation produces on the test surface a microscopic cut or scratch, the width of which is variable and governed by the relative hardness of the various structural components of the alloy. To ensure a uniform cut in any one constituent, the diamond is lubricated with a fine watch oil during the cutting operation.

FIG. 128.   The Microcharacter hardness tester, shown mounted on the stage of a table-type metallurgical microscope.   (*Courtesy of American Optical Co.*)

When the test is completed the lubricant is removed with a suitable solvent, such as xylol, and the cut width is measured by a suitable microscope arrangement, to be discussed later. From the numerical value of the cut width, the microhardness number may be computed.

Cutting Tool and Loads.   In order to produce an even and uniform cut on the test surface, the cutting tool must possess the following characteristics:[1]

[1] Conley, W. J., W. E. Conley, H. J. King, and L. E. Unger: The Microcharacter as a Research Tool, *Trans. ASM*, Vol. 24, 1936.

1. The point must be sufficiently sharp to produce a clear cut on both hard and soft materials.

2. The point must retain its cutting edge during continual use.

3. The shape must be such that the vertical pressure exerted upon the test surface is always greater than the horizontal cutting force. In addition, the depth of the indentation should not increase because of the vertical pressure, when the horizontal cutting force is removed.

It has been shown by Bierbaum that, to fulfill the above requirements, a diamond must be accurately ground to the form of a solid right angle (the corner of an imaginary cube), and it must be mounted on the balanced arm with the body diagonal of the cube normal to the surface of the specimen, with one edge of the solid right angle accurately in line with the cut. To increase the sensitivity of the diamond to abrupt changes in hardness, it is mounted elastically rather than rigidly to the arm.

In the use of the Microcharacter, extreme precautions must be taken in handling the diamond. It should not be subjected to blows of any kind and should not make contact with another diamond. An insensible pressure on the exceedingly sharp point will produce a stress far in excess of the crushing strength of the diamond, with accompanying disastrous results.

The standard vertical load applied to the diamond point is 3 g; it is applied by means of weights placed directly above the point. For special testing purposes, a 9-g load is frequently used.

Method of Measuring Cut Width. In determining the microhardness of the material under test, the width of the cut produced is measured at points of interest under a microscope. The microscope should be, preferably, one of high quality and provision should be made so that it may be rigidly mounted, to prevent vibration during manipulations associated with operation of the Microcharacter. Optical parts consisting of a 1.5- or 2-mm oil and a 3- or 4-mm dry objective, in combination with either a 15 or 25× Filar screw-micrometer eyepiece, will produce sufficient magnification and resolution so that measurements to 1 micron (0.001 mm) may be made.

Microhardness Scale. The hardness scale pertaining to the Microcharacter has been arbitrarily established to meet the requirements of the test and is therefore in no way related to any of the more common hardness scales described heretofore. The microhardness number K is equal to the reciprocal of the square of the cut width λ expressed in microns multiplied by an arbitrary constant of 10,000 for purposes of avoiding fractional hardness values. Thus, for tests made with the standard 3-g vertical load, the microhardness may be expressed as

$$K = \frac{10,000}{\lambda^2} \tag{29}$$

Table 36, page 465, gives the microhardness numbers for cuts of different widths, as determined by the above equation.

Condition of Test Surface. Inasmuch as the cut made in the test specimen by the Microcharacter is microscopic in size, it is evident that the condition of the surface is of primary importance in obtaining the best possible results. Without regard to the purpose of the test, the specimen surface must be highly polished and free from scratches and surface imperfections. Surface preparation by one of the methods described in Chap. 1 should be closely followed. In the event that differences in hardness between various constituents are to be determined, the specimen should be etched to reveal the structure. The roughening of the individual components due to the etching treatment will usually not affect the character or appearance of the cut.

THE EBERBACH HARDNESS TEST

The Eberbach hardness tester, illustrated in Fig. 129*a*, is a portable-type instrument designed particularly to measure the hardness of areas of microscopic dimensions. The microhardness tester consists of two essential units—the indenter assembly, shown in Fig. 129*b*, which attaches directly onto a table-type metallurgical microscope or, through the use of adapters, onto any standard metallograph; and a control box containing a pilot-light assembly, electronically controlled relays, and a load-calibrating stand to be described.

Penetrator and Loads. The indenter employed in the Eberbach hardness test is the standard 136-deg Vickers square-based pyramidal diamond described on page 225. The diamond penetrator is mounted on one end of a plunger rod, which is supported vertically in a carefully lapped bearing sleeve. The indenter assembly contains within it one of several calibrated coil springs that serves as the loading force to the indenter. In operation the plunger rod is raised very slightly when the diamond indenter makes contact with the test surface through manipulations of the coarse and fine focusing adjustments of the microscope. This slight vertical motion of the plunger rod actuates a normally closed electrical contact point in the indenter assembly, which, in turn, illuminates a pilot light in the control box. When this occurs, the operator is aware that the compressive force of the spring is effective and that the force opposing penetration of the indenter on the test surface is equivalent to that exerted by the coil spring. If the downward motion of the microscope tube is stopped the instant the pilot light is illuminated, the same load will be applied to the indenter in repeated hardness tests. The feature of the pilot light affords also a convenient means of timing precisely the time of load application, which normally is between 5 and 10 sec.

As mentioned heretofore, the load applied to the indenter is secured through the use of a coil spring appropriately mounted in the indenter assembly. The standard instrument includes six springs, which, in order, are capable of applying approximate loads of 7.5, 22.5, 64.5, 110.5, 185.5, and 550.0 g. At the time any one of the loading springs is to be used, it is

a. Control unit and fixture for calibrating the compression spring within the indenter assembly.

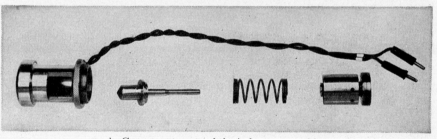

b. Component parts of the indenter assembly.

FIG. 129. The Eberbach microhardness tester. (*Courtesy of Eberbach and Son, Co.*)

necessary to calibrate it after it is placed within the indenter unit. Calibration of the spring is carried out by simply placing the indenter assembly on the calibrating stand of the control box (see Fig. 129*a*); and by loading the weighing pan temporarily attached to the plunger rod with sufficient calibrated balance weights to just overcome the compressive-loading factor of the coil spring.   The end point of balance is indicated by the pilot light, and the load exerted by the coil spring is then equivalent to the combined weight in kilograms of the weighing pan and the weights required.

Method of Measuring Impression.   When an impression on the test surface is to be measured, the indenter assembly is removed from the microscope and an appropriate objective lens substituted in its place.   The impression is then critically focused and the length of the two diagonals of the square surface impression (same shape as in the standard Vickers test) is accurately measured with a Filar eyepiece.   An average of the two measurements is then used in computing the diamond-pyramid hardness number.

Hardness Scale.   The numerical value of the Eberbach hardness number, expressed in terms of kilograms per square millimeter, is identical with that of the standard Vickers test, and may be computed from the following equation:

$$\text{DPH} = 1.854\,\frac{P}{D^2} \tag{30}$$

where DPH = diamond-pyramid hardness number
P = load, kg
D = diagonal length of square impression, mm

In the practical performance of the test, hardness numbers corresponding to different diagonal lengths may be determined by reference to Table 32, page 454.   Inasmuch as these data are based upon an applied load of 10 kg, appropriate computations must be made as described on page 229 to secure the hardness number corresponding to the actual load applied during test.

Condition of Test Surface.   Because the impression made by the diamond penetrator in the Eberbach test is microscopic in size, it is evident that the surface of the test specimen must be free from surface imperfections and other abnormalities.   It is essential that the surface be prepared in accordance with one of the procedures outlined in Chap. 1 for the preparation of metallographic specimens.   In most cases the specimen must be etched in order to distinguish one constituent from another.   The roughening of the individual structural phases, due to the etching treatment, generally does not affect the hardness measurements nor the appearance of the impression.

HARDNESS CONVERSION

Tables 37, 38, and 39, pages 466 to 470, give the approximate hardness number conversions from Vickers, Brinell, and Rockwell C numbers, respectively, to other hardness scales.   These data are applicable to steel and have been derived from extensive hardness tests on carbon and alloy steels, mainly in the heat-treated condition.   They have been found reliable, however, for practically all constructional alloy steels and tool steels in the as-forged, annealed, normalized, and quenched and tempered conditions, provided they are homogeneous. In the case of high-manganese steels and austenitic types, as well as constructional alloy steels and tool steels in the cold-worked condition, the hardness conversions may be less accurate than when applied to the precise steels for which they are intended.

Chart 4, page 471, shows the approximate relationship between various hardness scales as determined on cartridge brass. These data are valid only when the base hardness number is secured by standard test procedures and the specimen is sufficiently thick to avoid anvil effect during testing.

References

ASM: "Hardness and Hardness Measurements," Cleveland, Ohio, 1942.

ASM: "Metals Handbook," Cleveland, Ohio, 1948.

ASTM Standards, Part 1-A, 1946.

ASTM: "Symposium on the Significance of the Hardness Test of Metals in Relation to Design," Pittsburgh, Pa., 1943.

Clark, J. J.: Hardness of Metals, *Can. Min. J.*, May 22, 1925.

Conley, W. J., W. E. Conley, H. J. King, and L. E. Unger: The Microcharacter as a Research Tool, *Trans. ASM*, Vol. 24, 1936.

Dana, E. S.: "A Textbook of Mineralogy," 4th ed., John Wiley & Sons, Inc., New York, 1932.

Dohmer, P. W.: "Die Brinellsche Kugeldruckprobe," Verlag von Julius Springer, Berlin, 1925.

Holz, H. A.: On Hardness Testing of Metals and Metal Products, *Bull.* 27, New York.

Holz, H. A.: On Hardness Tests of Metals and Metal Products, *Special Bull.*, New York, 1937.

Knoop, F., C. G. Peters, and W. B. Emerson: A Sensitive Pyramidal-Diamond Tool for Indentation Measurements, *J. Research Nat. Bur. Standards*, Vol. 23, 1939.

O'Neill, H.: "The Hardness of Metals and Its Measurement," Chapman & Hall, Ltd., London, 1934.

Rockwell, S. P.: "The Testing of Metals for Hardness," *Trans. ASST*, Vol. 2, 1922.

SAE: "SAE Handbook," New York, 1946.

Shore Instrument and Manufacturing Co.: The Monotron, *Bull.* M-3, May 18, 1937, Jamaica, N.Y.

Tinius Olsen Testing Machine Co.: "The Herbert Pendulum and Cloudburst Machine," *Bull.* 4, Philadelphia, 1929.

Wilson Mechanical Instrument Co., Inc.: Rockwell Superficial *Catalogue Supplement* RS-4, New York.

CHAPTER 7

SPECIAL METALLURGICAL TESTS

In addition to the conventional mechanical tests to which metals and alloys may be subjected, there are a number of special-purpose tests designed primarily to reveal certain metallurgical properties not otherwise apparent. The usual tensile, impact, hardness, torsion, and similar tests are of considerable importance in determining engineering properties of a metal but by their very nature they fail as appropriate testing procedures to reveal such metallurgical characteristics as response to heat-treatment, austenitic grain size and hardenability in steel, internal defects within a metal that might arise from certain metallurgical operations, etc. Of the many special metallurgical tests that are currently in use, some are non-destructive in nature, as exemplified in the magnetic and supersonic methods of determining internal defects within a metal, whereas others are destructive, requiring either the whole section or appropriately selected specimens therefrom upon which to conduct the test.

The use of X rays and gamma rays in the testing of metals and alloys has become of considerable importance in the study of internal structures and in the inspection of certain semifinished and finished products. Radiographic techniques, for example, are particularly appropriate for detecting sub-surface discontinuities within a metal, such as casting, welding, and fabricating defects, which if undetected might initiate disastrous service failures. The application of X rays and associated diffraction techniques to the study of crystal structures has afforded clearer insight into the atomic arrangement and structure of metals and has solved many perplexing problems of this nature. Inasmuch as an adequate and thorough discussion of the subject is beyond the scope of this book, a selected textbook bibliography on the principles of X rays, radiography, and crystal structure is given on page 325.

METHODS FOR DETERMINING THE AUSTENITIC GRAIN SIZE IN STEEL

It is often observed that two heats of steel having similar chemical composition may respond quite differently to heat-treatment, and the properties of the steels secured by the same heat-treatment may likewise be different. Although these differences in behavior and ultimate properties cannot be satisfactorily attributed to the small variations in nominal composition, they can usually be correlated to some measure with the austenitic grain size established in the steels at heat-treating temperature and, to a lesser extent, with the ferritic grain size at ordinary temperature.

It is well known that any unhardened steel of more than approximately 0.04 per cent carbon is, at ordinary temperature, an aggregate of two phases—ferrite and iron carbide—which remains substantially unaltered during heating to a temperature of about 723°C (1333°F). At this tem-

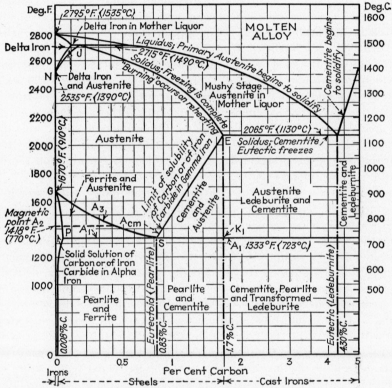

Fig. 130. The iron-carbon equilibrium diagram. (*Modified from Metal Progress, Vol. 52, 1947.*)

perature level, however, which is known as the A_1, or lower critical transformation temperature (line PSK_1 in Fig. 130), the two existing phases proceed to react to form a new phase, known as austenite. In terms of phase designations, austenite is a solid solution of carbon dissolved in gamma iron, which is stable at elevated temperatures but unstable at temperatures below the lower critical. As in the case of other homogeneous solid solutions, austenite exists in the form of equiaxed grains of some measurable size.

In the case of eutectoid steels, *i.e.*, steels containing approximately 0.83 per cent carbon, the ferrite–iron carbide reaction goes to completion at the A_1 transformation temperature and renders the steel fully austenitic. In those steels containing more or less than 0.83 per cent carbon (hypereutectoid and hypoeutectoid steels, respectively) the austenite formed at the lower critical temperature coexists with an excess of one of the reacting phases. As the temperature is increased above the A_1 temperature level, however, the excess constituent progressively dissolves into the austenite first formed which, as a consequence, alters the relative amount of austenite as well as its carbon content. For any of these steels, a temperature is soon reached upon further heating where the last traces of the excess phase disappears and the steel becomes completely austenitic. The temperature at which transformation is complete, *i.e.*, where the structure is 100 per cent austenitic, is known as the upper critical transformation temperature and, unlike the lower critical temperature, it varies with the carbon content of the steel. Specifically, in the case of hypoeutectoid steels, the upper critical is designated as the A_3 transformation temperature, and in hypereutectoid steels the designation is A_{cm}.[1] These transformation temperatures are shown in Fig. 130 as the lines *GS* and *SE*, respectively.

It is of interest to note that the newly formed austenitic grains at the transformation temperature are initially small regardless of the ferritic grain size and the size of the pearlite patches prior to heating. Precisely what the initial size will be, however, in terms of actual dimensions, is dependent upon a number of factors, the most important being the chemical composition of the steel, the manner by which the steel was deoxidized during manufacture, and the characteristics of the microstructure existing in the steel at room temperature. Although the size of the initial austenitic grains is small at the transformation temperature, an increase in temperature within the stable austenitic range will usually result in grain growth; in general, the higher the temperature the more rapid will be this growth and the larger will be the final grain size. At any particular temperature within the austenitic region, the size of the austenitic grains will be characteristic of that temperature in relation to the steel under consideration. The grain

---

[1] The notations A_1, A_3, and A_{cm} are used when the transformation temperatures which they designate, both upon heating and cooling, are considered collectively. Because of the dependency of transformation temperatures upon the rates at which steel is heated and cooled through the thermal critical range, the temperature of transformation upon heating may be different than that upon cooling. In order, therefore, to distinguish the critical points upon cooling and heating, the former are referred to as the Ar points, r designating cooling, from the French *refroidissement*. Upon heating, the critical points are referred to as Ac, c designating heating, from the French *chauffage*. Thus, upon heating, the respective A points are referred to as Ac_1, Ac_3, and Ac_{cm}.

growth of austenite due to a prolonged sojourn at a given temperature occurs at a diminishing rate and, because of this, grain growth may be considered, in a practical sense, to have ceased after a certain grain size has been established.

The austenitic grain size established at some maximum temperature (provided sufficient time has been allowed) will, therefore, remain substantially constant, and, as is well known, the attained grain size will not be influenced in any way by the rapidity with which the steel is subsequently cooled to room temperature. As a consequence, some of the mechanical properties of the steel and its behavior during subsequent heat-treatment may be associated in part with the austenitic grain size established at the temperature of interest, and the influence of that particular grain size may be reflected in the properties of the steel, whether it is furnace-cooled, normalized, or quenched.

In all steels, as mentioned heretofore, there is a definite tendency for grain growth to occur at temperatures in the austenitic region, but this tendency in some steels may be less than in others. Hence, as a means of distinguishing between these tendencies, steels are commonly referred to as being fine or coarse grained. A fine-grained steel is essentially one that resists grain growth over a wide range of temperatures in the austenitic region, whereas a coarse-grained steel is one that tends to coarsen near the transformation temperature and continues to coarsen steadily as the temperature is increased.

Although fine-grained steels usually resist coarsening in the normal heat-treating range of temperatures, they nevertheless often exhibit pronounced grain-growth tendencies above the so-called coarsening temperature range. The coarsening temperature, which pertains essentially to fine-grained steels, is not one that is definitely fixed for any given steel, but it is a temperature that varies within limits, depending primarily upon the prior heat-treatment. However, when the coarsening temperature range is reached or exceeded during heating, the austenitic grains of the so-called fine-grained steels very often grow at a faster rate and attain relatively larger sizes than the austenitic grains of a coarse-grained steel heated to the same temperature. Furthermore, at the coarsening temperature the influence of time at temperature with respect to uniformity of the final austenitic grain size becomes of importance. Because the grain-growth characteristics of a coarse-grained steel are more gradual over a wide range of temperatures, there is no well-defined coarsening temperature for these steels.

It should be evident from the past discussion that the austenitic grain size of a given steel is not inherently coarse or fine, but rather it is a variable characteristic that is influenced by the temperature to which the steel is heated and the time at that particular temperature.   Hence, in accurately describing the behavior of steel in terms of austenitic grain size, the description is incomplete unless the conditions of temperature and time are specified.   It is a common inference, however, unless otherwise stated, that the temperature and the time at temperature are those normally used when heat-treating the particular steel in question.

The importance of the prior austenitic grain size with regard to many of the characteristics of steel may be summarized and compared as follows:

Property	Coarse-grained austenite	Fine-grained austenite
1. In heating slightly above critical range.....	Tends to coarsen	Tends to stay fine
2. Hardenability...........................	Deeper	Shallower
3. Retained austenite......................	More	Less
4. Distortion during hardening..............	More	Less
5. Warpage in hardening....................	More	Less
6. Possibility of quenching cracks............	More	Less
7. Internal stress after hardening............	Higher	Lower
8. Possibility of soft spots in quenching......	Less	More
9. Behavior in carburizing..................	Deeper case	Shallower case
10. Machinability after normalizing...........	Better	Inferior
11. Embrittlement by cold-working...........	More	Less
12. Toughness.............................	Less	More

It is apparent, therefore, that, if the austenitic grain size is determined by methods to be described, it will be possible to predict, in certain respects, the behavior of a particular steel when subjected to different forms of heat-treatment, or when the steel is put into service under different existing conditions of stress.

Ascertaining the Austenitic Grain Size.   Inasmuch as austenite normally exists above the thermal critical range, it is of practical importance that the steel in question be so treated that the outlines of the prior austenitic grains will be apparent at room temperature upon subsequent microscopic examination.   The treatment to which the steel is subjected for this purpose may produce a change in the chemical composition of the specimen, such as is brought about by oxidation and carburization methods; or the method may be one in which the composition of the specimen remains substantially unaltered.   In most cases the latter method is preferred. Fortunately, by an appropriate cooling schedule from the significant heat-treating temperature, the austenite may be transformed in such a manner that the prior austenitic grain boundaries will be clearly revealed.   This may be brought about in one of the following ways:

1. By the rejection of either ferrite or iron carbide into the austenitic

grain boundaries, forming a complete or nearly complete network, in hypoeutectoid and hypereutectoid steels, respectively.

2. By the formation of a transformation-product network at the prior austenitic grain boundaries as secured by appropriate interruption of the austenite transformation.

3. By the etch contrast established between differently oriented grains of martensite, which have limits of orientation related to that of the parent austenite.

4. By high-temperature etching methods.

In certain respects the estimation of the austenitic grain size made from its decomposition products rather than upon the constituent itself is advantageous. This circumstance permits outlining the grains with constituents which may have appreciable visual contrast with the grains themselves. Plain carbon or low-alloyed steels with a carbon content remote from the eutectoid composition will reject, upon slowly cooling through the transformation range, either ferrite or cementite, depending upon the carbon content of the steel. The rejection of these constituents occurs largely at the austenitic grain boundaries and, under appropriate conditions, the constituents may be made to form a complete or nearly complete network around the original grains, after which the remaining austenite transforms into pearlite. Eutectoid steels, having no excess constituent to reject, begin to transform characteristically in the austenitic grain boundaries, forming under suitable conditions a network of fine nodular pearlite around the original austenitic grains. The fine pearlite outlining the grains may be preserved and the remainder of the austenitic grains may be transformed, by quenching, to light-etching martensite. Thus, by appropriate etching, the martensitic body of the grain may be distinguished from the dark-etching pearlite at the grain boundaries.

Appropriate Cooling Methods. As mentioned heretofore, pertinent information relative to quenching and tempering and to the final properties of the heat-treated steel may be obtained by determining the relative austenitic grain size that is established in heat-treating an appropriate sample at some definite temperature in exact accordance with the original heat-treating procedure.[1] With the austenitic grain size once established at the temperature of interest, the appropriate conditions of cooling to reveal the prior austenitic grain boundaries depend primarily upon the carbon content of the steel in question. Suitable cooling schedules

[1] It is common practice to establish the austenitic grain size under conditions mutually agreed upon between consumer and producer. For example, the agreement may specify that the steel possess a certain austenitic grain size when established by heating to a temperature not over 28°C (50°F) above the normal heat-treating temperature and for not over 50 per cent more than the normal heat-treating time. The normal values referred to are those mutually agreed upon.

recommended for steels of different carbon contents will be discussed separately.

HYPOEUTECTOID CARBON STEELS. The prior austenitic grain size in steels containing less than about 0.10 per cent carbon can be ascertained by the heat-etch method described on page 277, or more simply, by metallographic examination of an appropriately heat-treated specimen. The latter procedure involves austenitizing a small and very thin specimen at the temperature of interest for a short time (to prevent decarburization) and subsequently quenching the specimen into mercury, water, or brine. The metallographic structure resulting therefrom will consist of low-carbon martensite, and possibly a network of ferrite at the prior austenitic grain boundaries. In any event, the as-quenched specimen, or preferably one tempered for 5 to 10 min at about 200°C (392°F), is then prepared for metallographic examination and etched by immersion in a 5 per cent aqueous solution of ferric chloride. By this etching treatment, the prior austenitic grain size will be made apparent, owing to the etch contrast established between the differently oriented martensitic grains. The principles involved in this method of determining austenitic grain size are the same as those associated with a similar etching method applicable to higher carbon-content steels, as described on page 275.

The austenitic grain size of hypoeutectoid steels having a carbon content between 0.25 and 0.55 per cent carbon may be revealed by air cooling an appropriate specimen from the temperature of interest. The excess ferrite that is rejected at the austenitic grain boundaries during cooling will, upon appropriate etching at room temperature, reveal the prior austenitic grain-size pattern. This circumstance is brought about, as shown in Fig. 131, by the contrast developed between the lightly etching ferritic network and the more darkly etching areas of pearlite. However, if the carbon content of the steel exceeds about 0.55 per cent, the free ferrite available may be insufficient to form a complete network, and, if the carbon present is less than about 0.25 per cent, the ferrite will be in excess and may be rejected in large patches rather than as a thin network. Hence in both these cases the prior austenitic grain size will not be clearly revealed.

Carbon steels that are only slightly hypoeutectoid in composition may be cooled to about 730°C (1345°F) and held at this temperature before final cooling for approximately 10 or 15 min for small specimens (0.5 in. diameter) and proportionately longer for larger specimens. This treatment will generally produce a well-defined network of ferrite.

An alternate method for revealing the austenitic grain size in these steels, as well as in other steels to be described, is by gradient quenching. By this method of quenching it is possible to reveal the grain size in several ways in the same specimen. The specimen of interest should be at least 1.5 in.

long and about 0.25 in. thick. It is heated to the desired temperature, removed quickly from the furnace, and approximately 0.5 in. of its length is then quenched in brine. The remainder of the specimen is allowed to cool in still air above the surface of the quenching medium. In such a quenched specimen there will be a great variety of structures, ranging from martensite at the end of the specimen quenched in brine to a uniform carbide-ferrite aggregate at the end cooled in air. Along a longitudinal section of the gradient zone will be found an area wherein the cooling rate was appropriate to produce a structure consisting of martensitic grains surrounded by nodules of fine pearlite. Metallographic etching of this zone

Fig. 131.   The prior austenitic grain size in a 0.50 per cent carbon steel as revealed by a network of ferrite.   100×.   (*Courtesy of U.S. Steel Corp. Research Laboratory.*)

will reveal lightly etching martensitic grains (corresponding in size to the parent austenitic grains) surrounded, as shown in Fig. 132, by darkly etching nodules of fine pearlite. When the grain size is extremely small, the estimation of the grain size should be made in the fully hardened zone of the specimen in accordance with the procedure to be described for fully hardened steels.

In other parts of the gradient zone, particularly in the most slowly cooled regions of the specimen, the austenitic grain size may be revealed after proper etching by a network of well-defined ferrite, as shown in Fig. 131. This is especially true in the case of moderately hypoeutectoid steels. When the carbon content is close to the eutectoid composition and the specimen is relatively small in size, a continuous network of ferrite will not, in general, be formed even in the most slowly cooled portions of the specimen. Rather the austenitic grain size will be made apparent in other parts of the gradient-quenched specimen by a network of fine pearlite, as already described.   If for any reason a ferrite network is desired, it may be obtained in other specimens by cooling according to the method described for slightly hypoeutectoid steels.

EUTECTOID CARBON STEELS.   These steels are best treated by quenching a specimen (0.5 to 1 in. in diameter) in brine from the desired temperature. Provided the steel is not one that is exceptionally deep-hardening, this

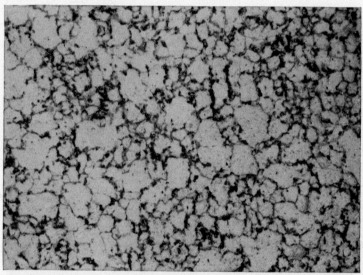

FIG. 132.   Region of a gradiant-quenched specimen of 0.50 per cent carbon steel wherein the prior austenitic grain size is revealed by a network of nodular pearlite surrounding martensite.   100 ×.   (*Courtesy of U.S. Steel Corp. Research Laboratory.*)

treatment will produce in the specimen an outside layer or case of martensite and a core of fine pearlite.   Between these two zones will be an intermediate region consisting of light-etching martensitic grains (corresponding in size to the parent austenitic grains), made apparent by contrast with the dark-etching fine pearlite formed at the grain boundaries.

Eutectoid steels that are of the very deep-hardening type must be quenched by the gradient quenching method described for hypoeutectoid steels, or be fully quenched and subsequently etched with the special grain-size reagent as described on page 276.   In the former case, an intermediate region will be found between the fully martensitic and fully pearlitic zones, where the prior austenitic grains will be outlined by a network of fine pearlite.

HYPEREUTECTOID CARBON STEELS. In general, the prior austenitic grain size of these steels is best determined by air or furnace cooling an appropriate specimen from the temperature of interest. Since transformation of the austenite during cooling begins predominately at the grain boundaries by rejection of excess iron carbide, the former austenitic grain boundaries will be made apparent at room temperature, after suitable etching, by a thin network of cementite in contrast to darkly etching pearlite. This is shown in Fig. 133. If the carbon content of the steel is below 1.10 per cent carbon, a complete network of cementite may not always be formed, and in these cases better results may be obtained by cooling the

FIG. 133.   The prior austenitic grain size in a hypereutectoid steel as revealed by a network of iron carbide.   100 ×.   (*Courtesy of U.S. Steel Corp. Research Laboratory.*)

specimen in accordance with the gradient-quench procedure described for hypoeutectoid and eutectoid steels.

An alternative cooling method for hypereutectoid steels consists in cooling the specimen to just above the lower critical temperature (Ar_1), holding the specimen at this temperature for several minutes, and subsequently quenching in water. The quenched specimen is then reheated to approximately 425°C (800°F) and air cooled. By appropriate etching, the former austenitic grains will be delineated by an iron carbide network surrounding tempered martensite.

FULLY HARDENED STEELS. In the case of fully hardened or quenched and tempered steels, the prior austenitic grain size may be determined by etching an appropriately prepared metallographic specimen with Vilella's

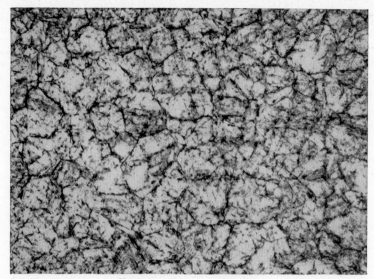

a. Untempered martensite.   100 ×.

b. Martensite tempered at 200°C (392°F) for 15 min.   100 ×.

FIG. 134.   The prior austenitic grain size in hardened steel as revealed by the contrast established between the martensitic grains through appropriate etching with grain-size reagent.   (*Courtesy of U.S. Steel Corp. Research Laboratory.*)

martensitic reagent.[1]   This reagent is quite selective in its action and excellent contrast is thereby established between the differently oriented martensitic grains whose orientations were not completely obscured during the austenite-to-martensite transformation.   As shown in Fig. 134, the best contrast is produced by the use of this reagent when the quenched specimen is tempered at 200 to 300°C (392 to 572°F) for approximately 15 min.

Heat Etching.   Heat etching is a simple and rapid method for revealing the austenitic grain size in any type of steel, and is particularly useful for grain-size determinations in steels containing less than 0.10 per cent carbon. The heat-etch method is based upon the fact that, when a polished specimen

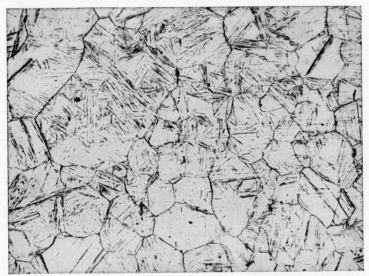

Fig. 135.   The prior austenitic grain size in a carbon − 2 per cent molybdenum steel as revealed by heat-etching techniques.   100 ×.   (*Courtesy of U.S. Steel Corp. Research Laboratory.*)

is heated to a temperature of interest in an inert atmosphere, such as purified hydrogen of low oxygen pressure, the metal at the grain boundaries of the austenite is preferentially vaporized, thus revealing the austenitic grains on the surface of the specimen.   Oxidation and tarnishing of the polished and etched surface are prevented by quenching the specimen in a mercury bath out of contact with air.   A typical heat-etched specimen is shown in Fig. 135.

Provided the partial pressure of the oxygen in the hydrogen atmosphere is sufficiently low, the surface of the specimen will remain substantially unaltered with respect to chemical composition, and the austenitic grain

[1] For composition, see Table 2, reagent No. 3, p. 412.

size as revealed on the surface of the specimen will then be characteristic of the entire specimen as heat-treated.   However, if the partial pressure of the oxygen is relatively high, elements such as aluminum may be oxidized on the surface of the specimen, with the result that the austenitic grain size so revealed may be smaller than that which is characteristic of the steel as a whole at that particular temperature.   This may be attributed to the effective retardation of grain growth by the oxides formed on and near the surface.

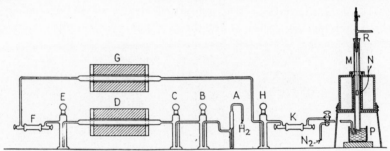

Fig. 136.   Diagrammatic arrangement of hydrogen-purification train (left) and heat-treating furnace (right) for heat-etching.   (*Courtesy of U. S. Steel Corp. Research Laboratory.*)

A. Sulphuric acid drying bottle through which commercially pure hydrogen is passed.
B. Soda-lime drying bottle.
C. "Drierite" drying bottle.
D. Catalyst of platinized asbestos at 540°C. (1000°F.) to convert much of the oxygen present into water vapor.
E. Activated alumina to absorb some of the water vapor formed in *D.*
F Phosphorus pentoxide to further absorb water vapor.
G. Second catalyst to further purify the hydrogen.   Same as *D.*
H. Activated alumina drying bottle.
K. Phosphorus pentoxide.
M Sintered pure-alumina tube within a platinum-wound furnace.   A silica tube may be used if induction heating is employed.
N. Furnace thermocouple.
N₂. To prevent an explosion at high temperature, the furnace tube should be flushed and filled with nitrogen before admitting the hydrogen.
P. Mercury quenching bath.
R. Rubber stopper or cap to support the specimen.

Adequate purification of the hydrogen to ensure successful heat etching may be obtained in an apparatus designed by Miller and Day,[1] shown in Fig. 136.

Fracture Method.   The austenitic grain size of steel established at some temperature of interest may be determined rapidly and effectively, particularly in medium- and high-carbon-content steels, by a method involving a comparison of fractures.   This procedure is possible because there is a direct relationship between the appearance of the fracture of fully hardened steel and the size of the austenitic grains prior to quenching and to fracturing of the sample.

The fracture method consists essentially in heating a previously notched

[1] Miller, O. O., and M. J. Day: Heat Etching as a General Method for Revealing the Austenitic Grain Size of Steels, *Trans. ASM*, Vol. 30, 1942.

specimen of appropriate dimensions to the heat-treating temperature of interest, holding the sample at temperature for a suitable length of time, and finally quenching it drastically in water or a brine solution.   The specimen is then broken at the notch, and the appearance of the fracture in the fully martensitic zone is compared and matched with one in a series of ten standard grain-size fractures.   When the steel under examination is one of low hardenability, the fracture of the pearlitic core that may be present should be disregarded as it in itself is not always indicative of the prior austenitic grain size.

The standard fractures most widely used in this country, and which originally were developed for tool-steel classification but which now have a

Fracture 1   2   3   4   5
number

Fracture 6   7   8   9   10
number

FIG. 137.   Appearance of the Shepherd grain-size fractures.   1 ×.

much broader application, are those known as the Shepherd standards. These standards, as shown in Fig. 137, consist of ten hardened-steel fractures, ranging from fracture No. 1, which is relatively coarse in appearance, to fracture No. 10, which is comparatively fine.   Intermediate are fractures that appear to the eye to increase in fineness in even increments from Nos. 1 to 10; these standards, when considered as a group, cover fully the range of coarseness and fineness usually encountered in the fractures of any steel.

Another set of fracture standards, known as the Jernkontoret standards, was developed in Sweden and is used extensively in European countries. These standards, consisting of ten fractures, are basically similar to the Shepherd fractures and the appearance of corresponding fractures is practically identical.   It is interesting to note, as shown in Fig. 138, that the fracture grain-size designation of both these standard sets is in remarkably close agreement with the ASTM austenitic grain-size rating as determined by metallographic examination.

In most cases little experience is necessary to rate grain size by the fracture method.   It has been shown that the eye is much more sensitive to slight changes in fracture appearance than to correspondingly small changes in grain size when determined by metallographic methods.   However, in cases where a mixed grain size is encountered, difficulty in rating may be experienced because the eye does not recognize immediately the coarse- and fine-appearing facets present, unless the fracture is relatively coarse over all.   Nevertheless, an experienced observer can identify the mixed grain sizes with a surprising degree of accuracy.

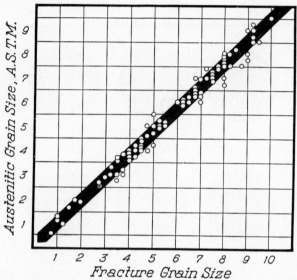

FIG. 138.   Relationship between the Jernkontoret and Shepherd grain-size designations and the ASTM austenitic grain-size numbers.   Within the limits of the band, the Jernkontoret and Shepherd standards are equivalent.   (*Courtesy of U.S. Steel Corp. Research Laboratory.*)

Carburizing Method.   The carburizing method (McQuaid-Ehn test) for ascertaining the prior austenitic grain size of steel is one not generally applicable for the purpose as it is a procedure that involves a change in chemical composition of the steel.   It is, however, a very appropriate method, as discussed on page 300, for determining the grain size in carburizing grades of steel that are normally heat-treated in a manner similar to the procedure of the McQuaid-Ehn test.

Oxidation Method.   Tobin and Kenyon[1] have suggested a procedure to ascertain the austenitic grain size of eutectoid steels by means of selective oxidation at elevated temperatures.   By providing a mildly oxidizing atmosphere at the temperature of interest, the austenitic grain boundaries

[1] Tobin, H., and R. L. Kenyon: Austenitic Grain Size of Eutectoid Steel, *Trans. ASM*, Vol. 26, 1938.

will be preferentially oxidized, and when the specimen is subsequently etched at room temperature to improve contrast, the prior austenitic grain boundaries will then be clearly revealed. Since this method is one that involves a change in the chemical composition of the steel at and near the surface, the austenitic grain size so determined must be interpreted with caution.

The oxidation procedure consists of heating to the temperature of interest a specimen of suitable dimensions and one that has previously been prepared metallographically (unetched), holding the specimen at temperature for an appropriate length of time, and then cooling it to room temperature either slowly or rapidly by quenching. Heating of the specimen is conducted in a mildly oxidizing atmosphere, and in most cases, when the temperature is only moderately high and the time at that temperature is relatively short, ordinary furnace atmospheres are satisfactory. If, however, the maximum temperature of heating is high, *e.g.*, in excess of 926°C (1700° F) and the time at temperature is relatively long, the specimen may be oxidized excessively and to a degree that would be unsatisfactory for clear delineation of the grain boundaries. In these cases, surface oxidation of the specimen may be limited by packing the sample in a suitable material that will limit the access of the furnace atmosphere; an alternative method is to heat the specimen in an appropriate salt bath or closed container and then, during the last few minutes of the heating period, to allow the specimen surface to become oxidized by coming into contact with the furnace atmosphere.

The outlines of the austenitic grains as revealed by this treatment alone will be faint, and because of this it is recommended that greater contrast be developed between the grain boundaries and the transformed grains proper by appropriate etching. If the specimen has been slowly cooled from the significant temperature, etching the repolished oxidized surface with either picral or nital will reveal the prior austenitic grain boundaries by white ferritic envelopes surrounding the darker etched areas of pearlite. Specimens that have been quenched may be satisfactorily etched in a solution consisting of 15 per cent hydrochloric acid in ethyl alcohol. This etching reagent will delineate the grain boundaries as a dark network in contrast to the lighter etched martensitic grains.

It is possible by this method to reveal the prior austenitic grain size established in a eutectoid steel sample by previous heat-treatment even though the specimen was not prepared by grinding and polishing prior to this heat-treatment. The technique employed is essentially one of metallographic preparation of the specimen, wherein considerable care is taken during the procedure so that the surface layer of metal on which the grain boundaries have been attacked is not completely removed.

Methods of Grain-size Designation. In the methods to be discussed for rating the austenitic grain size of an appropriately treated steel, estimating is facilitated by projecting the image of the structure on the focusing screen of a metallograph, or in some cases by obtaining a photomicrograph of the structure at some designated magnification. The prepared metallographic specimen in most cases must be suitably etched to produce good visual contrast between the grains and the grain-boundary constituent. Appropriate etching reagents for this purpose will be found in Table 2, pages 412 to 417.

The austenitic grain size may be reported in a number of ways, the following designations being acceptable and the first two being most common:

1. ASTM (American Society for Testing Materials) grain-size number—arbitrary exponential numbers with reference to the mean number of grains per square inch at a magnification of 100×.

2. The average number of grains intercepted by a line of some designated length. (Graff-Snyder intercept method.)

3. The average number of grains per square millimeter.

4. The average area of grains expressed in square millimeters.

The ASTM Comparative Method. The ASTM system of designating the austenitic grain size in steel is the preferred method in most cases, owing to the technical simplicity of the procedure, and the widespread acceptance of the basic indexing principles involved. The method is essentially one of comparison, in which the image of the structure of an appropriately treated specimen projected at a magnification of 100×, or a photomicrograph of the structure at the same magnification, is compared with a series of graded standard grain-size charts,[1] shown in the lower illustrations of Figs. 139 to 146, pages 283 to 290. By trial and error a match is secured and the grain size of the steel is then appropriately designated by a number corresponding to the index number of the matching chart. Steels showing a mixed grain size are rated in a similar manner, and it is customary in such cases to report the grain size in terms of two numbers, designating the approximate percentage of each size present.

The ASTM austenitic-grain-size standards have been designed to cover the range of grain sizes normally encountered in most grades of steel. The standard charts are indexed from No. 1 to No. 8, each index number representing some mean number of grains per square inch at a magnification of 100× according to the following relationship:

$$\text{Mean number of grains per square inch at } 100\times = 2^{n-1} \qquad (1)$$

where n = ASTM grain-size index number.

[1] Reproduced with permission of the ASTM.

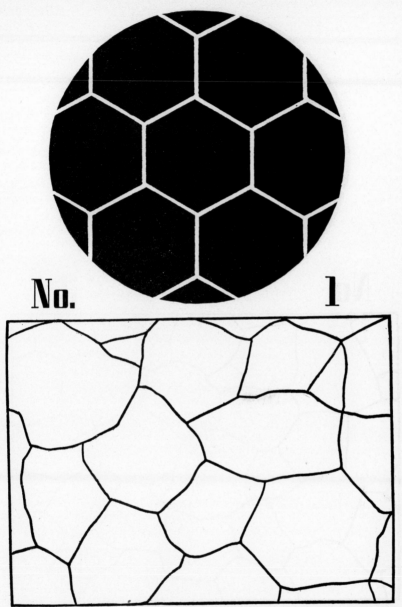

No. 1

Fig. 139. Upper, idealized hexagonal network for mean grain size No. 1, ASTM scale, 1 gr per sq in. Lower, ASTM standard grain size No. 1. Up to $1\frac{1}{2}$ gr per sq in. at 100 ✕.

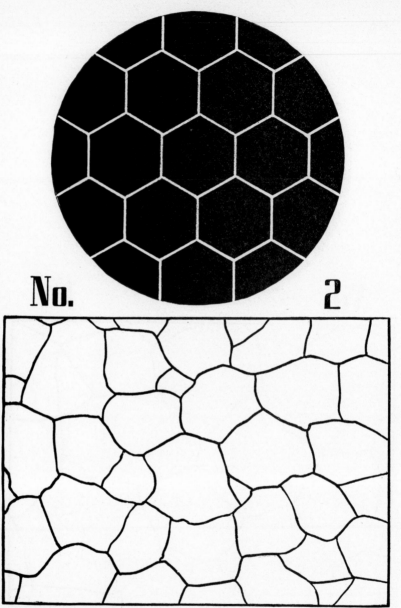

Fig. 140.   Upper, idealized hexagonal network for mean grain size No. 2, ASTM scale, 2 gr per sq in.   Lower, ASTM standard grain size No. 2, $1\frac{1}{2}$ to 3 gr per sq in. at 100 ×.

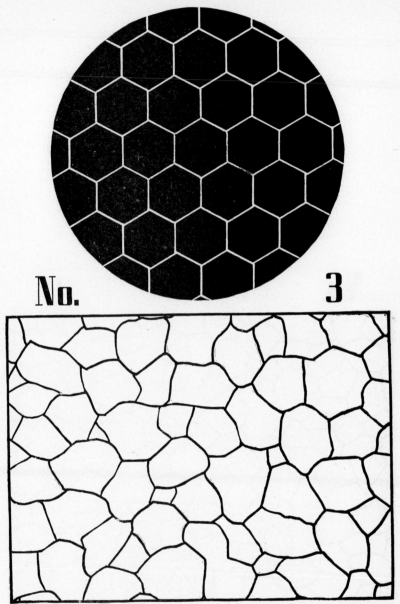

FIG. 141. Upper, idealized hexagonal network for mean grain size No. 3, ASTM scale, 4 gr per sq in. Lower, ASTM standard grain size No. 3, 3 to 6 gr per sq in. at 100 ×.

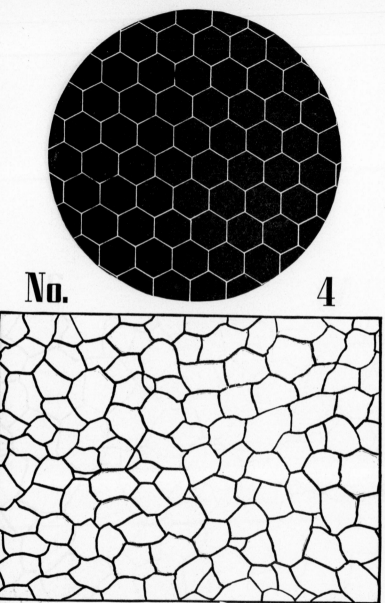

Fig. 142.   Upper, idealized hexagonal network for mean grain size No. 4, ASTM scale, 8 gr per sq in.   Lower, ASTM standard grain size No. 4, 6 to 12 gr per sq in. at 100 ✕.

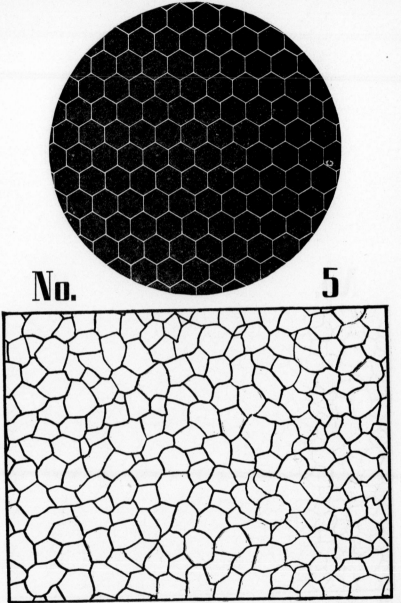

FIG. 143. Upper, idealized hexagonal network for mean grain size No. 5, ASTM scale, 16 gr per sq in. Lower, ASTM standard grain size No. 5, 12 to 24 gr per sq in. at 100 ×.

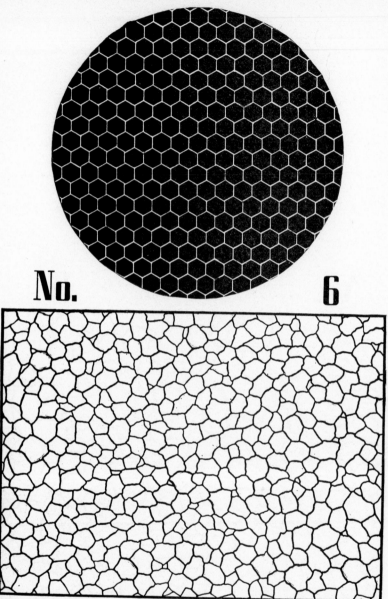

No.   6

FIG. 144.   Upper, idealized hexagonal network for mean grain size No. 6, ASTM scale, 32 gr per sq in.   Lower, ASTM standard grain size No. 6, 24 to 48 gr per sq in. at 100 ×.

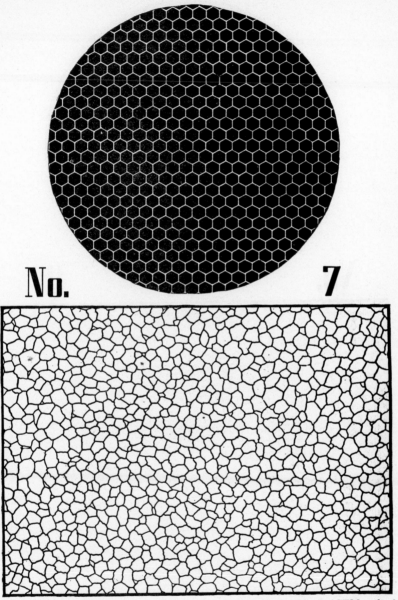

Fig. 145. Upper, idealized hexagonal network for mean grain size No. 7, ASTM scale, 64 gr per sq in. Lower, ASTM standard grain size No. 7, 48 to 96 gr per sq in. at 100 ×.

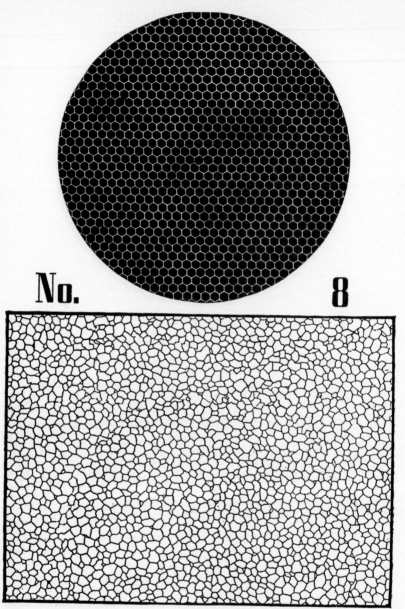

FIG. 146.   Upper, idealized hexagonal network for mean grain size No. 8, ASTM scale, 128 gr per sq in.   Lower, ASTM standard grain size No. 8, 96 to 192 gr per sq in. at 100 ×.

Pertinent data concerning the ASTM grain size standards and the inter-relationship between index number, mean number of grains per square inch at $100\times$, and actual existing grain sizes as determined by calculations, are shown in Table 40.

For all practical purposes, grain-size Nos. 1 to 5 may be considered to comprise the coarse-grained steels, whereas Nos. 5 to 8 may be considered as fine-grained steels. Size No. 5 may be considered as either coarse or fine, depending upon whether or not those relatively few grains which are outside the No. 5 range are coarser or finer than No. 5.

It is not uncommon to encounter austenitic grain sizes at $100\times$ that are either larger than ASTM No. 1 or smaller than No. 8. Larger grain sizes can be appropriately rated by using a projected magnification of $50\times$, instead of $100\times$, and securing a match with one of the standard charts in the usual manner. At this lower magnification, however, grain sizes Nos. 00 and 0 are reported when matched to standard charts Nos. 1 and 2, respectively. In the case of grain sizes smaller than No. 8, a magnification of $200\times$ may be used, and grain sizes Nos. 9 and 10 reported when rated against standards Nos. 7 and 8, respectively.

TABLE 40. DATA CONCERNING THE ASTM AUSTENITIC GRAIN-SIZE STANDARDS*

ASTM grain-size number	Number of grains per sq in. at $100\times$		Calculated diameter of equivalent spherical grain, not magnified		Calculated mean average of cross section of grain, sq in., not magnified
	Mean	Range	In.	Mm	
1	1	0.75– 1.5	0.01130	0.287	0.0001
2	2	1.5– 3	0.0080	0.203	0.00005
3	4	3 – 6	0.00567	0.144	0.000025
4	8	6 – 12	0.00400	0.101	0.0000125
5	16	12 – 24	0.00283	0.0718	0.00000625
6	32	24 – 48	0.00200	0.0507	0.00000313
7	64	48 – 96	0.00142	0.0359	0.00000156
8	128	96 –192	0.00100	0.0254	0.00000078

* From ASTM Standards, Part 1-A, 1946.

Grain-size Measuring Eyepiece. Rapid and accurate grain-size determinations may be made on the ASTM scale with a special microscope grain-size eyepiece,[1] thereby eliminating the necessity of first projecting the image and then comparing the grain size of the specimen with standard grain-size charts. Such an eyepiece is essentially of the Huygenian type ($7.5\times$ magnification) into which is constructed a revolving circular plate containing eight glass micrometer disks. On each disk is engraved an accurately ruled

[1] Bausch and Lomb Optical Co., Rochester, N.Y.

square, the size of the square on each disk corresponding respectively to the ASTM standards of 1, 2, 4, 8, 16, 32, 64, and 128 mean grains per square inch on the magnified image of the specimen at $100\times$. The disks are numbered from 1 to 8, corresponding to the grains per square inch as given above.

The micrometer square, as well as the corresponding engraved grain-size number, are brought into the central part of the field by revolving the circular plate. By construction of the eyepiece, the image of the ruled squares and that of the specimen are in sharp focus in the same plane. The different disks are rotated into the field of view and when a match is made between the grain size of the specimen and one of the ruled squares, the ASTM grain-size number is then read directly in the field.

The micrometer value of each disk is corrected for a magnification of $100\times$. This magnification may be obtained by using the eyepiece in conjunction with a 16-mm short-mounted objective at a mechanical tube length of 215 mm. For microscopes having a tube length of only 165 mm, special adapters must be used to extend the tube of the microscope to the proper distance.

The eyepiece may be adapted readily to photomicrography and, because the micrometer disks are in simultaneous focus with the specimen on the photographic plate, the grain-size classification and the matching square are made an integral part of the photographic print.

Graff-Snyder Intercept Method. The Graff-Snyder method of measuring and designating austenitic grain size is particularly well suited for high-alloy steels in general, and specifically for high-alloy tool steels including high speed grades. The method is applicable to metallographic samples of the steel of interest which have been appropriately processed to reveal at ordinary temperature by methods already described an outline of the prior austenitic grains; and it is advantageous, as compared to other measuring procedures, in that it is sensitive to small differences in grain sizes particularly when the over-all grain size is rather fine. This attendant sensitivity of the method, owing to the measurements being secured on an open and extended scale, is a necessary requisite in rating the grain size of the class of steels mentioned, since they usually possess a fine austenitic grain size as normally heat-treated and because slight differences in grain size are quite significant.

The principle of the Graff-Snyder method involves counting the number of grains intercepted by a theoretical line on the specimen surface, having a length of 0.005 in. The arithmetic average of ten such determinations randomly made over the surface of the specimen is the grain-size index. In the actual performance of the test, a line is drawn on the focusing screen of the metallograph of such length that it is equivalent, in terms of the projected magnification of the specimen image, to a line 0.005 in. inscribed on the face of the specimen. For example, at a projected magnification of 500×, the length of the line drawn on the focusing screen will be 2.5 in.; at a magnification of 750×, it will be 3.75 in.; and at 1000×, 5 in. Usually, 1000× is used principally because of the fineness of the austenitic grain sizes associated with the steels mentioned heretofore.

In a broad sense, intercept numbers of 8 and lower are considered to be representative of a coarse grain size; numbers 9 to 11 fairly coarse; 12 to 15 a moderately fine grain size; and above number 15, a very fine grain size.

METHODS FOR DETERMINING ACTUAL GRAIN SIZE IN NONFERROUS METALS AND ALLOYS

The ASTM Comparative Method. Similar to the ASTM method for indexing the austenitic grain size of steel, the grain size of nonferrous metals and alloys may be rated by comparing the microstructure of the unknown specimen, at a magnification of 75×, with standard grain size charts. The nonferrous grain size standards,[1] shown in Figs. 147 to 156, pages 294 to 298, consist of a series of ten photomicrographs of cartridge brass at a magnification of 75×. These standards are rated in terms of average grain diameters, not magnified, and range from 0.200 mm, as illustrated in Fig. 147, to 0.010 mm, as shown in Fig. 156.

Jefferies' Planimetric Method.[2] The planimetric method of determining the grain size in nonferrous metals and alloys, as modified by Jefferies, is one that is particularly accurate and precise, and relatively simple to carry out. Like the ASTM comparative method, the planimetric method is applicable only to materials possessing equiaxed grains, such as found in most cast and fully annealed metals and alloys.

A circle 79.8 mm in diameter (equivalent to a circular area of 5000 sq mm) is drawn with pencil on the rough side of the focusing screen of the metallograph and is so drawn that the center of the circle and the center of the rectangular focusing screen nearly coincide. The projected image of the etched specimen at a known magnification[3] is accurately focused upon the screen and the field-of-view diaphragm is adjusted so that the circumference of the circle is well within the image of the structure.

The grains that are intersected by the circumference of the circle are accurately counted. This procedure may be facilitated by carefully drawing short lines, perpendicular to the circumference of the circle, to correspond to the grain-boundary intersections. A soft crayon pencil is useful for this

[1] Reproduced with permission of the ASTM.

[2] Jefferies, Z.: Grain Size Measurements, *Chem. & Met. Eng.*, Vol. 18, 1918.

[3] Preferably at a magnification corresponding to one given in Table 41, page 299.

Fig. 147.   ASTM nonferrous grain-size standard.   Average grain diameter—0.200 mm.

Fig. 148.   ASTM nonferrous grain-size standard.   Average grain diameter—0.150 mm.

Fig. 149.   ASTM nonferrous grain-size standard.   Average grain diameter—0.120 mm.

Fig. 150.   ASTM nonferrous grain-size standard.   Average grain diameter—0.090 mm.

Fig. 151.   ASTM nonferrous grain-size standard.   Average grain diameter—0.065 mm.

Fig. 152.   ASTM nonferrous grain-size standard.   Average grain diameter—0.045 mm.

FIG. 153. ASTM nonferrous grain-size standard. Average grain diameter—0.035 mm.

FIG. 154. ASTM nonferrous grain-size standard. Average grain diameter—0.025 mm.

FIG. 155.   ASTM nonferrous grain-size standard.   Average grain diameter—0.015 mm.

FIG. 156.   ASTM nonferrous grain-size standard.   Average grain diameter—0.010 mm.

purpose. The grains that are completely included in the circular area are next checked and counted. From these experimental measurements or counts, the grain size may be expressed in grains per square millimeter by the following method of calculation.

One-half the number of grains intersected by the circumference of the circle, added to the number of completely included grains, gives the total number of equivalent whole grains included within the circle. Knowing the magnification of the projected image, the number of grains per square millimeter is then determined by multiplying the equivalent number of whole grains included in the circle by the corresponding magnification factor f given in Table 41.

TABLE 41. MULTIPLICATION FACTORS FOR DETERMINING THE NUMBER OF GRAINS PER SQUARE MILLIMETER

Magnification of the projected image	Magnification factor f
Full size	0.0002
10	0.02
25	0.125
50	0.5
75	1.125
100	2.0
150	4.5
200	8.0
250	12.5
300	18.0
500	50.0
750	112.5
1000	200.0
1500	450.0
2000	800.0

Thus, if the equivalent number of whole grains is found to be 40 at a magnification of $75\times$, then the number of grains per square millimeter is equal to 40×1.125, or 45.

Frequently it may be desirable to use a rectangle in the determination, in place of the conventional circle. In order to use the magnification factors as given in Table 41, to determine the number of grains per square millimeter, the area of the rectangle should be approximately the same as that of the circle. Convenient rectangle sizes conforming to the above requirements are as follows:

$$70.7 \times 70.7 \text{ mm}$$
$$65.0 \times 77.0 \text{ mm}$$
$$60.0 \times 83.3 \text{ mm}$$
$$55.0 \times 91.0 \text{ mm}$$
$$50.0 \times 100.0 \text{ mm}$$

The method of making a grain-size determination with a rectangle, in place of a circle, is similar to the procedure described. One-half the number of grains intersected by the perimeter of the rectangle is added to the number of completely included grains, the sum representing the number of equivalent whole grains within the rectangle. The number of grains per square millimeter is then calculated by use of the proper multiplication factor, corresponding to the magnification at which the measurements were made.

Heyn's Intercept Method. When grain-size determinations are to be made in metals that are in a strained condition or in cases where the grains are not equiaxed, the Heyn's intercept method is found particularly useful. The method consists of determining by count, at some appropriate magnification, the number of grains intercepted by two lines of known lengths. The lines referred to are inscribed on the focusing screen of the metallograph, perpendicular to one another, and so oriented with respect to the projected image that one of the lines is parallel to the direction of deformation or elongation of the grains. When the grain count has been made, the grain size may then be expressed as the average number of grains per linear unit in the two directions, or as the average number of grains per unit area, together with the ratio of length to breadth of grain.

THE McQUAID-EHN TEST

The McQuaid-Ehn test is principally a carburizing procedure wherein the specimen of interest is case carburized at a temperature of 925°C (1700°F) for a time sufficiently long to produce a case that is definitely hypereutectoid in nature. By appropriate cooling schedules, the austenite is made to transform to pearlite with attending rejection of excess iron carbide to the austenitic grain boundaries. The original austenitic grain size established at the test temperature will then be made apparent by a complete or nearly complete network of iron carbide. By appropriate metallographic examination of the structure so produced, it is possible, in certain respects, to determine within limits the furnace or melting practice followed in production of the steel. As associated with the prior austenitic grain size established at the carburizing temperature, it is possible to predict the expected response of the final carburized product to subsequent quenching, and it enables, to a limited extent, the steel to be classified according to the use for which it is best suited.

When the importance of austenitic grain size in steel, with its attending ramifications, was first realized, the McQuaid-Ehn test procedure served as the method for revealing the grain size. It is now known, however, that the austenitic grain size established according to the McQuaid-Ehn test schedule cannot with any degree of certainty be extended to, and correlated with, the grain size established in the same steel by a different mode of heat-treatment either at the same or at a lower temperature. This circumstance arises from the fact that in the McQuaid-Ehn test both carbon and oxygen are introduced into the steel and thus radically change the chemical composition of the steel, at least near the surface. The austenitic grain size as then revealed in the altered steel may be entirely different from that which is characteristic of the original metal at the same temperature as revealed by noncarburizing methods. Thus it is often observed that some steels which show evidence of being coarse-grained when heat-treated in the absence of a carburizing atmosphere and at a temperature considerably below 925°C (1700°F) may actually appear to be relatively fine-grained according to the McQuaid-Ehn test procedure. This anomaly may be attributed in part to the introduction of oxygen into the steel during carburizing, since under appropriate circumstances the formation of oxides in this manner will inhibit the growth of austenite grains.

Since the McQuaid-Ehn test is conducted at a temperature of 925°C (1700°F) for 8 or more hours, and knowing the behavior of austenitic grain growth with respect to temperature and time, it is obvious that the grain size revealed may be very much different than that ascertained under more normal heat-treating conditions, *i.e.*, at a lower temperature and for a shorter length of time at the temperature. To illustrate, a steel having a coarsening temperature below 925°C (1700°F) but above that usually reached during a normal heat-treating procedure may, by the McQuaid-Ehn test, appear to be a comparatively coarse-grained steel; on the other hand, if this steel is heat-treated in the absence of a carburizing atmosphere at a temperature below the coarsening range, it may show evidence of being fine-grained and possess all the attributes of a fine-grained steel insofar as the normal heat-treating procedure is concerned.

The McQuaid-Ehn test is further limited in general application because the procedure, being one of carburizing, produces a case having a higher carbon content than is found in the core. Because of this, two different austenitic grain sizes are frequently found in the same specimen (case and core, respectively), causing, as would be expected, considerable confusion in determining the true austenitic grain-size characteristic of the specimen.

It is evident from the illustrations cited that the McQuaid-Ehn test does not disclose the really significant grain size of the steel, *i.e.*, the austenitic grain size established at the precise heat-treating temperature and in exact accordance with the appropriate heat-treating schedule for that steel. However, the test is admirably suited for revealing the austenitic grain size in carburizing grades of steel, as the test procedure bears close similarity to commercial carburizing practice, and hence the grain size ascertained may be rated and may justifiably be used to estimate the characteristics of the carburized product.   Unlike the noncarburizing methods of revealing grain size, the McQuaid-Ehn test yields additional information concerning depth of case, presence of soft spots, and the degree of normality or abnormality of the steel.

Method of Carburizing.   Specimens that are to be subjected to the test are packed into a suitable solid carburizing compound and heated to a temperature of approximately 925°C (1700°F).   The carburizing material should contain a sufficient amount of energizer so that the case produced will have a carbon content well above the eutectoid range (greater than 0.83 per cent carbon).   The specimens are allowed to carburize at the designated temperature until a carburized case is produced that is approximately 0.05 in. in thickness.   This thickness of case will be sufficient to permit a microscopic examination over the hypereutectoid zone at the requisite magnification of 100×.   The exact length of time required to produce this case is best determined by trial and may take anywhere from 8 to 15 hours, depending upon the original composition of the steel.

When carburizing is completed, the steel must be properly cooled to develop a satisfactory pearlitic structure in the hypereutectoid zone.   The rate of cooling depends upon the composition of the steel and upon such factors as the size of the specimen, size of carburizing container, and the heating furnace.   Obviously, the proper rate of cooling is best determined by experience, unless certain of the influencing factors are standardized. Satisfactory structures may be obtained by relatively rapid cooling for low-carbon, low-manganese steels containing no alloys.   Very slow cooling, such as furnace cooling, is required for high-alloy air-hardening steels of the high-nickel and chromium-nickel-molybdenum types.   Plain carbon steels may be satisfactorily cooled by air cooling in the carburizing box.   Where the carburizing is done in a laboratory furnace, it is quite common to cool all test specimens in the furnace and, in the case of high-alloy steels, it has been found necessary to cool them in the furnace at a very definite maximum rate.

Interpretation of Results.   The austenitic grain size of steel as established according to the McQuaid-Ehn test procedure may be assigned an ASTM grain-size number in the manner described before.   Such a procedure provides a suitable and convenient means for properly selecting steels according to grain size, so that the finished product after carburizing will possess certain desirable, and sometimes highly important, physical and structural properties.

By careful examination of the structure in the hypereutectoid zone, much useful information may be obtained regarding the normality or abnormality of the steel.   In strictly normal steels the pearlitic areas are well defined, with a continuous and fairly uniform network of iron carbide surrounding the pearlite patches. In strictly abnormal steels, however, the pearlite is generally coarse and irregular, with excess ferrite and globular iron carbide present in either or both the regions of the prior austenitic grain boundaries and the pearlite areas proper. The structural appearance of a normal and an abnormal steel is shown in Fig. 157.

In general, the coarse-grained steels are more normal than are the fine-grained steels; it is not uncommon, however, for coarse-grained steels to exhibit, at times, pronounced abnormality. The cause of abnormality in fine-grained steels is not known with certainty, but it is believed to be associated with thorough deoxidation during manufacture, as, for example, by relatively large additions of aluminum. In the case of coarse-grained steels exhibiting abnormality after carburizing, the cause may be related to dissolved oxygen in the steel prior to carburizing or, more likely, to the introduction of oxygen during the process.

The coarse-grained normal steels, as compared to the finer grained abnormal steels, tend to respond better to quenching, but with accompanying greater tendency for distortion. The inherent properties of fine-grained abnormal steels—such as lower hardenability, less tendency for distortion upon quenching, and greater toughness—are primarily attributed to the fineness of grain and not to the circumstance of abnormality.

It is quite evident that the grain size of steel and the degree of normality are independent factors, as either fine-grained normal or abnormal steels may be made, as well as coarse-grained normal and abnormal steels. These factors are, undoubtedly, related to the furnace practice used in making the steel.

THE JOMINY HARDENABILITY TEST

Of the many methods that have been proposed for measuring the hardenability of steel, the Jominy hardenability test is perhaps the one most widely used, primarily because the test procedure is relatively easy to carry out and because it is one that is well suited to furnish the fundamental relationship between cooling rate and corresponding hardness for most any given steel. The data obtained from this test are relatively easy to interpret, and by appropriate plotting the hardenability of different steels may be readily compared. Furthermore, the Jominy test data afford a means of predicting readily after appropriate heat-treatment the hardness that may be expected at any location within an object of given shape, regardless of the steel from which the object might be made.

The Jominy hardenability test is particularly applicable to steels having a critical cooling rate not exceeding about 38°C (100°F) per sec. Although the standard test procedure describes a shape of test specimen, known as

a. Normal structure.   100 ×.

b. Abnormal structure.   100 ×.

FIG. 157.   Typical appearance of normal and abnormal steels as revealed by pack carburizing at 945°C (1725°F) for 16 hr.   (*Courtesy of U.S. Steel Corp. Research Laboratory.*)

the type "L" bar, to meet the requirements of low-hardenability steels, it is one not too satisfactory, owing to the difficulties in machining the bar to shape and the tendency for the specimen to crack when quenched. In view of this, Shepherd[1] has proposed a hardenability test, known as the P-V test, that is particularly applicable for measuring the hardenability of those shallow-hardening steels for which the Jominy test is not particularly suited.

The procedure of the Jominy test consists of heating a specimen of standard dimensions to the temperature of interest; holding at that temperature for about 20 min; and then end-quenching the specimen with water, at a temperature of 4.5 to 29°C (40 to 85°F), in a specially designed quenching fixture. End-quenching is secured by directing a stream of water against the bottom face of the test specimen at a velocity defined in terms of a column of water from a $\frac{1}{2}$-in.-diameter orifice that rises to a free height of 2.5 in. above the opening. The water stream is regulated from a $\frac{1}{2}$-in.-diameter orifice, the opening of which is located $\frac{1}{2}$-in. below the bottom face of the specimen. After quenching for not less than 10 min, the test specimen is prepared in a manner to be described and hardness measurements made along the prepared areas.

Standard Test Specimens. The preferred Jominy test bar, which is particularly suited for measuring the hardenability of relatively deep-hardening steels, *i.e.*, those whose critical cooling rates are lower than about 38°C (100°F) per sec., is illustrated in Fig. 158a. The specimen consists of a 1-in.-diameter solid bar of the steel of interest, approximately 4 in. in length, on one end of which is a machined shoulder (or a fitted detachable collar ring) to permit the specimen to be suspended vertically in the quenching fixture.

As mentioned heretofore, the type "L" specimen bar, illustrated in Fig. 158b, is suggested for use with shallow-hardening steels whose critical cooling velocities are near or above 38°C (100°F) per sec. The design of this test specimen, particularly with respect to the shape of the hole in the lower end, provides a more gradual change in the relatively fast cooling rates near the quenched end of the bar than is possible to obtain with the conventional standard test bar. Because of the shape of the type "L" specimen the quenching procedure is slightly modified. Instead of a rise in free height of the water stream to 2.5 in., as in the case of the standard test specimen, the height is increased to 4 in.

Hardness Measurements. After the test bar has been end-quenched from the significant temperature according to the specifications of the test

[1] Shepherd, B. F.: Hardenability of Shallow Hardening Steels Determined by the P-V Test, *Trans. ASM*, Vol. 38, 1947.

procedure, two flat surfaces, 180 deg apart, are ground to a depth of 0.015 in. along the entire length of the specimen. During grinding it is essential that the specimen be kept sufficiently cool to prevent tempering of the hardened end. To determine whether or not tempering has occurred, the specimen may be etched with a 5 per cent aqueous solution of nitric acid until the specimen surface is blackened, followed by a thorough wash in hot water, and finally etched by immersion for about 3 sec in a 50 per cent aque-

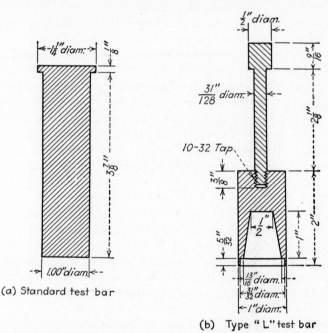

(a) Standard test bar

(b) Type "L" test bar

F ig. 158. The standard and type L Jominy hardenability-test specimens.

ous solution of hydrochloric acid. Any darkened areas at or near the quenched end, revealed by this treatment, indicate regions that have been tempered.

Along the longitudinal center line of each flat surface, Rockwell hardness measurements are made at intervals of $\frac{1}{16}$ in. from the quenched end.[1] At identical distances from the quenched end of the bar the hardness measurements on the two flat surfaces are averaged and plotted on coordinates of hardness vs. distance in inches from the quenched end. Such typical

[1] The Wilson Mechanical Instrument Co. manufactures a fixture, known as the *Equitron*, that is designed especially to facilitate in the hardness measurements. The fixture appropriately supports the Jominy test bar, and by means of a hand-actuated micrometer lead screw, the test specimen may be advanced beneath the Brale indenter at precisely $\frac{1}{16}$-in. intervals.

curves for a number of SAE steels are shown in Fig. 159. When considering a hardenability curve, such as any one of those illustrated, the interpretation must be associated only with the particular conditions under which the data were obtained; and any changes that may affect the hardenability of the steel, such as changes in composition, austenitic grain size, or quenching temperature, make it necessary to redetermine the hardenability characteristics of the steel under the new conditions.

Interpretation of Test Results. Appropriate plotting of the Jominy hardness data, *i.e.*, measured hardness vs. distance from the quenched end,

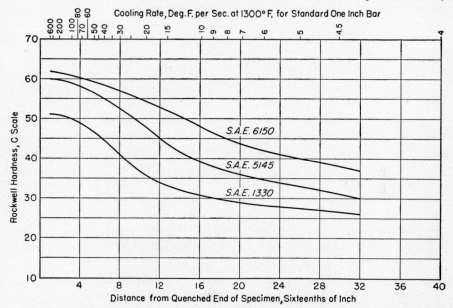

FIG. 159. Jominy hardenability curves for three different SAE steels, end quenched from their appropriate austenitizing temperatures. Data secured from standard Jominy test bars.

affords a means for readily distinguishing between the relative hardenability of steels when quenched from their appropriate austenitizing temperatures. This is illustrated in Fig. 159, and for the steels shown it will be noted that SAE 1330 steel is of a more shallow-hardening variety than are the other two, whereas the deepest hardening propensities are shown by SAE 6150 steel.

As mentioned heretofore, the Jominy test, together with appropriate data, is ideally suited for predicting the hardness at a particular location within a section of any given size regardless of the steel from which the section is made. This procedure comprises making the object of some relatively low-hardenability steel, heat-treating it in accordance with the production specifications for such a piece, sectioning it after heat-treatment, and finally determining the hardness at various locations throughout the cross section.   The hardness of the object at the particular location of interest is then referred to the appropriate Jominy hardenability curve, *i.e.*, one for the same steel from which the object was made and similarly austenitized, and the distance from the quenched end noted which corresponds to that particular hardness.   The hardness that may then be expected in this particular location of the object when it is made of some other steel, but similarly heat-treated, may be found by referring to Jominy curves of different steels and by noting the hardness that corresponds to the same distance from the quenched end as that first determined.   In like manner, any given steel may be selected which will give a desired hardness in the object when heat-treated in a manner identical to the heat-treatment of the test object.

Inasmuch as the cooling rates along the longitudinal surface of the Jominy test bar are nearly independent of composition, and dependent only on the standardized conditions of the test (size and shape of test bar, quenching medium, quenching procedure, etc.), it is, therefore, possible to correlate definite cooling rates at some designated temperature with the various $\frac{1}{16}$-in. positions on the test specimen.   This relationship, which is a fundamental one and remains substantially unaltered so long as the test specifications are strictly adhered to, has been determined experimentally for both the standard and type "L" Jominy test bars, and is illustrated in Fig. 160.   The average cooling rate at different $\frac{1}{16}$-in. locations is expressed as the instantaneous rate at 705°C (1300°F), since at this temperature level the cooling velocities are free from influence of heat evolution during transformation of austenite and are not significantly affected by variations in the quenching temperature from 815 to 900°C (1500 to 1650°F). This correlation, therefore, permits the Jominy hardenability curves to be referred to coordinates of hardness vs. cooling rates at 705°C (1300°F), as shown in Fig. 159, and enables a prediction to be made concerning the expected hardness within any given-shaped object once the cooling rate, under a given heat-treating schedule, is known at the location of interest.   Such cooling rates at 705°C (1300°F) within the object may be determined experimentally, or in the case of steel plates of different thicknesses and round bars of different diameters, either water or oil quenched, such data pertaining to different distances beneath the surface may be obtained by reference to the literature.[1,2]

By reference to a group of hardenability curves referred to coordinates of hardness vs. cooling rates, it is not only possible to predict the expected

[1] American Iron and Steel Institute: "Hardenability of Alloy Steels," New York, 1947.

[2] *Metal Progress:* "Cooling Rates of Plates and Rounds," Data Sheet, January, 1947.

hardness at a location of interest within a given-shaped object on the basis of cooling rate, but also on the same basis a steel may be selected to render some given hardness at the location of interest. Obviously, after such determinations are made and the object is changed in any way that may affect the cooling rate—*e.g.*, a change in size or shape, quenching medium,

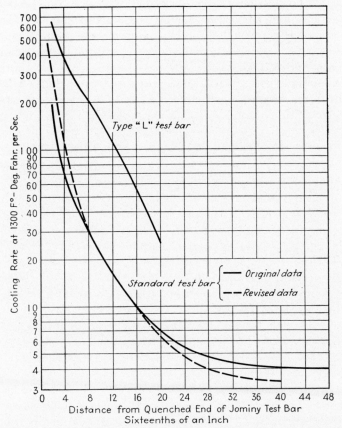

FIG. 160. Experimentally determined cooling rates at different distances from the quenched end of the standard and type L Jominy test bars. (*The revised cooling-rate curve for the standard test bar as determined by Boegehold and Weinman, Metal Progress, Vol.* 52, 1947.)

austenitizing temperature, temperature of quenching medium, etc.—it then becomes necessary to redetermine experimentally the cooling rate within the object at the location of interest.

From Jominy hardenability data, the hardenability of a steel may be designated by a code which will indicate the distance or distances from the quenched end of the specimen within which the designated hardness is obtained. For example, a steel containing 0.50 per cent carbon could be specified to have a hardenability of $J_{45} = 12$; *i.e.*, the minimum requirement for this steel would be a Rockwell hardness of C 45 at a distance of $1\frac{2}{16}$ in. from the quenched end.   If both minimum and maximum limits are specified, the index of hardenability might be specified as $J_{45} = 6$ to 15.

AISI "H" Steels.   A group of commercial steels are now available, known as "H" steels, that are manufactured to meet certain hardenability requirements as defined by the Jominy end-quench test.   These steels, designed by the letter H following the standard SAE and AISI notations, are manufactured with greater latitude in specified chemical composition than the conventional standard steels.   This greater latitude in composition was found necessary in order to secure heats of steel that would exhibit hardenability characteristics within limits defined by tentatively established maximum and minimum hardenability curves for each of the different "H" steels.[1]

THE SHEPHERD PENETRATION-FRACTURE TEST

The Shepherd penetration-fracture test, more commonly referred to as the P-F test, is primarily a method for determining the quality of tool steels and classifying the heats into definite types with regard to melting practice, hardening sensitivity, and physical properties of the hardened tool after appropriate heat-treatment.   The test affords a means whereby the "timbre" of the steel may be intelligently evaluated, *i.e.*, the evaluation of that property of tool steels which governs the grain size and hardness penetration developed by quenching.   The P-F test consists essentially of subjecting four samples of the tool steel in question to a brine quench from temperatures of 788, 815, 843, and 871°C (1450, 1500, 1550, and 1600°F), respectively.   Where time and expense of the test are to be considered, the test may be modified to include only two quenching temperatures, 788 and 843°C (1450 and 1550°F).   Each specimen is then notched and broken and, from a comparison of one of the fractures with a set of standard fractures, the fracture grain size may be determined.   The depth of hardening produced by quenching is then determined by macroetching the other half of the broken specimen.   From a correlation of these data, as will be discussed later, the P-F values may be designated.

Preparation of the Test Specimens.   The specimens to be used in the P-F test should be machined accurately to 0.75 in. $\pm$ .0001 in. diameter, and be approximately 3 in. in length.   For purposes of identification, each specimen should be stamped on both ends with a number corresponding to the temperature from which it is quenched.   When the samples have been accurately machined to size they are subjected to a preliminary heat-treatment.   This treatment is for the purpose of normalizing, or making the

---

[1] See current issue of "Hardenability of Alloy Steels," American Iron and Steel Institute, for latest tentative hardenability bands of various "H" steels.

structural conditions of the steel uniform, and is carried out by heating the specimens to 871°C (1600°F), and holding at this temperature for at least 40 min. The specimens are then quenched in oil.

Hardening Treatment. The oil-quenched specimens are uniformly heated to temperatures of 788, 815, 843, and 871°C (1450, 1500, 1550, and 1600°F), respectively, and held at each respective temperature for 30 min. It is important that the temperature-time cycle be strictly adhered to, as both the hardenability and the austenitic grain size tend to increase with time and temperature on all except very stable heats.

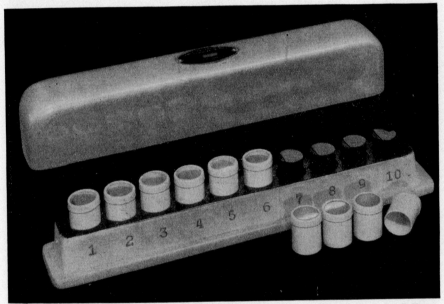

Fig. 161.   A complete set of Shepherd grain-size fractures for rating grain size in the P-F test. (*Courtesy of B.F. Shepherd.*)

The specimens are then quenched in a standard vertical jig flushed with a 10 per cent aqueous brine solution.   When the specimens are sufficiently cooled to be handled, they are notched midway between the ends with a thin cutting wheel, $\frac{1}{16}$ to $\frac{3}{32}$ in. in thickness, to a depth not greater than $\frac{1}{16}$ in.   The actual fracturing after notching is made by transverse impact.

Determination of Grain Size.   The grain size of the fracture is determined by comparing the hardened-case fracture with fracture grain-size standards.   The standards,[1] as shown in Figs. 137 and 161, consist of ten

[1] In the actual performance of the test, illustrations of the fracture standards are not suitable.   A complete series of fracture grain-size standards, as well as a set containing the most frequently used standards (Nos. 6 to 10), may be procured from The Thermist Company, Phillipsburg, N.J.

fractures numbered from 1 to 10 in order of decreasing grain size, in accordance with the standard ASTM metallographic method of grain-size designation. When judged by an experienced observer, classification of each fracture can be made to fractional values. Thus, a fracture having a grain size between standard Nos. 4 and 5 would be designated as 4.5, and between this and No. 5 would be 4.75.

Determination of Penetration. The second half of the fractured specimen is cut at least 1 in. from the original end and the cross section is prepared for macroetching. A final grinding on No. 0 or 00 emery paper is recommended. The specimen is etched in a 50 per cent aqueous solution of hydrochloric acid for 3 min at a temperature of 82°C (180°F). The depth to which the specimen has hardened is then measured to the nearest half of $\frac{1}{64}$ in. and the hardness penetration expressed in terms of the number of sixty-fourths of an inch of hardened case.

P-F Designation. The P-F characteristics are recorded as eight numbers, for example,

$$4, 5, 9\tfrac{1}{2}, 12 \;/\; 9, 7\tfrac{3}{4}, 6, 3$$

The first group of four numbers represents the depth of hardness in sixty-fourths of an inch for each of the four specimens, respectively. Thus, in the above designation, 4 is equivalent to a hardness penetration of $\frac{4}{64}$ in., 5 is equivalent to $\frac{5}{64}$ in., etc. The last group of four numbers give the fracture grain-size numbers for the four specimens, respectively, as determined by visual comparison with the standards. The first number in each group represents the P-F characteristics for the specimen quenched from a temperature of 788°C (1450°F), the second number of each group represents the P-F characteristics for the specimen quenched from a temperature of 815°C (1500°F), etc. Thus, for 788°C (1450°F), the P-F characteristic is 4–9; for 815°C (1500°F), the P-F value is 5–7$\frac{3}{4}$, etc.

THE SPARK TEST

The spark test, when conducted by an experienced observer, is a quick and effective method for the classification of steels according to their chemical composition. The test is to be considered neither as a substitute for a chemical analysis nor as an unequivocal method for identifying unknown steels. However, for definite-type steels, a skillful operator can classify carbon steels to within 0.20 per cent carbon content and can identify alloyed steels with regard to chromium, vanadium, tungsten, etc., within approximately 1 per cent.

The test consists essentially of holding the steel in contact with a high-speed abrasive wheel and visually noting the characteristics of the spark stream that is thrown off.

Principles. When a piece of steel is held in contact with a high-speed grinding wheel, the friction between the metal and the revolving wheel is so great that particles of both the metal and the wheel are torn loose. Because of the intense frictional heat evolved, the temperature of the particles is raised to incandescence. As these glowing particles are hurled through the air the trajectory is easily followed by the unaided eye, especially when noted against a dark background or under subdued lighting conditions. This trajectory is called a carrier line; the characteristics of it are different for steels of different composition. The incandescent particles from steel contain combined carbon that reacts with the oxygen in the air to form gaseous carbon dioxide. This reaction is supported by the evolution of heat attending the oxidation of iron or ferrite. The transition of the solid carbide phase to gaseous carbon dioxide is accompanied by an increase in volume, and as a consequence, an internal pressure is built up within each particle. This pressure is released only by an explosion of the particles, causing what is known as spurts, or bursts, *i.e.*, forking or deviations from the line of travel of the normal iron spark. This forking effect occurs in increasing amounts as the carbon content of the steel increases.

Abrasive Wheels. The spark test is best conducted by holding the revolving wheel of a small portable grinder against the steel to be tested. This type of grinder is particularly useful in testing pieces of steel that are too large to be handled conveniently or that are racked and shelved. If portable grinders are not available, the test may be carried out on any of the conventional, stationary types of grinders.

The particle size of the abrasive in the wheel is relatively unimportant. A 30-grain wheel will generally produce shorter carrier lines and fewer sparks than a 60-grain wheel under identical test conditions, but the important characteristics of the spark stream, in general, will not be affected.

The rate of travel of the sparks as they are hurled from the wheel depends upon the speed at which the wheel is revolving. It is recommended that for best results the wheel speed be maintained between 3600 and 5000 rpm.

Pressure between the Wheel and the Test Section. The position and pressure of the test section against the wheel, or vice versa, should be such that the spark stream is approximately 12 in. in length and at right angles to the line of vision. Wheel pressure is important, as too heavy a pressure will raise the temperature of the spark stream and bursts, and will increase the volume of the sparks, with the result that the test will indicate to the inexperienced observer a steel of higher carbon content than is actually present. Experience will assist the observer in determining the correct length and volume of the spark stream for the various types of steel.

Examination of the Spark Stream. The conditions under which the spark stream is examined determine to some extent the accuracy of the

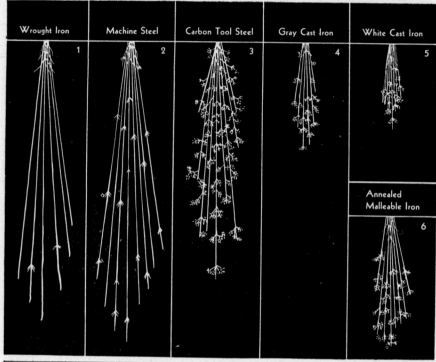

Line	Metal	Volume of Stream	Length of Stream (a)
1	Wrought iron	Large	65 in.
2	Machine steel	Large	70
3	Carbon toolsteel	Moderately large	55
4	Gray cast iron	Small	25
5	White cast iron	Very small	20
6	Annealed malleable iron	Moderate	30
7	High speed steel	Small	60
8	Manganese steel	Moderately large	45
9	Stainless steel	Moderate	50
10	Tungsten-chromium die steel	Small	35
11	Nitrided nitralloy	Large (curved)	55
12	Stellite	Very small	10
13	Cemented tungsten carbide	Extremely small	2
14	Nickel	Very small (d)	10

(a) Figures obtained with 12-in. wheel on bench stand, and are relative only. Actual length in each instance will vary with grinding wheel and pressure.

Fig. 162.   Illustrating schematically the typical spark characteristics of some metals and alloys

High Speed Steel	Manganese Steel	Stainless Steel	Tungsten-Chromium Die Steel	Nitrided Nitralloy
7	8	9	10	11
		Stellite	Cemented Tungsten Carbide	Nickel
		12	13	14

Color of Stream		Quantity of Spurts *(b)*	Nature of Spurts *(b)*	Line
Close to Wheel	**Near End**			
Straw	White	Very few	Forked	1
White	White	Few	Forked	2
White	White	Very many	Fine, repeating	3
Red	Straw	Many	Fine, repeating	4
Red	Straw	Few	Fine, repeating	5
Red	Straw	Many	Fine, repeating	6
Red	Straw	Extremely few	Forked	7
White	White	Many	Fine, repeating	8
Straw	White	Moderate	Forked	9
Red	Straw *(c)*	Many	Fine, repeating *(c)*	10
White	White	Moderate	Forked	11
Orange	Orange	None		12
Light orange	Light orange	None		13
Orange	Orange	None		14

(b) "Spurts" are the sparks (seemingly small explosions)
that occur at intervals on the carrier lines.
(c) Spurts are blue-white.   *(d)* Some wavy streaks may be observed.

associated with spark testing. (*From Grits and Grinds, June, 1940, copyrighted by Norton Co.*)

test. Subdued lighting conditions are to be preferred, so that the characteristics of the bursts are apparent. The spark stream should be protected from drafts of air, to prevent hooking of the tail sparks, which may lead to confusion in interpreting the results.

The components of the spark stream to be noted during the test are:

1. The carrier lines, which will vary in length, breadth, number, and color, depending upon the type of steel under test.

2. The spark bursts, which will vary in size, number, shape, intensity, and distance from the wheel or ends of the carrier lines.

In plain-carbon steels, the variation in the number and intensity of the bursts indicates changes in the carbon content. Low-carbon steels produce sparks that are long and tend to follow straight lines, growing more luminous and broader as they advance. As steels of increasing carbon content are tested, more and more forking of the carrier lines is evident, with a proportional increase in the intensity of the bursts. The characteristic spark streams from different irons and steels and some nonferrous metals and alloys are shown in Fig. 162.

In alloyed steels the masking effect on the spark stream, due to the alloying elements, makes classification of such steels somewhat difficult. For the same carbon-content steel, the effect of an alloying element may retard or accelerate the carbon bursts and alter the brightness of the carrier lines. Whenever classification of a particular steel is doubtful, it is recommended that their spark characteristics be compared with those from standard samples. Such standards are merely a group of steels of known chemical composition.

MAGNETIC TEST METHODS

Magnetic testing is ordinarily applied to the ferromagnetic metals, such as iron, nickel, cobalt, and their alloys. It affords a convenient means of detecting surface and subsurface defects in a metal, such as cracks, seams, blowholes, etc. (called continuity testing), and of determining comparatively the physical properties of a metal (called magnetic analysis). This method of testing is advantageous in that it is rapid and nondestructive and permits the test to be conducted on the finished product instead of a representative specimen. A number of factors, such as variation in analysis, structure, and stress, influence magnetic measurements, with the result that the choice of a method of magnetic testing depends upon which of the several magnetic characteristics shows most variation with change in the property of interest.

The use of direct current in magnetic testing is most appropriate when the fundamental magnetic properties of a metal are to be determined, when it is necessary in other cases to secure data at relatively high induction values, and when the specimen of interest is of rather large cross-sectional dimensions. In contrast to the use of alternating current in continuity testing to be described, direct current is preferred when detection of deep-seated defects within the specimen, as well as surface discontinuities, are of primary interest.

In the alternating-current methods, the sample serves essentially as the core of a transformer. Characteristics of the core, such as permeability and watt-loss values, provide shifts in the phase angles between the current and the voltage of the induced current, with the result that harmonics are set up in the alternating-current wave. By measurement of these harmonics, variations in the magnetic characteristics of the core can be distinguished. Alternating-current methods are useful in determining uniformity of analysis, structure, and physical properties, and afford great sensitivity to small differences.

Magnetic Analysis. The correlation of the magnetic properties of a metal with its physical properties is somewhat complicated. Two specimens that are identical in their magnetic properties are usually identical in their physical properties, but the converse is not true. Often specimens that show like physical properties by the usual test methods are entirely different with regard to magnetic properties. This is often due to the fact that magnetic measurements are sensitive to variables, such as internal stresses and slight temperature differences, that normally have little if any effect on the physical properties. The effect of stress variation is most noticeable in cases where high frequencies or low magnetizing forces are used.

Magnetic analysis serves to compare the material under test with a standard of known physical properties. Care must be taken to eliminate variables due to unimportant differences in analysis and structural conditions that will affect the magnetic measurements. Because the relation between magnetic and physical properties is complicated by such factors as chemical composition, heat-treatment, flaws, etc., each application must be considered separately and the testing procedure must be carefully coordinated with mechanical tests.

The comparative test, in its simplest form, is carried out on an alternating-current bridge, in which the impedance of a coil surrounding the test specimen is balanced against either the standard specimen in a similar coil or a synthetic standard of equal impedance. The unbalance of the bridge is indicated by an alternating-current galvanometer, or by an oscilloscope when additional information is desired on the wave form of the magnetizing current. Good results are obtained by this method in detecting small structural changes in steel due to variations in quenching temperatures and in determining structural abnormalities and decarburized surfaces.

Continuity Tests. A widely used magnetic test method, known as the Magnaflux test, is particularly useful for detecting in ferromagnetic materials the presence of invisible surface or subsurface discontinuities, such as incipient fatigue cracks, surface seams, shrinkage cracks in weldments, etc. The test is nondestructive in nature, and large test specimens as well as small can be tested, provided the equipment is of sufficient capacity to handle the larger sections of interest.

The principle of operation of the Magnaflux test consists of inducing a magnetic flux in the specimen of interest by either passing an electric current of appropriate amperage through it or by the influence of an external magnetic-yoke coil surrounding the specimen. Any abrupt discontinuities located in the path of the magnetic flux will induce local flux leakage fields, which may be detected by the application of finely divided particles of a ferromagnetic material to the surface of the test section. The ferromagnetic powder will offer paths of low reluctance to the leakage fields, and, as a consequence, the particles will tend to outline the effective boundaries of the discontinuities which disrupted originally the induced magnetic flux.

Inasmuch as the leakage fields diffuse and decrease in intensity with increase in distance from the discontinuities which caused them, the detection of subsurface defects will depend upon the strength of the magnetic field, the distance from the surface at which the defect is located, the ratio of the height of the defect to the thickness of the test section, and the width of the discontinuity. Under favorable conditions of testing, defects may be detected that are located about 2 in. from the test surface, and as close to the surface as 0.0002 in.

In the practical performance of the Magnaflux test, the specimen of interest may be magnetized with either alternating or direct current. The use of alternating current limits the test to detection of only those discontinuities that are open at the surface—a circumstance associated with the "skin effect" common to alternating current. However, direct-current magnetization makes possible the detection of large and small subsurface discontinuities, as well as those located directly on the test surface.

Either simultaneously with, or subsequent to, magnetization of the test section, the inspection medium may be applied to the piece in one of two ways. In the so-called dry method of application, a special ferromagnetic material in finely divided form is dusted onto the surface or surfaces of the test specimen by means of a hand shaker, vibrating screens, or other methods of application that will uniformly distribute the powder. In those regions on the surface of the test section where local flux-leakage fields exist, the ferromagnetic particles will tend to concentrate and effectively outline the boundaries of the defects causing the flux leakage. This circumstance is illustrated in Fig. 163. The dry method of application is more sensitive to revealing deep-seated discontinuities within the test section than is the wet method to be described. Furthermore, the dry method is less messy to carry out and is better suited to unmachined castings, weldments, forgings, heavy sections, etc.

In the wet method of application, appropriately sized magnetic oxide particles in the form of a paste, as supplied by the manufacturer, are suspended in a suitable vehicle such as kerosene, petroleum oils of appropriate grades ("Bayol D," "Ultrasene," "Base Oil C"), etc. The suspended particles are then flowed over the magnetized test section or applied by other suitable methods such as dipping or spraying. The wet method is used principally to detect defects on finished and bright parts, such as bearings, ground and polished engine parts, or on other sections where the

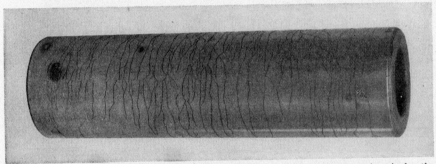

FIG. 163. Invisible grinding cracks revealed on the surface of an engine wrist pin by the Magnaflux test and dry-powder method. (*Courtesy of Magnaflux Corp.*)

primary interest is in locating nonmetallic stringers and very small discontinuities.

For the purpose of establishing good contrast between the powder pattern and the surface of the test section, Magnaflux powders are available in different colors. Dry powders may be obtained in colors of gray, black, and red; and pastes for use in the wet method in colors of black and red. For maximum contrast, however, or in cases where the test surface of interest may be somewhat obscured (interior of tubes, borings, etc.) and the powder patterns difficult to see, special fluorescent powders are available, known as Magnaglo. The Magnaglo material is applied to the test section by the wet-method techniques described, and final examination of the test section is conducted in a darkened enclosure under ultraviolet radiation of wave lengths between 3500 and 4000 Å. The powder pattern so revealed appears as greenish-yellow lines against a rather dark background of the surface of the test section, and is rendered in excellent contrast, as illustrated in Fig. 164.

FLUORESCENT TEST METHODS

In addition to the Magnaglo fluorescent process described for magnetic-test sections, certain types of invisible surface defects in nonmagnetic materials, such as austenitic stainless steels, nonferrous metals and alloys, plastics, ceramics, etc., may be detected by a fluorescent penetrant inspection process, known as Zyglo.

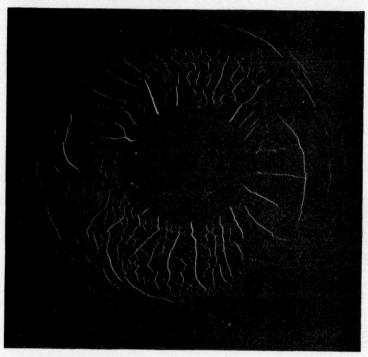

FIG. 164.   Invisible grinding cracks on the surface of a steel die revealed by the Magnaglo fluorescent process.   (*Courtesy of Magnaflux Corp.*)

The Zyglo process, licensed by the Magnaflux Corporation, consists of applying a specially prepared fluorescent penetrant to the test section by means of dipping, brushing, spraying, or other convenient methods.   After appropriate wetting, the test section is allowed to drain of excess penetrant for a time sufficiently long to permit the penetrant to seep into the surface discontinuities.   Depending upon the type of defect to be located, the drainage period may be anywhere between a few seconds and several hours. The test section is then washed in a water-spray and subsequently dried in a stream of warm air.

The test section is then dusted with a special developing powder that absorbs some of the penetrant from the surface defects and covers any general fluorescent background that may be retained on the surface. An alternate method of applying the developer is by immersing the test section into an aqueous suspension of the developer and subsequently removing the test section and drying it in a drying oven. By this latter method of application, the developer is deposited on the test section in a uniformly thin layer.

After the test section has been appropriately treated with developer, the detection of surface defects is made by visual observations under ultra-violet radiation, as described for the Magnaglo process. An example of grinding cracks, as revealed by the Zyglo method, is illustrated in Fig. 165.

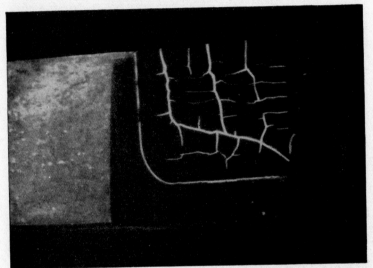

Fig. 165.   Invisible grinding cracks on the surface of a tungsten carbide tool tip revealed by the Zyglo fluorescent process.   (*Courtesy of Magnaflux Corp.*)

SUPERSONIC TEST METHODS

The Sperry Supersonic Reflectoscope, illustrated in Fig. 166, is an instrument designed to locate discontinuities within a metal or other materials through the use of supersonic waves, *i.e.*, sound waves having frequencies well above the upper limit to which the human ear is sensitive (about 16,000 cycles per sec). Owing to the characteristics of supersonic waves, as will be described, the principle of supersonic testing involves a measure of the time required for supersonic vibrations to penetrate the material of interest, reflect from the opposite side or from an internal discontinuity, and return to the point where the waves were first introduced. The behavior of the waves through such a cycle of travel with regard to time can be appropriately recorded on a cathode-ray oscilloscope screen, and by visual observation of the wave pattern the presence of defects and their location within the test section can be readily noted.

Supersonic testing is practically an instantaneous process of indication, since the time of travel of supersonic waves is in the order of microseconds, regardless of the length of a practical test section.   The test method is surprisingly sensitive to revealing small internal defects and is capable of locating discontinuities having a projected lateral area of about 0.1 per cent of

FIG. 166.   The Sperry supersonic reflectoscope unit.   (*Courtesy of Sperry Products, Inc.*)

the distance that the defects exist from the test surface.   In practical terms, a defect having a projected lateral area of about 0.006 in. (roughly equivalent to a projected circular area of this diameter) can be detected within 6 in. from the test surface, and one about 0.125 in. at a distance of 10 ft.

Supersonic waves, unlike audible sound waves, are propagated rectilinearly owing to their relatively short wave lengths, and can be beamed from an appropriate source with little tendency for the waves to be refracted around small obstacles.   They are strongly reflected by interfaces involving a change in densities or elasticities, and particularly so at air-metal interfaces. These circumstances are responsible for the test method being applicable for detecting discontinuities within a metal, since the supersonic waves are reflected equally as well from a defect, like a crack, whose surfaces are closely pressed together as from one whose surfaces are widely separated.

Principles of Operation. The Sperry reflectoscope consists essentially of two components—a transmitter and a receiver. The transmitter is a supersonic oscillator or pulse generator that is capable of generating supersonic energy in the frequency range of 0.5 to about 12 megacycles and specifically in the commonly used frequencies of 0.5, 1, 2.25, and 5 megacycles. The supersonic energy from the transmitter, in the form of electrical energy, is transformed into mechanical vibrations by a quartz transducer crystal, the frequency of the vibrations being dependent upon the crystal and the frequency of the pulse generator. These supersonic vibrations are radiated to the test section when the specially prepared (usually flat) quartz crystal makes intimate contact with one of the surfaces. Usually the coupling medium between the test-section surface and the crystal is a thin oil or glycerine to ensure maximum transfer of energy through

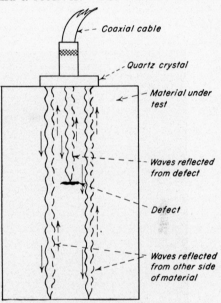

FIG. 167. Illustrating schematically the behavior of supersonic waves when transmitted through a section of metal containing an open defect within.

the interface. The pulse generator is so designed that the quartz crystal sends out supersonic waves in pulses of a few microseconds duration, at the rate of 60 pulses per second. Between the transmitting periods, the crystal acts as a receiver for the supersonic waves as they are reflected back to the origin. The returning waves oscillate the crystal mechanically, and owing to the inverse piezo effect, electrical impulses are generated that are amplified by the receiving unit and recorded on the oscilloscope screen. The receiving unit is capable of a resolution of about 0.25 microseconds, which is sufficiently great to indicate simultaneously discontinuities in steel or aluminum spaced about 0.03 in. apart, at the minimum.

As illustrated in Fig. 167, the supersonic waves transmitted to the test section by the crystal transducer travel unhindered through sound unaffected metal, reach the opposite side, and are reflected back through the section to the crystal.   This gives rise to two peaks in the oscilloscope pattern as illustrated in Fig. 168, defining the initial pulse and the reflected pulse from the opposite side.   Actually the distance between the initial and reflected pulses is an exact measure of the time required for the supersonic waves to travel from the test surface to the opposite side, and back again.   The transmitted waves, however, that meet a defect within the test section are reflected back to the crystal within a shorter interval of time than those waves reflected from the opposite side.   This circumstance

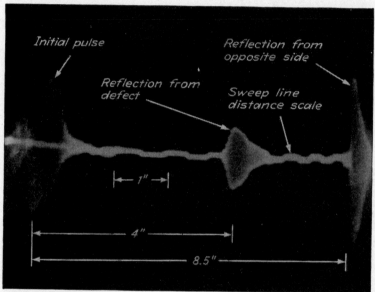

Fig. 168.   A typical oscilloscope record of the behavior of supersonic waves in passing through a mass of steel 8.5 in. thick that contains an internal defect about 4 in. from the test surface.   (*Courtesy of Sperry Products, Inc.*)

gives rise to an intermediate peak in the oscilloscope pattern, defining the location of the internal discontinuity.   To assist in locating more precisely the depth to which the defect exists, a timing circuit superimposes a simple distance scale on the oscilloscope sweep.   This timing wave is in the form of a square wave and can be adjusted to represent distances of 0.5 in. to 2 ft (1 in. as illustrated in Fig. 168).   The timing wave is calibrated by merely placing the transducer crystal upon a test section of known length and the appropriate number of scale marks set between the initial pulse and one occurring by reflection from the point of known distance.

Having once located the defect, and noting the distance from the test surface at which it exists, it is a simple matter to trace the projected outline of the flaw by merely moving the crystal about on the test surface and observing the oscilloscope indications.

Depth of Penetration.   The depth to which supersonic waves will effectively penetrate a material and be appropriately reflected depends upon a number of variables.   In general, the depth of penetration will be greater the more uniform and smaller the existing grain size in the metal; the smoother the surface which contacts the crystal transducer; the larger the transducer crystal; and the lower the frequency (longer the wave length) of the supersonic waves.   On the basis of wave frequencies, Table 42, page 472, shows the depth of penetration of supersonic waves in various materials.

Selected References on the Principles of X Rays and Crystal Structure

ASTM, "Symposium on Radiography," Philadelphia, 1942.

Barrett. C. S.: "Structure of Metals," McGraw-Hill Book Company, Inc., New York, 1943.

Bragg, W. H.: "Introduction to Crystal Analysis," D. Van Nostrand Company, Inc., New York, 1929.

Bragg, W. L.: "The Crystalline State," The Macmillan Company, New York, 1934.

Buerger, M. J.: "X-ray Crystallography," John Wiley & Sons, Inc., New York, 1942.

Clark, G. L.: "Applied X-rays," 3d ed., McGraw-Hill Company, Inc., New York, 1940.

Compton, A. H., and S. K. Allison: "X-rays in Theory and Experiment," D. Van Nostrand Company, Inc., New York, 1935.

Davey, W. P.: "A Study of Crystal Structure and Its Application," McGraw-Hill Book Company, Inc., New York, 1934.

Eastman Kodak Co.: "Radiography in Modern Industry," Rochester, New York, 1947.

Evans, R. C.: "An Introduction to Crystal Chemistry," The Macmillan Company, New York, 1939.

General Electric X-ray Corporation: "X-ray Studies in Advanced Radiographic Technique," 3d ed., Chicago, 1935.

Hilger, A.: "X-ray Metallurgical Crystallography," Adam Hilger, Ltd., London, 1932.

Hume-Rothery, W.: "The Metallic State," Oxford University Press, New York, 1931.

Hume-Rothery, W.: "The Structure of Crystals," Institute of Metals, London, 1936.

Hume-Rothery, W.: "The Structure of Metals and Alloys," Institute of Metals, London, 1945.

International Conference on Physics: "The Solid State of Matter," Vol. II., Physical Society, London, 1935.

Joffe, A. F.: "The Physics of Crystals," McGraw-Hill Book Company, Inc., New York, 1928.

Pullin, V. E. A.: "Engineering Radiography," George Bell & Sons, Ltd., London, 1934.

Pullin, V. E. A.: "X-rays and Radium," Ernest Benn, Ltd., London, 1929.

Randall, J. T.: "Diffraction of X-rays and Electrons by Amorphous Solids, Liquids, and Gases," John Wiley & Sons, Inc., New York, 1934.

Seitz, F.: "The Modern Theory of Solids," McGraw-Hill Book Company, Inc., New York, 1940.

Sproull, W. T.: "X-rays in Practice," McGraw-Hill Book Company, Inc., New York, 1946.

St. John, A., and H. R. Isenburger: "Industrial Radiography," John Wiley & Sons, Inc., New Hork, 1934.

Stillwell, C. W.: "Crystal Chemistry," McGraw-Hill Book Company, Inc., New York, 1938.

Taylor, A.: "An Introduction to X-ray Metallography," John Wiley & Sons, Inc., New York, 1945.

Terrill, H. M., and C. T. Ulrey: "X-ray Technology," D. Van Nostrand Company, Inc., New York, 1930.

Von Hevesy, G. : "Chemical Analysis by X-rays and Its Application," McGraw-Hill Book Company, Inc., New York, 1932.

Worsnop, B. L.: "X-rays," E. P. Dutton & Co., New York, 1930.

Wyckoff, R. W. G.: "The Structure of Crystals," ACS Monograph Series 19, Reinhold Publishing Corporation, New York, 1931.

Wyckoff, R. W. G.: "The Structure of Crystals," ACS Monograph Series 19A, supplement to second edition, Reinhold Publishing Corporation, New York, 1935.

Zachariasen, W. H.: "Theory of X-ray Diffraction in Crystals," John Wiley & Sons, Inc., New York, 1945.

References

American Iron and Steel Institute: "Hardenability of Alloy Steels," No. 11, New York, June, 1947.

ASM: "Electronic Methods of Inspection of Metals," Cleveland, Ohio, 1947.

ASM: "Metals Handbook," Cleveland, Ohio, 1948.

ASTM: Standards, Parts I-A and I-B, 1946.

ASTM: "Tentative Method of End-Quench Test for Hardenability of Steel," ASTM Designation A255-46T, January, 1947.

Bullens, D. K.: "Steel and Its Heat Treatment," Vols. 1 and 2, John Wiley & Sons, Inc., New York, 1948.

Doan, F. B.: "Principles of Magnaflux Inspection," Photopress, Inc., Chicago, Ill., 1942.

Enos, G. M.: Notes on Spark Testing, *Trans. ASST*, Vol. 12, December, 1927.

Epstein, S.: "The Alloys of Iron and Carbon," Vol. I, "Constitution," The Engineering Foundation, McGraw-Hill Book Company, Inc., New York, 1936.

Jefferies, Z.: Grain Size Measurements, *Chem. & Met. Eng.*, Vol. 18, 1918.

Jominy, W. E.: A Hardenability Test for Shallow Hardening Steels, *Trans. ASM*, Vol. 27, 1939.

Jominy, W. E., and A. L. Boegehold: A Hardenability Test for Carburizing Steel, *Trans. ASM*, Vol. 26, 1938.

McQuaid, H. W., and E. W. Ehn: Effect of Quality of Steel on Case-carburizing Results, *Trans. AIME*, Vol. 67, 1922.

Metal Progress: Metal Progress Data Sheet, January, 1947.

Miller, O. O., and M. J. Day: Heat Etching as a General Method for Revealing the Austenitic Grain Size of Steels, *Trans. ASM*, Vol. 30, 1942.

Palmer, F. R.: "Tool Steel Simplified," Carpenter Steel Company, Reading, Pa., 1937.

Shepherd, B. F.: Hardenability of Shallow Hardening Steels Determined by The P-V Test, *Trans. ASM*, Vol. 38, 1947.

Shepherd, B. F.: The P-F Characteristic of Steel, *Trans. ASM*, Vol. 22, 1934.

Tobin, H., and R. L. Kenyon: Austenitic Grain Size of Eutectoid Steel, *Trans. ASM*, Vol. 26, 1938.

Vilella, J. R.: The Grain Size of Steel, *Mech. Eng.*, Vol. 62, 1940.

Vilella, J. R., and E. C. Bain: Revealing the Austenitic Grain Size of Steel, *Metal Progress*, September, 1936.

Wilson, R. L.: Grain Size in Steel, *Metal Progress*, August, 1934.

CHAPTER 8

THE PRINCIPLES OF PYROMETRY
AND PYROMETRIC PRACTICE

The successful operation of many metallurgical processes, such as smelting, refining, casting, and heat-treating, depends largely upon the accuracy with which temperatures are measured and controlled. Often, a variation of 10 to 15°C (18 to 27°F) from the correct process temperature will result in products of inferior quality. The temperatures encountered in many of these processes are generally very high and the ordinary glass thermometer, needless to say, cannot be used. One of the simplest methods for estimating the temperature of a metal, particularly of steel when enclosed in a heated furnace, is that of noting the color or tint of the hot body. There is an apparent variation between the temperature of a metal and its color, the temperature varying approximately as follows:

Color	Degrees centigrade	Degrees Fahrenheit
Faint red......................	500	930
Blood or dark red..............	550–625	1022–1157
Dark cherry...................	635	1175
Full cherry red................	700	1292
Bright cherry.................	800	1472
Dark orange...................	900	1652
Orange........................	950	1742
Full yellow or lemon...........	950–1000	1742–1832
Light yellow..................	1100	2012
White.........................	1150 and up	2102 and up

The greatest objection to this method of measuring temperatures is the variation of individual judgments. Nevertheless, careful observation of the color of heated metal can be used to advantage for rough temperature estimations, and when they are made by a skilled and experienced observer the results are surprisingly accurate.

Scientific temperature-measuring devices that are used most extensively in metallurgical work are known as pyrometers and may be divided into two general groups. The first of these are thermocouples and resistance pyrometers, which, in order to operate, must be brought into contact with the object or locality of which the temperature is to be measured.   Optical and radiation pyrometers comprise the second group and operate at a distance from the heated body, utilizing the heat and light radiation emitted by a high-temperature source.

THERMOELECTRIC PYROMETERS

A simple thermoelectric pyrometer, shown in Fig. 169, consists essentially of three distinct units: (1) the thermocouple, composed of two different metals or alloys; (2) two lead wires connecting the free ends of the thermocouple to a suitable indicator; and (3) an indicator for measuring the emf developed by the thermocouple.

Thermocouples.   A thermocouple is a system composed of two homogeneous metal wires, dissimilar in chemical composition, that are soldered,

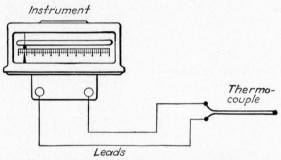

FIG. 169.   A thermoelectric pyrometer, consisting of a thermocouple, extension lead wires, and an emf-indicating instrument.

fused, or welded together at one end, forming a closed circuit when the free ends are connected to an indicating instrument.   The welded end, because it is that part of the thermocouple in contact with the hot medium, is called the hot junction.   The connections that the free ends of the thermocouple element make with the binding posts of the indicating instrument, with copper lead wires, or any electrical conductors having thermal emf characteristics different from those of the respective thermocouple wires, are known as the cold, or reference, junctions.

The wires comprising a thermocouple are electrically insulated from one another by some appropriate insulating and refractory material, such as porcelain beads or special tubular shapes of porcelain as illustrated in Fig. 170a, asbestos tubing, etc.

A specially constructed iron-constantan thermocouple, illustrated in Fig. 171, consists of one wire element within a hollow tube of the other. The tubular element serves both as the second component of the thermocouple and as a protection tube to the element enclosed.   The two elements are fused together at one end, as illustrated, and are electrically insulated from one another by an appropriate insulating material.   When a thermocouple of this type is used at relatively high temperatures, the outer tubular ele-

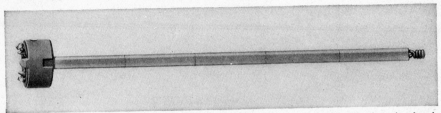

a. Thermocouple elements separated by porcelain insulators and attached to junction head. Primary mounting.

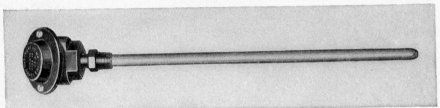

b. Thermocouple elements as in *a* above enclosed within a porcelain protection tube and metal junction-head enclosure.   Secondary mounting.
Fig. 170.   Typical construction of a base-metal thermocouple.   (*Courtesy of the Brown Instrument Co.*)

ment is usually protected from excessive oxidation by a conventional protection tube.

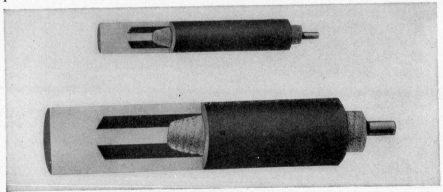

Fig. 171.   An iron-constantan thermocouple constructed so that the inner constantan element is enclosed within the iron element of tubular form.   (*Courtesy of Foxboro Co.*)

On the basis of the kind of metals and alloys that comprise the elements of a thermocouple, thermocouples may be divided into two general groups: (1) the base-metal thermocouples, which are composed of the more common and relatively inexpensive metals and alloys, such as iron, copper, chromel, alumel, etc.; and (2) the noble-metal thermocouples, composed of platinum and platinum alloys. With each general group of couples there are associated certain technical advantages and disadvantages, including such factors as resistance to oxidation, maximum operational temperature limits, resistance to contamination, etc. Notwithstanding the many desirable characteristics of noble-metal thermocouples, they are not used as extensively as base-metal couples in routine metallurgical practice, owing to their relatively high cost. The characteristics of the more common thermocouple combinations are given in Table 43, page 473.

Principles. The measuring of temperature by an appropriately combined thermocouple is possible only because an emf is developed in the thermocouple circuit when the fused junction is at a temperature different from that of the cold junctions. If the cold junctions are maintained at a constant and reproducible temperature level, say the melting point of ice, then the measurable emf developed in the thermocouple circuit will be some definite function of the temperature of the hot junction. By appropriate calibration at a number of fixed temperature points or by comparison with a working-standard thermocouple so standardized (see page 339), it will be possible to establish a precise relationship between developed emf and true temperature of the hot junction. For some of the metallic combinations used in practice, this function will be approximately linear, and for the degree of accuracy required for most ordinary temperature measurements, the calibration need only be made at a relatively few temperature levels over the working range of the thermocouple of interest.

Under the circumstances described above, the emf developed in the system and the flow of electrical current arising therefrom may be associated with two independent sources of emf as defined by the Peltier and Thomson effects.

Peltier Effect. When two homogeneous metallic wires of different chemical composition are brought into electrical contact with one another, a difference of potential will exist between the two at the junction point. The magnitude of the emf so developed will be governed by the temperature of the contact junction and the precise chemical composition of the two elements. Usually, the emf will increase proportionally with increase in temperature of the junction, but in some cases—*e.g.*, iron vs. copper—the emf may increase to some maximum value and then decrease with further increase in temperature of the function point.

Thomson Effect. If a temperature gradient is established in a homogeneous metallic wire by heating one of its ends, a potential difference will be established between the heated and unheated ends of the wire. Usually the end at the higher temperature will be at the greater potential, and the magnitude of the emf developed will be a function of the composition and homogeneity of the wire and of the temperature difference between the two ends. In the case of some metals and alloys, the Thomson emf may actually reverse at certain elevated temperatures and cause the direction of current flow along the wire element to reverse also.   Metals and alloys that behave in this manner are, of course, unsuitable for thermocouple construction.

From the past discussion, it is evident that the net emf developed in a simple thermoelectric system is equal to the sum of the following four emfs, taking into account the algebraic sign of each:

1. Peltier emf at the hot junction
2. Peltier emf at the cold junction
3. Thomson emf along one of the wire elements
4. Thomson emf along the second wire element

Thermoelectric Potentials.   It is apparent from the discussion of the Peltier and Thomson effects that different combinations of metals will develop different emfs under the same conditions of hot- and cold-junction temperatures.   In order, therefore, to evaluate the potential of each metal, it must be referred to some metal designated as a standard.   Because lead appears to have a zero Thomson effect, it is referred to as the standard reference source, although any metal would serve equally as well for this purpose.   In practice, platinum is usually used as the standard, a potential value of zero at all temperatures being arbitrarily assigned to it.

In order to obtain potential data of a metal, a couple is first made between pure platinum and the metal in question.   An extension wire of platinum is then attached to the free end of the metal.   If the temperature of one of the metal-platinum junctions is maintained constant, say 0°C, and the other metal-platinum junction is heated to some predetermined temperature, an emf will be produced between the free ends of the platinum wires.   This difference in potential is due to the temperature difference between the hot and cold junctions of the metal.   If a series of such measurements is made at different temperatures, and if the same procedure is followed for different metals, emf temperature data will be obtained, so that a family of potential curves may be plotted.   Such a group of curves for the common thermocouple metals and alloys is shown in Fig. 172.

It will be noted that some of the metals are positive, others negative to the platinum reference standard.   The direction in which the current flows at the hot junctions distinguishes a positive metal from one which is negative.   A metal is considered positive with respect to the platinum standard if the current flow is from platinum to the metal and negative if the flow of current is in the opposite direction.   The polarity may be quickly ascertained during the experimental procedure by noting the order in which the lead wires are connected to the indicating instrument.

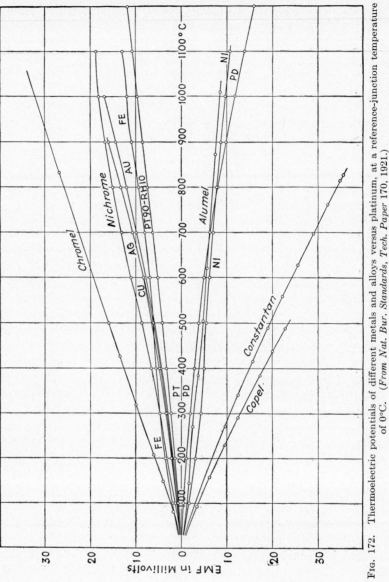

Fig. 172. Thermoelectric potentials of different metals and alloys versus platinum, at a reference-junction temperature of 0°C. (From Nat. Bur. Standards, Tech. Paper 170, 1921.)

If platinum wire is not available for carrying out the actual experimental measurements, copper wire may be used instead. The measured potential value so obtained may then be corrected and referred to platinum. The potential value between copper and platinum is calculated first by the following formula:

$$E = 0.00313t + 0.0000123t^2 \tag{1}$$

where E = emf, mv, when the cold junction is at 0°C

t = temperature of hot junction, °C

The potential of the metal with respect to platinum is subsequently determined by adding the measured emf of copper versus the metal (with regard to polarity) to the value of E as obtained from the above equation.

A family of potential curves is useful in determining the emf temperature relation between any two metals. The emf developed between two metals at a particular temperature is equal to the difference of the individual potentials of the metals, the positive element being the one having the highest potential. Thus, at 800°C a chromel-alumel combination will produce an emf equal to $26 - (-7.3)$ or 33.3 mv. The values of 26 and -7.3 for chromel and alumel, respectively, are obtained from Fig. 172.

Cold-junction Correction. As was previously shown, the magnitude of the emf developed by the hot junction of a thermocouple depends upon the temperature of the hot junction, the composition of the thermocouple elements, and the temperature of the cold junction. If the cold-junction temperature is allowed to vary from a constant value, there will be a small variable Peltier effect at this junction and, in addition, a variation in the Thomson emf along the wires. Both of these variations will introduce errors in the temperature measurements. In order, therefore, to maintain a reproducible and direct relationship between developed emf and the temperature of the hot junction, either the cold junction of the thermocouple must be kept at constant temperature, or appropriate methods must be used to compensate for the error introduced.

Inasmuch as thermocouples are generally calibrated at a cold-junction temperature corresponding to the melting point of ice (0°C), it is desirable to use the thermocouple subsequently at the same cold-junction temperature. By so doing, calibration data may be used directly without having to correct the temperature measurements by computation, or setting the emf-measuring instrument to compensate for the involved error. Appropriate cold-junction temperatures may be maintained constant and at the calibration temperature by placing the insulated cold junction in a cracked ice-water mixture contained within a Dewar flask or Thermos bottle. Because this procedure is at times inconvenient to carry out, particularly in large metallurgical installations, it is common practice in such cases to correct the temperature measurements to correspond to the constant cold-junction temperature actually used, or to compensate for variations of the actual cold-junction temperature with time through appropriate choice of an emf-measuring instrument.

As mentioned heretofore, the calibration curves of most base-metal thermocouples are approximately linear. If a thermocouple exhibiting such calibration characteristics is used at a constant cold-junction temperature different from the temperature at which it was calibrated, the correct hot-junction temperature may be computed by merely adding to the observed temperature the following correction factor:

$$(t_1 - t_2) K \qquad (2)$$

where t_1 = actual temperature of cold junction, °C

t_2 = calibrated cold-junction temperature, °C

$K = \begin{cases} \text{approximately 1.0 for base-metal couples} \\ \text{approximately 0.5 for noble-metal couples} \end{cases}$

As a rule, this correction can be made directly, without computation, by setting the indicator of the millivoltmeter to correspond to the cold-junction temperature at which the thermocouple is used. In the case of a potentiometer, provision usually is made in the form of an adjustable resistance which can be set manually at the emf corresponding to the junction temperature involved. New settings of either indicator are obviously necessary whenever the cold-junction temperature is altered.

Compensating extension lead wires, owing to their relatively low cost as compared to thermocouple alloys, are widely used for moving the cold junction of a thermocouple from surroundings where the temperature varies to some location where the temperature remains more constant; or for any other reasons where it is desirable to move the reference junction from the junction-head of the thermocouple. Compensating leads are merely inexpensive metal or alloy wires that are of different composition than the thermocouple wires to which they are attached, but which have practically indentical thermal emf charactersitics as those of the element wires.

Constant-temperature sources to which the cold junction may be referred by means of compensating lead wires may be a constant-temperature room where the measuring instrument is located; an ice bottle as already mentioned; or a hole in the ground, about 10 ft deep, into which the cold-junction connections may be buried. It is common practice, however, to ignore the constancy of the reference-junction temperature and, by means of lead wires as described, move the cold junction of the thermocouple to the connecting binding posts of an emf-measuring instrument equipped with automatic compensating devices to correct for changes in temperature that may occur at this location.   In millivoltmeters, these compensating devices are usually bimetallic strips that, in effect, automatically adjust the zero position of the indicator pointer as the temperature varies.   In a potentiometer, small coils of special wire are located near the connecting-binding posts, which automatically affects the balance of the potentiometer circuit by an amount related to the change in resistance of the compensating coils arising from variations in temperature.   Thus, the temperature of the hot junction of the thermocouple is always correctly indicated by the slide-wire dial of the instrument, regardless of the temperature changes occurring at the reference junction.

Requirements and Preparation of Thermocouples.   Although any two dissimilar metals will develop a thermal emf when appropriately joined together, it is only those combinations possessing additional requisite characteristics that will function efficiently as thermoelectric pyrometers.   It is therefore of general importance, when considering the appropriateness of any two metallic elements for the purpose intended, that:

1. The thermal emf developed by the thermocouple increases uniformly and continuously with increase in temperature, and at any temperature the corresponding emf is sufficiently great to afford accurate emf measurements.

2. The two dissimilar metals are reasonably resistant to corrosion, contamination, and oxidation at elevated temperatures.

3. The thermoelectric characteristics of the thermocouple are not changed appreciably during calibration or during subsequent use.

4. The melting temperatures of the two metallic elements are above any temperature at which the thermocouple will be used.

5. The metals comprising the element are reproducible with regard to quality and thermal emf characteristics.

Thermocouples are very often made in the laboratory from a supply of stock wire.   If the thermocouple is to be used at relatively low temperatures, the junction of the two wires may be readily affected by soldering with a silver solder of appropriate melting-point.   It can be demonstrated that the introduction of a third metal at the hot junction, which is the case when the junction is soldered, has no effect on the developed emf.   For use at relatively high temperatures, the thermocouple junction is best made by fusing the ends of the two wires in an electric arc or by means of an oxygen-gas flame.   For the purpose of increasing the mechanical strength of the joint, the two wires may be twisted together several times before the ends are fused.   In joining together noble-metal thermocouple wires, fusion of the two ends in an electric arc is preferred over other methods.   When using gas flames, particularly an oxyacetylene flame, there is considerable danger of contaminating the platinum element and rendering it rather brittle.

Before calibration of noble-metal thermocouples by methods to be described, it is appropriate to homogenize the element wires by annealing at a temperature of about 1500°C (2732°F) for approximately 1 hr. This annealing treatment is best carried out by suspending the uninsulated thermocouple element between two binding posts in a 110-volt circuit with adjustable resistances to supply a current of 10 to 14 amp. Temperature measurements may be made during annealing with an optical pyrometer.

In the case of base-metal thermocouples, stock wire supplied by most manufacturers is sufficiently well homogenized so that it is unnecessary to anneal the complete thermocouple after it is made.

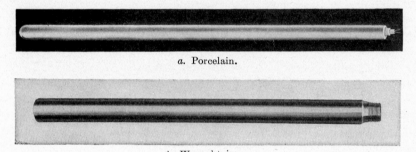

a. Porcelain.

b. Wrought iron.

Fig. 173.   Thermocouple protection tubes.   (*Courtesy of the Brown Instrument Co.*)

Thermocouple Protection Tubes. For purposes of retaining constancy of calibration and of prolonging the useful life of a thermocouple, it is essential during use at relatively high temperatures or in corrosive media to protect the thermocouple wires adequately from excessive oxidation and corrosion. Protection is best afforded a thermocouple by the use of an appropriate protection tube. As illustrated in Fig. 173, a protection tube is merely a refractory or metal tube, closed at one end, into which the insulated thermocouple may be placed. As a general rule, a thermocouple should have as little protection as possible, consistent with life characteristics of the element. Protection tubes that are of heavy wall thickness will reduce the sensitivity and delay the response of the thermocouple to small changes in temperature.

Protection tubes suitable for thermoelectric pyrometry should be constructed of materials having the following characteristics:

1. Relatively high heat conductivity so that by the presence of the tube the response of the thermocouple to small changes in temperature will not be significantly retarded.

2. The ability to withstand high temperatures for an appreciable length of time without cracking or seriously deteriorating.

3. High rigidity at elevated temperatures so that the tube does not suffer deformation if inadequately supported.

4. Sufficient strength at elevated and ordinary temperatures to withstand mechanical shock and normal handling.

The characteristics and appropriate uses of the more common thermocouple protection tubes are given in Table 44, page 474.

Calibration of Thermocouples. Thermocouples may be standardized by one of two methods, namely, (1) calibration at known fixed temperature points, which is designated as a primary calibration, or (2) calibration by comparison with a standarized instrument, such as a primary calibrated thermocouple, resistance thermometer, etc., which is known as a secondary calibration.

Primary Calibration. Temperatures in the range of 660 to 1063°C on the International Temperature Scale[1] are defined by a platinum–platinum-10 per cent rhodium thermocouple, one junction of which is kept constant at a temperature of 0°C. In this range, the true temperature–emf relationship is given by the equation

$$e = a + bt + ct^2 \tag{3}$$

where e = emf, mv
t = true temperature, °C
a, b, and c = constants of the thermocouple

The constants a, b, and c are determined by simultaneous solution of Eq. (3) after measuring the emf developed by the thermocouple element at the freezing points of gold, silver, and antimony. By appropriate use of this equation, the true temperature on the International Temperature Scale may be computed for any measured emf.

Other noble-metal thermocouples, as well as base-metal thermocouples, may be calibrated at known fixed freezing points of metals whose precise temperatures of freezing have been accurately determined by designated primary temperature-measuring instruments. Metals suitable for this

[1] The International Temperature Scale was adopted in the year 1927 by 31 nations. "The experimental difficulties incident to the practical realization of the thermodynamic scale have made it expedient to adopt for international use a practical scale designated as the International Temperature Scale. This scale conforms with the thermodynamic scale as closely as is possible with present knowledge, and is designed to be definite, conveniently and accurately reproducible, and to provide means for uniquely determining any temperature within the range of the scale, thus promoting uniformity in numerical statements of temperature. Temperatures on the international scale will ordinarily be designated as '°C'. ... " (Excerpt taken from Burgess, G. K.: The International Temperature Scale, *Nat. Bur. Standards, Res. Paper* 22, 1928.)

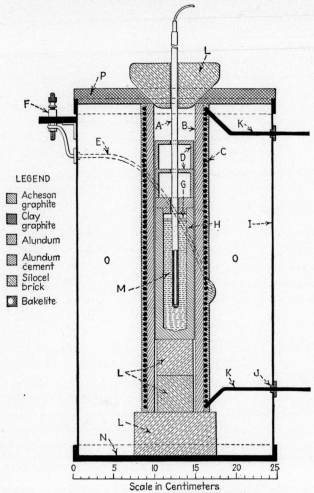

LEGEND

░ Acheson graphite

▓ Clay graphite

▨ Alundum

▧ Alundum cement

▦ Silocel brick

■ Bakelite

Scale in Centimeters

A. Thermocouple to be calibrated.
B. Alundum heater tube.
C. Heater unit.
D. Acheson graphite diaphragms.
E. Thermocouple to measure temperature of the furnace.
F. Binding posts.
G. Graphite powder.
H. Graphite crucible.
I. Sheet-metal furnace shell.
J. Gasket.
K. Heating-wire leads.
L. Silocel supports and cover.
M. Metal of known melting point.
N. Furnace bottom.
O. Sliced powder.
P. Furnace cover.

Fig. 174. A design of electric furnace suitable for calibration of thermocouples at fixed freezing points. (*From Nat. Bur. Standards, Res. Paper* 99, 1929.)

purpose are given in Table 45, page 476, and an electric-furnace design appropriate for the purpose is illustrated in Fig. 174.

When the calibration of a thermocouple (other than the primary standard described) by the primary method is to be of high order of precision, it is customary to determine the temperature–emf relationship at a sufficiently large number of fixed temperature points over the working range of the thermocouple to enable subsequent precise interpolation of the data or to permit accurate definition of the temperature–emf calibration curve, as illustrated in Fig. 175. Determination at a relatively large number of fixed points is necessary because the equations defining the temperature–emf relationship for most thermocouples is not known with the same degree of certainty as in the case of the standard platinum–platinum-10 per cent rhodium thermocouple [Eq. (3)].

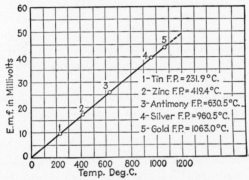

Fig. 175. Typical calibration curve of a chromel-alumel thermocouple standardized at fixed freezing points. Cold-junction temperature 0°C.

Where the accuracy of interpolation need not be more precise than about 10°C, the thermocouple may be calibrated at about four fixed temperature points over its working range. From these data, a smooth calibration curve may be constructed on coordinates of temperature and emf.

Secondary Calibration. Owing to the inconvenience associated with the primary calibration method described, and the relatively large number of determinations required in some cases, thermocouples may be calibrated to an accuracy sufficient for most purposes by direct comparison with (1) a standard platinum–platinum-10 per cent rhodium thermocouple calibrated at fixed points; (2) a primary-calibrated thermocouple of the same type as the one to be calibrated; or (3) a primary-measuring instrument designated to define the temperature scale in the appropriate temperature range of exploration. The standard platinum–platinum-10 per cent rhodium thermocouple, although serving only to define the temperature scale in the range 660 to 1063°C, is often used both above and below this range as a working standard.

The success of the comparison method of calibration depends mainly upon the ability of the observer to bring the hot junction of the unknown thermocouple to precisely the same temperature as the actuating element of the comparison standard. The methods by which this equality in temperature is secured depends upon the type of thermocouple involved, the type of comparison standard, and the method of heating. When a standard thermocouple is used as the comparison source, the junctions of the standard and the unknown thermocouples may be brought to approximately the same temperature by inserting the thermocouples into a bath of molten metal; or by inserting the two junctions into drilled holes in a large block of metal having relatively good thermal conductivity and which may be appropriately heated in an electric furnace. Details of these methods, as well as others, are fully described in publications of the National Bureau of Standards.

Owing to the rapid decrease of the thermal emf per degree of the standard platinum–platinum-10 per cent rhodium thermocouple at temperatures below about 300°C, this type of thermocouple is usually never used as a calibration standard below 0°C. In the temperature range of 300°C to about −150°C, thermocouples are best calibrated against a standard resistance thermometer (see page 351). The element of the resistance thermometer and the junction of the thermocouple are brought to the same temperature in an agitated liquid bath. Baths appropriate for this purpose consist of various combinations and proportions of inorganic and organic liquids, each of which will remain liquid over some given temperature range.

Millivolt-temperature Equivalents. Tables 46 to 53, pages 477 to 486, give the millivolt-temperature relations for the common base- and noble-metal thermocouples. These tables have been computed specifically for thermocouples manufactured by the Brown Instrument Company. Since the composition of base-metal thermocouples made by different manufacturers may vary, it is recommended that, where extreme accuracy is required, equivalents be used which correspond exactly to the specific make of thermocouple used. However, for ordinary purposes the variation in equivalents due to slight composition differences is small and can be disregarded.

Emf Measuring Instruments. The emf developed by a thermocouple is generally measured by a potentiometer or a millivoltmeter, calibrated either in millivolts or directly in degrees of temperature. Instruments that indicate temperatures directly must be used only with the type of thermocouple for which they were calibrated, as the emf developed at a particular temperature is different for different types of thermocouples.

Potentiometers. A potentiometer is one of the most accurate instruments available for the measuring of small emfs. Essentially, the emf developed by a thermocouple is balanced against a known emf and is measured in terms of this standard. This principle is admirably applicable for measuring the emfs developed by a thermocouple, as the measurements are independent of varying resistance in the circuit, resistance of the extension-lead wires and switches, and corroded thermocouples. However, large changes in resistance of the thermocouple circuit are to be avoided, since the precision by which the potentiometer can be balanced is greatly influenced by such changes. With increase in resistance, the sensitivity of response of the galvanometer decreases, requiring, as a consequence, larger changes in emfs to produce a perceptible deflection of the galvanometer indicator.

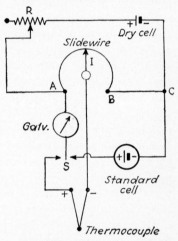

Although potentiometric instruments suitable for this purpose are available in a great many designs, they nevertheless are all basically the same in principle and may be classed under two general types—one type that is portable and requires manual manipulation of the instrument when a temperature measurement is to be made, and another general type that is designed for more permanent installation and which is semiautomatic or fully automatic in operation.

Fig. 176. Illustrating the principle of the simple direct-indicating potentiometer. (*Courtesy of Leeds & Northrup Co.*)

The principle of the direct-reading potentiometer as applied to the measuring of emfs is illustrated in Fig. 176. Current from the dry cell is passed through a main circuit consisting of a slide-wire and an adjustable resistance R. The slide-wire AB is a uniform resistance wire which may be considered as divided into an equal number of divisions. This division is effected by a graduated dial directly connected to the slider I. With the polarity of the dry cell as shown, there is a drop of potential along the slide-wire from A to B, the magnitude of which depends upon the current flowing through it from the dry cell. Since the slide-wire is of uniform resistance, there are equal drops of potential across each division.

In order to standardize the drop between A and B to correspond to the fixed markings on the indicating dial, a standard cell of known voltage is connected into the circuit by means of switch S. It will be noted that the polarity of the standard cell is such that the current flowing from it opposes the current flowing from the dry cell. By adjustment of the resistance R these currents may be made equal, with the net result that no current flows through the circuit, as indicated by zero deflection of the galvanometer.

The circuit is now standardized so that the potential drop across each division of the slide-wire corresponds to the indications on the slide-wire dial. For example, if the slide-wire is divided into 100 equal divisions and the standard cell has a voltage of 1.0185, the drop across each division of the slide-wire is equivalent to 0.010185 volt. Usually, in the construction of potentiometers, a fixed resistance is connected between B and C, thereby permitting smaller potential drops across each division without the necessity of an abnormally long slide-wire.

When the emf of the thermocouple is to be measured, it replaces the standard cell in the circuit by means of switch S. Similar to the polarity of the standard cell, the thermocouple is appropriately connected so that the current flowing from it opposes the flow of current from the dry cell. The circuit is balanced, not by adjusting the resistance R, but by adjusting the resistance of that portion of the slide-wire which is contained in the thermocouple circuit. This adjustment is made by turning the indicator dial. When a balance is attained, no current flows through the circuit, as indicated by zero deflection of the galvanometer, and the emf of the thermocouple may then be read directly from the dial. With continual use of the potentiometer, the current from the dry cell will decrease, and because of this, the circuit will depart from its original standardization. Therefore, it is necessary to check the circuit frequently with the standard cell for the purpose of restandardization.

As was previously pointed out, some instruments are equipped with either manual-adjusting or automatic compensating devices for cold-junction temperatures. Figures 177 and 178 illustrate, respectively, the manner in which these devices are incorporated into the circuit.

The manual-adjusting reference-junction compensator consists of a variable resistance in the form of a slide-wire which is connected in parallel with the main slide-wire of the potentiometer, forming with the galvanometer a typical Wheatstone-bridge circuit. The slider of the reference-junction compensator is adjusted, by means of a separate graduated dial, to correspond to the departure from the calibrated cold-junction temperature of the thermocouple. It can be seen in Fig. 177 that any adjustment of the cold-junction compensator will require a new setting of the slider on the main slide-wire of the potentiometer to effect a balance of the circuit. Settings of the slider on the cold-junction compensating slide-wire are so calibrated with the main slide-wire that the true temperatures of the hot junction of the thermocouple will be indicated directly.

The automatic cold-junction compensator, shown in Fig. 178, consists of two fixed resistances, X and Y. Resistance X is of a material having a high temperature coefficient of resistance, and it is located within the instrument very near to the connecting binding posts onto which the thermocouple wires are attached, *i.e.*, very near to the cold junction of the thermocouple. Hence, variations in temperature of the surroundings will affect equally the temperature of the cold junction and the temperature of resistance X. Resistance Y has a temperature coefficient of resistance equal to about zero at ordinary temperature.

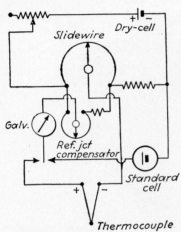

FIG. 177. Illustrating the principle of the manual-adjusting reference-junction compensator in the potentiometric circuit. (*Courtesy of Leeds & Northrup Co.*)

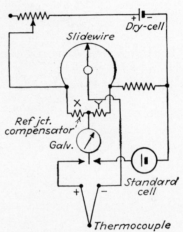

FIG. 178. Illustrating the principle of the automatic-adjusting reference-junction compensator in the potentiometric circuit. (*Courtesy of Leeds & Northrup Co.*)

If the temperature of the cold junction of the thermocouple increases, the temperature of resistance X increases the same amount, with an accompanying increase in the electrical resistance of X. This increase in the resistance of X has an effect on the balance of the circuit which is similar to the effect produced by moving the slider of the manual-adjusting compensator (Fig. 177) in a counterclockwise direction. When the temperature of the cold junction decreases, the circuit will be automatically readjusted by similar principles, but in an opposite manner.

Millivoltmeters. A millivoltmeter, shown in Fig. 179, consists essentially of a movable coil which is mounted on pivots between the poles of a permanent magnet. A pointer, which moves over a graduated scale, is attached to the moving coil. The deflection of the pointer depends upon the thermoelectric current passing through the coil, this current being proportional to the developed emf of the thermocouple. Millivoltmeters may be obtained that are self-recording and that automatically control furnace temperatures. These features operate on principles that are similar to the operation of like devices on potentiometers.

The emf at the terminals of the millivoltmeter is always less than the actual emf developed by the thermocouple. This difference is due to the external resistance of the couple and the leads. Generally, however, the scale of the instrument is calibrated to read the correct emf for a particular fixed external resistance. It can readily be shown from Ohm's law that the

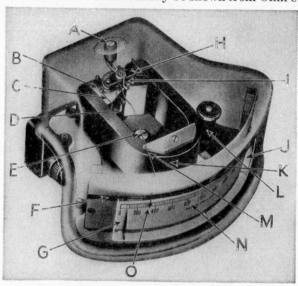

FIG. 179. A typical thermoelectric millivoltmeter, calibrated to indicate degrees of temperature directly when used with a given type of thermocouple. (*Courtesy of the Brown Instrument Co.*)

A. Adjusting screw for setting the pointer to zero.
B. Magnetic-pole pieces.
C. Resilient-pointer stops.
D. Soft-iron cores on pole pieces.
E. Base screw.
F. Mirror to avoid errors due to parallax when reading the scale.
G. Reference-junction compensator.

H. Support for movable coil.
I. Copper coil spring.
J. Scale cover glass.
K. Permanent magnet.
L. Series resistance of manganin wire.
M. Millivoltmeter pointer.
N. Temperature scale.
O. Pointer dot.

relation between the emf across the terminals of a millivoltmeter (e_o) and the emf developed by the thermocouple (e) may be expressed as follows:

$$e_o = \frac{R_m}{R_m + R_c + R_1} e \qquad (4)$$

where R_m = resistance in ohms of the millivoltmeter
R_c = resistance in ohms of the thermocouple
R_1 = resistance in ohms of the lead wires

This equation is particularly useful in determining the error involved in using instruments of different resistances, as well as correcting for known variations in the resistance of the lead wires.

Temperature Controllers. A temperature controller is an instrument consisting of a potentiometric circuit (sometimes a Wheatstone bridge for use with a resistance thermometer, or a millivoltmeter circuit) and auxiliary electrical or mechanical equipment whose purpose as a unit is to control and to maintain within very narrow limits the preset temperature of some external heat source, *e.g.*, the temperature of an electric furnace. This control of temperature is achieved by either semiautomatic or fully automatic mechanical methods; by photoelectric methods; or by electronic circuits.

In principle, a controller regulates the input of energy to a heat source whose temperature is to be controlled by appropriately opening and closing one or more contacting relays in the input circuit. Actuation of the relays can be made to correspond to any desired emf that is impressed across the controlling circuit by a thermocouple located within the heat source of interest.

Depending upon the design of the instrument, a temperature controller may regulate the temperature of only one heat source (single-point controller), or through suitable automatic-operating commutating switches, the temperature of several sources may be controlled.

Mechanical-operating Types. The principle of operation of all mechanical-operating controllers is basically the same, differing only in mechanical features by which the instrument functions. In the potentiometric type of controller, the galvanometer and its attendant deflections is the main controlling element of the system. In the use of a controller of this type, the slide-wire contactor of the potentiometer is set at some predetermined emf which is equivalent to that which will be developed by the furnace thermocouple at the temperature of interest. On many instruments the slide-wire scale is calibrated directly in degrees, facilitating this adjustment.

The galvanometer in the potentiometer circuit, in one type of instrument, is of the vertically suspended type, permitting the galvanometer needle or boom to move in a horizontal plane. Mounted above and to one side of the zero deflection position of the boom is a depressor bar which is slowly and continuously raised and lowered by a motor-driven cam arrangement. The depressor bar, depending upon the extent of its movement, controls a relay within the controller, which in turn may actuate an external contactor in the power-input circuit of the furnace; or if the controller relay is of sufficiently high power rating, this switch may be connected directly into the furnace circuit.

During the heating cycle of the furnace, the potentiometer circuit is unbalanced, causing the galvanometer boom to be deflected beneath the depressor bar. This position of the boom prevents the continuously moving depressor bar from completing a full downward cycle of movement.

The controller relay then remains closed and the furnace is allowed to heat. However, as soon as the emf developed by the thermocouple within the furnace is equal to that for which the potentiometer was preset, the galvanometer boom swings out from beneath the depressor bar and comes to rest at either zero deflection or slightly to the other side of zero.   This position of the boom now allows the depressor bar to complete a downward cycle of movement, which in turn actuates and opens the relay and breaks the power circuit of the furnace.   As the furnace cools, the boom again swings into position beneath the depressor bar and the heating cycle is repeated.

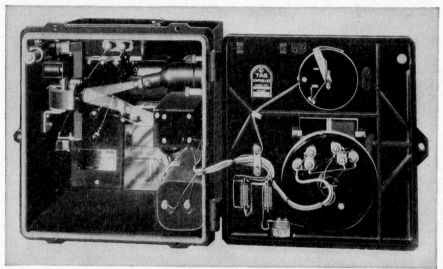

Fig. 180.   A photoelectric-type of potentiometric temperature controller.   (*Courtesy of Tagliabue Mfg. Co.*)

During the use of a controller, as in the case of any potentiometer, it is necessary to standardize periodically the potentiometer circuit with the standard cell.   This is usually accomplished by manual adjustment, but in some instruments a balance of the circuit is effected automatically.

Photoelectric-operating Types.   The principle of operation of the photoelectric-type controller, one model of which is shown in Fig. 180, is similar to that of the mechanical type since the galvanometer is the main controlling element, but it is different in that the relays are actuated by photo electric means rather than by a mechanical mechanism.   In the photoelectric controller the galvanometer boom is replaced by a small circular mirror onto which a beam of light is directed from a light-projection tube. The reflected beam of light from the mirror is directed toward a photoelectric tube of the cesium-coated vacuum type.   When the potentiometer circuit is unbalanced, *e.g.*, when the furnace is cold, the galvanometer is deflected and the beam of light reflected from the mirror is incident on the plate of the photoelectric tube. The small electric current induced by light excitation of the photoelectric cell is amplified by an amplifying circuit, which in turn closes a relay and in effect closes the external heating circuit of the furnace. When the preset temperature of the furnace is reached, the galvanometer comes to rest at nearly zero deflection, with the result that the beam of reflected light is moved off the sensitized photoelectric plate. This permits the controller relay to open and in effect opens the power-input circuit of the furnace.

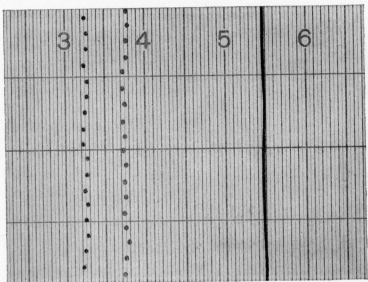

FIG. 181. Section of chart paper from a temperature recorder illustrating the unbroken temperature curve from a single-point recorder (right), and printed-dot curves from two points of a multiple-point recorder (left).

Temperature Recorders and Controllers. By the addition of auxiliary electrical circuits, reversing motors, and other equipment, all types of controllers described may be made to record automatically the temperature indicated by the temperature-indicating circuit. Single-point recorders, or recorder-controllers, trace an unbroken temperature curve upon a roll of graduated chart paper moving at a definite rate beneath an inked stylus attached to the temperature-indicating pointer of the instrument. This is illustrated in Fig. 181.

In multiple-point instruments, one model of which is shown in Fig. 182, the individual temperatures of several sources are recorded on chart paper by printed dots of various colors, numbered dots of the same color, or distinguishing symbols. The printing mechanism consists of a revolving printing wheel that is positioned along the scale to correspond to the temperature of the particular thermocouple whose temperature is to be recorded.   Each printing symbol or colored dot on the periphery of the wheel corresponds to one of the thermocouples, serving as a means of identification if more than one thermocouple is involved.   When the symbol is printed, an automatically operated commutating switch actuates a mechanism that revolves the printing wheel to the next symbol, corresponding to another thermocouple that is now connected into the recording circuit.   The printing wheel then moves along the scale to the temperature

Fig. 182.   A potentiometric type of temperature controller and recorder.   Multiple-point instrument.   (*Courtesy of Leeds & Northrup Co.*)

corresponding to that of the thermocouple now involved, and the printing cycle repeated.   This sequence of operations is repeated for as many thermocouples as the instrument is designed to handle.

For applications where fast-changing temperatures must be recorded or controlled with split-second speed or in cases where temperatures from several sources must be recorded on one chart in very rapid succession, special high-speed recorders and controllers are available.   These instruments, one model of which is illustrated in Fig. 183, are designed as either single- or multiple-point recorders, and are appropriately scaled to indicate and record degrees of temperatures directly from conventional types of thermocouples.   The main controlling circuit in these instruments is the conventional potentiometer circuit; and the high response to temperature changes of the thermocouple and the high over-all sensitivity of these instruments is brought about by an electrical balancing system consisting of an electronic converter-amplifier unit and a two-phase balancing motor.

SUGGESTIONS ON THE PROPER USE OF THERMOELECTRIC PYROMETERS

Thermocouples

1. When thermocouples are to be used in a gas-fired furnace, they should never be installed in the direct path of the flame.

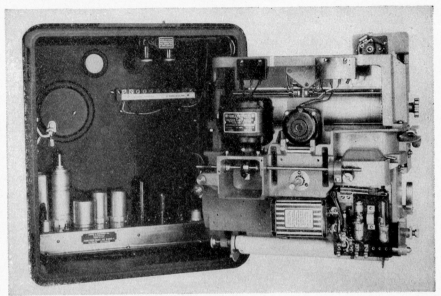

FIG. 183. A multiple-point, high-speed temperature controller and recorder. (*Courtesy of Leeds & Northrup Co.*)

2. Always insert control thermocouples into a furnace so that they will measure the average temperature of the furnace. A second thermocouple, placed with its hot junction on the charge of interest within the furnace, should be used for precise temperature check purposes.

3. Horizontally installed base-metal thermocouples should be supported at their ends if inserted more than 12 in. into the furnace. This is to prevent sagging. Noble-metal couples should be supported, for the same reason, if inserted more than 6 in.

4. Be sure the thermocouple is adequately protected with a protecting tube if used at high temperatures or in a corroding atmosphere. Porcelain-protected thermocouples must be heated slowly to prevent cracking.

5. Errors in temperature measurements due to thermocouples may result from

    a. Short circuit of the thermocouple wires
    b. Broken hot-junction connection
    c. Corroded thermocouple elements
    d. Loose cold-junction terminals
    e. Improper immersion into the heated medium
    f. Varying cold-junction temperatures not compensated for
    g. Large lag in temperature indications may be due to improper type of protection tube

Indicating Instruments

1. Have the instrument mounted where it will not be affected by vibrations.

2. Be sure thermocouple lead wires are tightly connected to the appropriate binding posts of the instrument.

3. If a potentiometer is used, be sure that the circuit is properly balanced.

4. Adjust the cold-junction compensator if the instrument is not equipped with an automatic compensating device.

5. Never allow the instrument to become heated. Always use extension leads long enough so that the instrument may be well away from the heated source.

6. Errors in temperature measurements due to the indicating instrument may arise from

    a. Calibration of the instrument having changed
    b. Loose terminal connections
    c. Short circuit in the instrument
    d. Cold-junction compensator having been accidentally moved
    e. Parts of the instrument functioning improperly, such as the indicating needle dragging on the scale, etc.

Extension Lead Wires

1. Be sure that appropriate extension leads have been chosen; *i.e.,* the leads should have approximately the same thermal emf characteristics as the thermocouple elements.

2. Tighten and clean electrical connections, both at the thermocouple and at the indicating instrument (see Item 4).

3. Be sure that lead wires are not partially broken, as this condition will increase the resistance of the thermocouple circuit.

4. All connections should be soldered, except where binding posts are provided for connecting purposes.

RESISTANCE THERMOMETERS

A resistance thermometer, illustrated in Fig. 184, is a temperature-measuring device well suited for precise temperature measurements from about −200 to 600°C. With appropriate resistance-measuring instruments, it is not uncommon to measure temperatures with this device to an accuracy of 0.001°C. By international agreement, the platinum resistance thermometer has been designated as the instrument to define the International Temperature Scale (thermodynamic centigrade scale) from −190 to 660°C.

A resistance thermometer consists of a high-purity, uniform, resistance wire wound into the form of a coil and supported usually on mica insulating supports. The resistance coil, referred to as the bulb, or thermometer, element, is appropriately mounted within a closed protection tube of either metal, glass, quartz, or porcelain. Lead wires from the element are extended through the protection tube and terminate at binding posts mounted on a junction head. Appropriate lead wires can then be connected from the junction-head terminals to a suitable resistance-measuring instrument.

Fig. 184. Illustrating schematically the construction of a resistance thermometer with three lead wires for compensation purposes (see page 356).

The wire from which the thermometer element is made must be of high purity and possess a constant and reproducible electrical resistance value at any given temperature. As will be shown later, the purity of the material and the over-all quality of thermometer element is best defined by numerical values deduced from certain constants of the temperature-resistance relationship. Thermometer elements are usually constructed of platinum wire, owing to the desirable characteristics of this metal for the purpose intended, although at times the element is made from nickel, copper, or palladium. The approximate temperature limits within which each of these thermometers may be safely used are as follows:

Type of thermometer	Degrees centigrade	Degrees Fahrenheit
Platinum...................	−250 to 600	−418 to 1112
Nickel.....................	−100 to 200	−148 to 392
Copper.....................	−100 to 200	−148 to 392
Palladium..................	−200 to 600	−328 to 1112

In the use of any of the above resistance thermometers, it is important not to exceed by very much the upper limit of temperature given.   Often platinum thermometers are inadvertently used at temperatures approaching 1000°C (1832°F).   Exposure to such high temperatures, even for relatively short times, will shorten the useful life of the thermometer and, owing to some volatilization of the platinum, will invalidate the initial calibration of the instrument.

Principles.   The principle of the resistance thermometer is based upon the well-known fact that the electrical resistance of a metal changes with change in temperature.   If this relationship between resistance and temperature is calibrated and appropriately corrected to conform to the thermodynamic centigrade scale, it is then possible to secure an indication of the true temperature of the thermometer element by measuring critically its electrical resistance at the temperature of interest.

By determining the resistance of a platinum thermometer at two known fixed points—the normal melting point of ice (the ice point) and the normal boiling point of water (the boiling point)—the two resistance values will define the slope of a temperature-resistance curve.   The relationship will be linear, and may be expressed by the equation

$$R_t = R_o(1 + \alpha_o t_{Pt}) \tag{5}$$

where R_t = resistance in ohms of the platinum element at any temperature t_{Pt}

R_o = resistance in ohms of the platinum element at the normal melting temperature of ice

α_o = zero temperature coefficient defining the fundamental temperature interval

t_{Pt} = temperature, °C, as indicated by this linear function between temperature and resistance.   This temperature is referred to as the platinum temperature.

The temperature coefficient α_o may be expressed as

$$\alpha_o = \frac{R_{100} - R_o}{100\,R_o} = 0.003915 \text{ for high-purity platinum}$$

where R_{100} = resistance in ohms of the platinum element at the normal boiling point of water

R_o = resistance in ohms of the platinum element at the normal melting point of ice

Thus, Eq. (5) may be written

$$t_{Pt} = \frac{100\,(R_t - R_o)}{R_{100} - R_o} \tag{6}$$

Unfortunately, the temperature as related to measured resistance in Eqs. (5) or (6) does not correspond to the true, or thermodynamic centigrade, temperature scale, except at 0 and 100°C. Between 0 and 100°C, the platinum temperature (t_{Pt}) is greater than the true temperature, and above 100°C the platinum temperature is less. It is evident, therefore, that if temperature indications are to be referred to the thermodynamic centigrade scale, these equations are not sufficiently critical for defining the resistance-true temperature relationship.

A closer approach to a relationship between measured resistance and true temperature is given by the following equation, which defines the International Temperature Scale from the normal melting point of ice (0°C) to 660°C:

$$R_t = R_o (1 + At + Bt^2) \tag{7}$$

where R_t = resistance in ohms of the platinum element at any temperature t

R_t = resistance in ohms of the platinum element at the ice point

t = true temperature, °C

A and B = constants related to each individual platinum thermometer

The value of R_o is determined by measuring the resistance of the platinum element at the ice point; and the constants A and B by simultaneous solution of Eq. (7) after measuring the resistances at the normal boiling points of water and sulphur, respectively. This calibration then permits Eq. (7) to be used for determining the true temperature at any measured resistance, and for any given platinum resistance thermometer such data may be plotted in the form of a calibration curve.

To define adequately the purity and physical condition of the platinum wire of which the thermometer element is composed, the ratio R_t to R_o [Eq. (7)] should not be less than 1.391 at the normal boiling point of water, nor less than 2.645 at the normal boiling point of sulphur. By appropriate calibration as described, the values of A and B for high-purity platinium are very close to 0.00392 and -0.000000588, respectively.

To correct for the departure of the platinum temperature as defined by Eq. (5) from the true temperature [Eq. (7)], at temperatures other than the normal ice point and normal boiling point of water, Callendar[1] proposed an equation whereby the true temperature could be calculated by a series of approximations. The constants of the Callendar equation fully define the over-all quality of the thermometer element in relation to purity, construction, etc., which is not so apparent from the constants of Eq. (7).

[1] Callendar, H. L.: On the Practical Measurement of Temperature, *Phil. Trans.* (*London*), Vol. 178, 1887.

Callendar Equation for Temperatures above 0°C. The Callendar equation, which is related mathematically to Eq. (7), may be expressed in one form as

$$t - \frac{100\,(R_t - R_o)}{R_{100} - R_o} = \delta\left[\left(\frac{t}{100}\right)^2 - \frac{t}{100}\right] \tag{8}$$

In terms of platinum temperature,

$$t - t_{Pt} = \delta\left[\left(\frac{t}{100}\right)^2 - \frac{t}{100}\right] \tag{9}$$

where t = true temperature, °C

t_{Pt} = platinum temperature, °C

R_o and R_{100} = resistance in ohms of the platinum element at the normal ice point and boiling point of water, respectively

δ = thermometer constant determined at the normal boiling point of sulphur. The value of δ may be expected to be between 1.49 and 1.50 for high-purity platinum

In order to use Eq. (9), the value of t_{Pt} as indicated by the resistance-measuring instrument is substituted for t in the right-hand side of the equation, and the correction factor $(t - t_{Pt})$ computed. This correction factor is then added to the original t_{Pt} value, and the corrected temperature is again substituted in Eq. (9) for temperature t. This redetermination will give a correction factor, which when added to t_{Pt} will be closer to the true temperature than that arrived at by the first computation. About three such substitutions will be sufficient to render a temperature value very close to the true temperature.

Equations for Temperatures below 0°C. Although Eqs. (7) and (9) are intended only for use in computing temperatures above the ice point, the Callendar equation is nevertheless used quite frequently for determining temperatures approaching −40°C. This extended use of the Callendar equation is not without error, but in the range of 0 to −40°C the error involved is rather small and may, in general, be neglected. However, to determine the temperature more precisely below 0°C, and in particular, to define the thermodynamic temperature scale from 0 to −190°C, the following equation is appropriate:

$$R_t = R_o\,[1 + At + Bt^2 + C\,(t - 100)t^3] \tag{10}$$

where R_t and R_o are the same as in Eq. (7) and A, B, C are constants of the thermometer which are determined at the normal boiling points of water, sulphur, and oxygen, respectively.

To define the purity, physical characteristics, and suitability for use of the platinum element, the ratio of R_t to R_o at the boiling point of oxygen should be less than 0.250.

On the basis of the original Callendar equation, Van Dusen[1] has proposed the following equation for temperatures below that for which Eq. (9) cannot be applied with certainty:

$$t - t_{Pt} = \delta \left[\left(\frac{t}{100} \right)^2 - \frac{t}{100} \right] + \beta \left[\left(\frac{t}{100} \right)^4 - \left(\frac{t}{100} \right)^3 \right] \qquad (11)$$

where t, t_{Pt}, and δ are the same as defined in Eq. (9), and β is a thermometer constant determined at the normal boiling point of oxygen.

To determine the true temperature by the use of Eq. (11), after the constants have been determined, appropriate substitutions are made in the right-hand member of the equation as described in the use of Eq. (9).

Resistance-measuring Methods. Perhaps the most widely used instrument for measuring the resistance of a thermometer element is the Wheatstone bridge. Bridges appropriate for this purpose must be capable of a precision of at least one part in a million, and the accuracy of the measurement must be sufficiently great so that differences of any two resistances, each observed to one part in a million, will have real significance. Wheatstone bridges usually indicate resistance in ohms, although some bridges designed especially for resistance thermometry indicate the temperature directly in degrees.

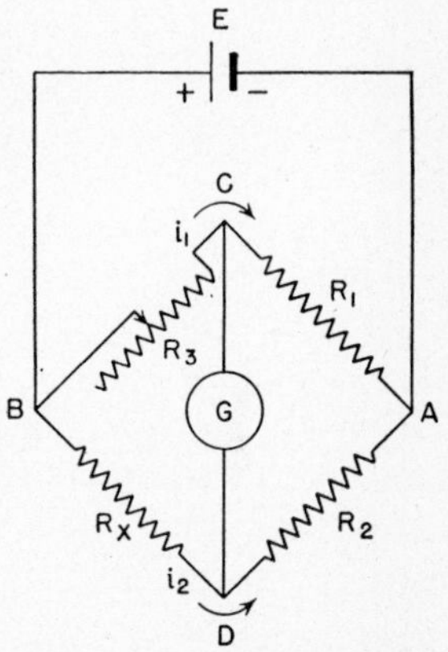

Fig. 185. Illustrating the principle of the simple Wheatstone bridge.

Wheatstone-bridge Principle. The principle of the simple Wheatstone bridge is illustrated in Fig. 185. Resistances R_1 and R_2 are of some fixed value, generally in the ratio of 1:1. R_x is the unknown resistance to be measured, corresponding to the resistance thermometer, and R_3 is a variable resistance. Two corners of the bridge, A and B, are connected to a battery E.

It is evident from Fig. 185 that the fall of potential along branch BCA is the same as that along BDA. Since the potentials of points C and D are intermediate between those of B and A, it is possible to adjust resistance R_3 so that the potential of point C is made to equal the potential of point D. When this equality in potentials is established, the bridge circuit is balanced and no current flows through the galvanometer, as indicated by zero deflection. Thus

$$E_{R_3} = E_{R_x}$$

and

$$E_{R_1} = E_{R_2}$$

[1] Van Dusen, M. S.: Platinum Resistance Thermometry at Low Temperatures, *J. Am. Chem. Soc.*, Vol. 47, 1925.

Let the current flowing through R_3 and $R_1 = i_1$ and the current through R_x and $R_2 = i_2$.

Then by Ohm's law,

$$R_3 i_1 = R_x i_2$$

and

$$R_1 i_1 = R_2 i_2$$

Whence

$$\frac{R_3}{R_1} = \frac{R_x}{R_2} \quad \text{or} \quad R_x = R_3 \frac{R_2}{R_1} \tag{12}$$

Since by adjustment or by construction of the instrument, $R_2 = R_1$ or the two are maintained in a known ratio, and R_3 is known by reading the instrument dial, the unknown resistance R_x is readily computed.

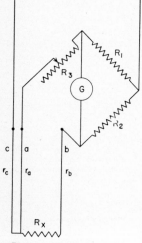

It is evident that if the thermometer is located at a considerable distance from the bridge, correct measurements will not be obtained unless the resistance of the lead wires are adjusted to a fixed value or compensated for by other means. The varying resistance of the lead wires, due to temperature changes along their lengths, may be compensated for by the three- or four-wire compensation methods to be described.

Three-wire Compensation Method. The three-wire compensation method for variation in resistance of the lead wires due to temperature is applicable to thermometers whose lead wires from the thermometer element to the junction head, and those from the junction head to the terminals of the Wheatstone bridge, are matched, *i.e.*, they are identical with regard to resistance. As shown in Fig. 186, two of the matched leads are connected to one end of the thermometer element; the third is connected to the other end of the element.

Fig. 186. Illustrating the principle of the three-wire compensation method in resistance thermometry.

If r_a, r_b, and r_c are the resistances of lead wires a, b, and c, respectively, it is evident that r_b is in series with R_x (resistance of thermometer element) and r_a with R_3. When the bridge is balanced by appropriate adjustment of R_3,

$$\frac{R_3 + r_a}{R_1} = \frac{R_x + r_b}{R_2} \tag{13}$$

By construction of the instrument, or through adjustment, $R_1 = R_2$. Hence

$$R_x = R_3 + r_a - r_b \tag{14}$$

Since the resistance characteristics of the lead wires are identical, then

$$R_x = R_3 \tag{15}$$

It is evident that by this method of compensation any variation in resistance of the lead wires produces no errors, since any change in one lead is compensated for by a similar change in the other.

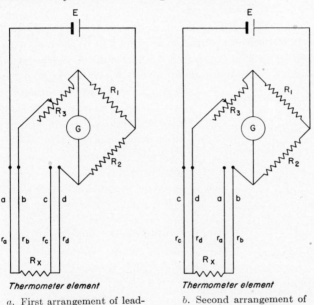

a. First arrangement of lead-wire connections.

b. Second arrangement of lead-wire connections.

Fig. 187. Illustrating the principle of the four-wire compensation method for precision measurements in resistance thermometry.

Four-wire Compensation Method. The four-wire compensating method is used mainly for precision measurements, and in cases where matched lead wires for any reason are not used. The method requires two resistance measurements of the thermometer element with the lead wires connected differently to the instrument for each measurement. The change in order of connections can be readily achieved by means of a multiple-pole, double-throw switch.

As illustrated in Fig. 187*a*, the four-wire method consists of attaching two lead wires of unmatched or unknown resistances to each side of the thermometer element R_x. In the first order of connections as indicated, only three lead wires a, b, and d are introduced into the bridge circuit.

Lead wire c is left idle and is only used in the second arrangement of connections to be described.

If r_a, r_b, r_c, and r_d are the resistances of lead wires a, b, c, and d, respectively, then when the bridge is appropriately balanced by adjustment of R_3,

$$\frac{R_3 + r_b}{R_1} = \frac{R_x + r_d}{R_2} \tag{16}$$

By construction of the instrument, $R_1 = R_2$. Hence, Eq. (16) may be simplified to

$$R_x = R_3 + r_b - r_d \tag{17}$$

After the first determination, or appropriate recording of R_3, the thermometer element is then connected to the bridge as indicated in Fig. 187*b*. This second order of connections involves the use of wire c instead of a as a battery lead, and arranges r_b in series with R_x and r_d with R_3—a state of affairs directly opposed to the first arrangement illustrated in Fig. 187*a*.

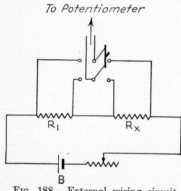

When the bridge is again balanced by adjustment of R_3,

$$\frac{R_3' + r_d}{R_1} = \frac{R_x + r_b}{R_2} \tag{18}$$

Simplified as before,

$$R_x = R_3' + r_d - r_b \tag{19}$$

Adding Eqs. (17) and (19),

$$R_x = \frac{R_3 + R_3'}{2} \tag{20}$$

Fig. 188. External wiring circuit for measuring resistances with a potentiometer.

It is evident, therefore, that by this method of compensation the value of R_x (resistance of the thermometer element) is independent of the resistance of the lead wires.

Potentiometer Method. By the use of a standard calibrated external resistance appropriately connected in the thermometer circuit, a potentiometer may be used to measure the resistance of the thermometer element. This potentiometric method is equally as precise as other measuring methods, but it is generally less convenient to use. The method is limited, however, to only such cases where the heating effects, arising from passing a small current through the thermometer element when a measurement is made, are negligible and insignificant.

As illustrated in Fig. 188, a standard resistance of known value is connected in series with the thermometer element R_x. The same current, by this circuit arrangement, passes through both resistances from an external battery B. If the fall of potential across each resistance is individually measured with the potentiometer, then from Ohm's law

$$E_1 = IR_1 \tag{21}$$

and

$$E_x = IR_x \tag{22}$$

whence

$$\frac{E_1}{R_1} = \frac{E_x}{R_x} \tag{23}$$

Since E_1 and E_x are known by measurements and R_1 is known by previous calibration, R_x may be readily computed. It is evident that by this method any variation in the resistance of the lead wires will not introduce errors into the final determination of R_x.

OPTICAL PYROMETERS

An optical pyrometer is essentially a photometer which singles out one particular narrow band of visible radiation that is being emitted by an incandescent body and compares the intensity of this radiation with the intensity of similar radiation from a standard calibrated source. The radiations so compared are generally confined to a narrow band of nearly monochromatic red light having an average wave length of 0.65 micron. The purpose in selecting red light upon which to make the comparison is that optical filters which transmit red produce a more nearly perfect monochromatic radiation, without appreciable loss in intensity, than can possibly be obtained from filters transmitting other colors. The importance of using monochromatic radiation when comparing intensities will become apparent.

It was shown in the temperature scale of colors, page 327, that as the temperature of a metal increases the tint or color of the radiating source becomes more brilliant and intense, owing mainly to the presence of shorter wave lengths of radiation. The intensity of different glowing materials in the open, however, is not the same even at the same temperature; it is a circumstance that depends largely upon the particular kind of material constituting the radiating source. For example, if carbon and platinum are heated to the same high temperature, the brightness of the carbon in the open will appear to be about three times that of the platinum. In other words, the rate at which radiant energy is being emitted by carbon is greater for all wave lengths of radiation than the rate of emission by platinum at the same temperature. In terms of comparing these rates of emission to some standard radiating source—a black body to be described —it may be said that the emissivity of carbon is greater at all wave lengths of radiation than is the emissivity of platinum.

It is apparent, therefore, that in order to determine temperatures of hot bodies by comparison of intensities of radiation, the emissivity of each particular material of interest must be appropriately considered. If the comparison is based on a given wave length of radiation, as in optical pyrometry, then the so-called partial or spectral emissivity of any given material must be known for the wave length of radiation involved. On the other hand, if the basis of comparison involves all wave lengths of radiation, as in total radiation pyrometry, then the total emissivity must be known. The numerical values of spectral and total emissivities are different for the same material at the same temperature.

In order, therefore, to evaluate the emissivity of a material, it is necessary to refer the emissivity characteristics to a selected standard source. An acceptable standard for this purpose is a black body, which at all wave lengths of radiation has maximum emissivity and to which an emissivity value (both spectral and total) of 1 has arbitrarily been assigned. All other materials, or radiating conditions, which are not black bodies have emissivity values of less than 1.

Black Body. The basic principles associated with optical and total radiation pyrometry involves the establishment and appropriate conception of a standard radiating source, known as a black body. A black body is defined as a body that absorbs radiation of all wave lengths made incident upon it, none of which is transmitted or reflected. The radiation from such an ideal source is due to temperature only and is not affected by outside radiation or by the material of which the black body is composed. All other bodies, known as nonblack bodies, or gray bodies, absorb only a part of the incident radiation, the remainder being either transmitted or reflected, or both.

A black body may be produced experimentally by uniformly heating the walls of a hollow enclosure and noting the radiation through a very small hole in the side. Although the interior of the enclosure may be incandescent and may not appear black in the common sense of the word, nevertheless black-body conditions prevail because all the radiation that is being emitted is absorbed by the walls. Approximate black-body conditions are attained in many technical processes requiring high operating temperatures. This may be demonstrated by viewing the interior of a glowing furnace through a small peephole. If black-body conditions are effected within the furnace the details of the interior, such as corners and edges, will be indistinguishable from the general glowing background.

Inasmuch as black bodies are perfect absorbers (from definition), they are, therefore, perfect radiators. If this were not so, then the black bodies would become hotter than their surroundings, which, of course, would be contrary to the second law of thermodynamics.

Optical pyrometers generally are calibrated to indicate black-body temperatures, *i.e.*, true temperatures when sighted on black bodies. Therefore, the apparent temperatures which are indicated when sighted on nonblack bodies, such as glowing materials in the open, will always be lower than the true temperatures. This difference, however, is due mainly to the emissivity of the material and, as will be shown later, true temperatures may be computed once the emissivity of interest is known.

These same nonblack-body materials, however, when placed within a black body or a glowing furnace and allowed to reach temperature equilibrium, will, in all respects, simulate black-body conditions. That is, the total intensity of radiation leaving the material will be the same as that emitted by the black body. If a pyrometer is now sighted on the material, the true temperature will be indicated directly. This apparent paradox is due to the fact that, although the material emits only a certain proportion of the total radiation of a black body, the radiation falling upon it from the furnace walls is reflected, with the net result that the material appears to be of the same radiating intensity as its surroundings.

Emissivity. The emissivity of an opaque material may be defined as the ratio of the rate at which radiant energy is emitted from the material as a consequence of temperature only, to the corresponding rate for that emitted from a black body at the same temperature. That is,

$$E_t = \frac{\delta}{\delta'} \tag{24}$$

where E_t = total emissivity
δ = rate of emission of radiant energy at all wave lengths
δ' = rate of emission of radiant energy at all wave lengths for a black body

In terms of monochromatic radiation,

$$E_\lambda = \frac{\delta_\lambda}{\delta'_\lambda} \tag{25}$$

where E_λ = partial or spectral emissivity at a wave length of radiation λ
δ_λ = rate of emission of radiant energy at a wave length λ
δ'_λ = rate of emission of radiant energy at a wave length λ for a black body

The radiation which is made incident upon an opague body may be either completely absorbed, as in the case of a black body, or it may be partially absorbed and the remainder reflected, as in the case of a nonblack body. The proportion of all wave lengths of radiation absorbed A, and the proportion reflected R, may be expressed as

$$A + R = 1 \tag{26}$$

For monochromatic radiation of wave length λ,

$$A_\lambda + R_\lambda = 1 \tag{27}$$

From definition of a black body, R and R_λ are equal to zero. Therefore, for a black body,

$$A' = A_\lambda = 1 \tag{28}$$

The absorbing power of an opaque body for all wave lengths of radiation may be defined as the ratio of the rate at which radiant energy is absorbed by the body to the rate at which radiant energy is made incident upon it. For all opaque bodies, the ratio of the rate of emission of all wave lengths of radiation to the absorbing power of the body is equal to some function of temperature. That is,

$$\frac{\delta}{A} = f(T) \tag{29}$$

where δ = rate of emission of radiation at all wave lengths
A = rate of absorption of radiation at all wave lengths
For monochromatic radiation at a wave length λ,

$$\frac{\delta_\lambda}{A_\lambda} = f(T, \lambda) \tag{30}$$

In terms of a black body, Eqs. (29) and (30) may be expressed as

$$\frac{\delta'}{A'} = f(T) \tag{31}$$

and

$$\frac{\delta_\lambda'}{A_\lambda'} = f(T, \lambda) \tag{32}$$

For a black body, Eqs. (29) and (31) and (30) and (32) may be equated to give

$$\frac{\delta}{\delta'} = \frac{A}{A'} \tag{33}$$

and

$$\frac{\delta_\lambda}{\delta_\lambda'} = \frac{A_\lambda}{A_\lambda'} \tag{34}$$

and Eqs. (24) and (33), and (25) and (34) may be equated and expressed as

$$E_t = \frac{A}{A'} \tag{35}$$

and

$$E_\lambda = \frac{A_\lambda}{A_\lambda'} \tag{36}$$

Since A' and A_λ' for a black body are equal to 1, then the relationship between emissivity and absorbing power for any body, either at all wave lengths of radiation or at a wave length λ, may be expressed, respectively, as

$$E_t = A \tag{37}$$

and

$$E_\lambda = A_\lambda \tag{38}$$

It is evident, therefore, that the absorbing power of any body is equal to its emissivity at the same temperature and for either all wave lengths of radiation or some selected wave length. Since the absorbing power of a black body represents a maximum (value of 1), then its emissivity must be a maximum as compared to any other body.

It has been observed that the spectral emissivity of a material changes slightly with temperature, and often greatly with wave length of radiation. The exact variation with temperature is not known with certainty, but because the change is small, the influence of temperature can generally be disregarded within limits for all practical purposes. Because the spectral-emissivity values of a material differ at different wave lengths of radiation, it is essential to select emissivity values in terms of the wave length of interest. In optical pyrometry, the wave length of radiation upon which intensities are matched is about 0.65 micron, equivalent to that of visible red radiation. In total radiation pyrometry, the interest lies in total emissivity values, since all wave lengths of radiation are involved.

The spectral-emissivity values for several materials, at a wave length of radiation of 0.65 micron, are given in Tables 55 and 56, pages 490 and 491; and the total emissivity values at different temperatures in Tables 57 and 58, pages 492 and 493.

Wien's Law for the Distribution of Energy. The temperature scale of all optical pyrometers up to 3000°C (5432°F) is based on Wien's distribution law for the energy of a black body. It has been shown that Planck's distribution law is more rigid and exact, but Wien's law is more commonly used because it is the simpler of the two to apply. Furthermore, in the region of monochromatic red, the difference between the Planck and Wien equations is small and, ordinarily, can be neglected.

Wien's distribution law for black bodies as applied to optical pyrometry may be formulated as

$$J_\lambda = c_1 \lambda^{-5} e^{-\frac{c_2}{\lambda T}} \tag{39}$$

where J_λ = intensity of radiation of wave length λ

c_1 = constant which disappears from the final working equation

λ = wave length of radiation—generally red for all optical pyrometers $\cong$ 0.65 micron

e = base of the Napierian logarithm system

c_2 = constant of a value between 14,320 and 14,360 micron deg

T = true absolute temperature of the black body

Wien's equation has no practical significance in the calibration of an optical pyrometer below 1500°C (2732°F), but above this temperature, where the true melting points of materials are unknown, the equation affords a means of extrapolation.

The rate at which radiant energy is emitted from a source at constant temperature and wave length depends upon the emissivity of the material. Therefore, for nonblack bodies, the emissivity of the material for a particular wave length must be included in Wien's equation. Thus for a nonblack body

$$J_\lambda = c_1 E_\lambda \lambda^{-5} e^{-\frac{c_2}{\lambda T}} = c_1 \lambda^{-5} e^{-\frac{c_2}{\lambda T_a}} \tag{40}$$

The term T_a is the apparent absolute temperature of the nonblack body and it is that temperature which is measured by the optical pyrometer. The value of T_a is always less than the true temperature T for any material except for a black body, where, in this case, it is equivalent to T.

Using a value of c_2 equal to 14,320, Eq. (40) may be rearranged and put into the following useful form:

$$\frac{1}{T} - \frac{1}{T_a} = \frac{\lambda \log E_\lambda}{6219} \tag{41}$$

Thus, if the emissivity of a material of interest is known, or determined by reference to Tables 55 or 56, the true temperature of the material may be readily computed from the apparent temperature T_a, as indicated by the optical pyrometer. Alternate to computing the true temperature after each measurement, and particularly so when the accuracy of interpolation need not be too precise, the data in Table 59, page 494, shows the true temperatures corresponding to various observed or apparent temperatures at different spectral emissivity values.

Types of Optical Pyrometers. Optical pyrometers may function on either of the following two principles:

1. Where the intensity of radiation from a calibrated source is varied and made to equal the intensity of radiation from the source whose temperature is to be measured

2. Where the intensity of radiation from the radiating source is varied in the pyrometer by means of an optical wedge and made to equal the fixed intensity of radiation from a calibrated source

There are numerous pyrometers which operate by each of these principles, but only one of the more important types in each group will be discussed.

Leeds and Northrup Milliammeter Type. The Leeds and Northrup pyrometer, a schematic section of which is shown in Fig. 189, operates by the first principle presented before. The pyrometer is sighted upon the object whose temperature is to be measured and is pointed towards the

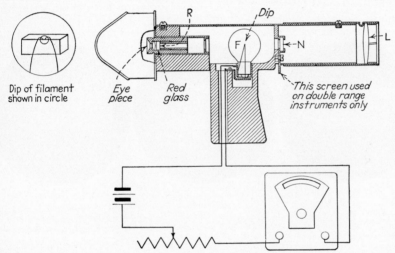

Dip of filament shown in circle   Eye piece   Red glass   This screen used on double range instruments only

FIG. 189. Illustrating the functioning principle of the milliammeter-type optical pyrometer.
(*Courtesy of Leeds & Northrup Co.*)

source so that the hot body is brought into the field of view. The visible radiation from the source is brought into focus by lens L, in the same plane as the dip of the tungsten lamp filament F. By adjustment of the current flowing through the lamp filament, the intensity of the filament dip is made to equal the intensity of the source. This comparison is made with visible red radiation ($\lambda = 0.65$ micron), secured by means of a light filter R set in the eyepiece of the instrument. When the intensity of the filament dip is appropriately matched to that of the radiating body, as shown in Fig. 190c, the filament current required to produce this matched condition is noted on the milliammeter. The black-body temperature is then determined subsequently by reference to calibration data or a calibration curve of milliamperes vs. black-body temperatures. If the radiating source departs from black-body conditions, then the true temperature may be computed by the use of Eq. (41), page 364, or, more simply, by reference to the data in Table 59, page 494.

At temperatures above 1400°C (2552°F), the intensity of radiation from the source is too blinding for direct observation, even though the radiation is first passed through the red filter R. Under these circumstances, a neutral filter N is inserted between the lens L and the image in the plane of the lamp filament. This neutral filter diminishes the intensity of radiation from the source by a known amount and, in effect, decreases by a definite proportion the value of J_λ in Wien's equation for black bodies. Since at these high temperatures the pyrometer cannot be calibrated by known melting points, calibration is made by extrapolation with Wien's equation. In the actual operation of the pyrometer, this extrapolation is made by

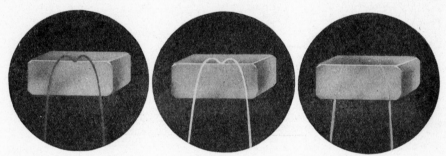

a. Temperature indication　　b. Temperature indication　　c. Temperature correctly
     too low.　　 too high.　　 indicated.

FIG. 190.　Appearance of the photometric field in the Leeds and Northrup milliammeter-type optical pyrometer (Fig. 189) when the instrument is adjusted to indicate a temperature higher, lower, or equivalent to the temperature of the radiating source.

including a special indicating scale on the milliammeter for use when the filter N is used.

Leeds and Northrup Potentiometric Type.　The potentiometric-type optical pyrometer, illustrated in Fig. 191, functions on the same "disappearing-filament" principle as the milliammeter type already described. It differs, however, in certain design features, and particularly in the method by which the lamp-filament current is measured, *i.e.*, by a potentiometric method rather than by a milliammeter. The matching lamp consists of a thin ribbon filament enclosed within an evacuated glass envelope having flat windows, instead of the former round, dipped filament within a spherical bulb. This improved lamp design has certain advantages for the purpose intended, among them being lower requirements in filament current; a perfect disappearance of the filament when a match is obtained with a radiating body; and negligible time required for the filament to become incandescent.

The diagram of the electrical and measuring circuit of the pyrometer is illustrated in Fig. 192. By means of a spring switch located on the hand-grip of the telescope unit, contacts a, b, and c are closed in the order given when the switch is actuated during sighting the pyrometer on the hot body of interest. This manipulation closes a simple series circuit, consisting of a 6-volt battery E, the lamp filament, and resistances R_1 and R_3 and parallel resistances R_4 and R_2. By adjustment of R_1, through rotation of the larger control knob illustrated in Fig. 191, the current flowing through the lamp filament may be critically adjusted so that the brightness of the filament corresponds to that of the hot body onto which the pyrometer is sighted. Through a mechanical clutch device, the slider of R_1 and R_2 (the slide-wire

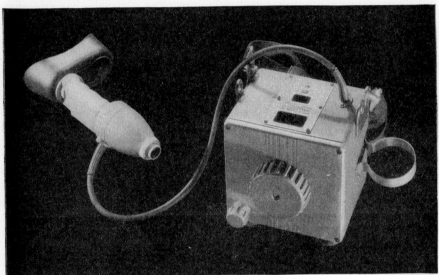

FIG. 191. The Leeds and Northrup potentiometric-type optical pyrometer. (*Courtesy of Leeds & Northrup Co.*)

of a potentiometer circuit) are so connected that when R_1 is adjusted to regulate the filament current, R_2 is simultaneously adjusted so that the drop in potential across R_3 and the included portion of R_2 (along st), when connected later into the circuit, is about equal to the opposing emf of the standard cell E'. This unique arrangement, brought about by appropriate resistance values of fixed resistances R_3 and R_4, always assures that the potentiometer circuit is roughly balanced as the filament current is altered.

When a brightness match has been secured, the larger control knob is left undisturbed. If, now, the smaller control knob, shown in Fig. 191, is pushed inwardly, the potentiometer circuit is closed through switch e (Fig. 192). This smaller control knob, in addition to actuating switch e, is coupled directly, and only, to the slider of the potentiometer slide-wire R_2.

By appropriate and precise adjustment of R_2 from its initial rough adjustment, the fall of potential across R_3 and that portion of the slide wire included in the circuit may be made equivalent to that of the standard cell E', as indicated by zero deflection of the galvanometer G. The fall of potential across R_3 and that portion of the slide wire included arises from the current flowing in the lamp-filament circuit, and will, of course, remain constant and unaltered so long as R_1 is not readjusted. The slider of R_2 is coupled to an indicating dial which is calibrated directly in degrees of temperature.

Standardization. Calibration of optical pyrometers that indicate the observed temperature in terms of milliamperes or corresponding emf, may be carried out by sighting the pyrometer on an incandescent source, not

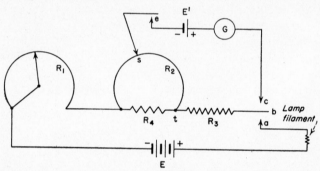

Fig. 192.   Illustrating the principle and electrical circuit of the potentiometric-type optical pyrometer.

necessarily a black body, and comparing its temperature measurements with those obtained by a calibrated pyrometer, utilizing the same wave length of radiation.

When a calibrated pyrometer is not available, calibration of the lamp filament may be made by comparing its intensity with the intensity of radiation at the known freezing points of four or more metals, experimentally arranged so that black-body conditions are approximated. The current required by the filament to produce a match is noted, and from a number of such determinations at different known freezing points, a calibration curve of current vs. black-body temperature may be plotted. The curve follows very closely the empirical relation

$$I = a + bt + ct^2 + dt^3 \qquad (42)$$

where I = current at temperature $t°C$
  t = black-body temperature, $°C$, of the freezing points of the metals used
a, b, c, d = constants

The constants may be determined by four standardization points. The metals generally used in this method of calibration are as follows:

Metal	Freezing point, °C
Aluminum	660.0
Copper-silver eutectic	779.0
Silver	960.5
Copper	1083.0

Another laboratory method of calibration is by the use of a broad-filament tungsten lamp that serves as a source of constant radiation. Such lamps may be obtained from pyrometer manufacturers. Each is calibrated so that the effective black-body temperature of the filament is known for all values of current required to make it incandescent. To carry out the calibration, the pyrometer is sighted on the filament of the lamp, which is already at a particular temperature, and the current is noted that is required by the pyrometer lamp to produce an intensity match. A series of such comparisons over the range of temperatures of the source filament will permit the plotting of a calibration curve, pyrometer lamp current in milliamperes vs. black-body temperature.

Because degrees of temperature are indicated directly by the potentiometric-type pyrometer, these instruments are best maintained in calibration by changing to a new filament lamp whenever it is suspected that the characteristics of the old lamp have changed. Each lamp is calibrated by the manufacturer with a separate shunt resistance (R_4 in Fig. 192) and with a definite setting of the temperature-indicating dial. Adjustment of the position of this dial about its axis can be made without disturbing the setting of the slider of resistance R_2. When a change of lamps is made, the appropriate shunt resistance (R_4) is installed in the circuit, and the setting of the temperature-indicator dial is made to correspond to that designated for the new lamp.

Fig. 193. The Pyro optical pyrometer. (*Courtesy of Pyrometer Instrument Co.*)

Pyro Optical Pyrometer. The Pyro optical pyrometer, shown in Fig. 193, is unique in that all the essential parts of the pyrometer are embodied in one compact unit, thus eliminating the ammeter, batteries, and connecting leads as separate auxiliary parts. The pyrometer operates on the principle of obtaining an intensity match by varying the intensity of radiation from the source to correspond to the fixed intensity of a filament lamp. As shown in Fig. 194, radiation from the source is collimated by an objective lens; is passed through an optical wedge of a density dependent upon its position in the optical axis; and finally through an optical prism and red filter, to the eye. The intensity of the source is made to match the intens-

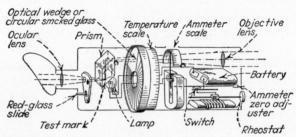

FIG. 194.   Illustrating the principle of the Pyro optical pyrometer. (*Courtesy of Pyrometer Instrument Co.*)

ity of the calibrated lamp filament by rotation of the optical wedge. By indications similar to those in the Leeds and Northrup pyrometer, a correct photometric match is obtained when the wedge is in such a position that the entire field of view is uniform, as shown in Fig. 195*a*. When such a condition is attained, the temperature of the source may be read directly on an indicating dial located on the side of the instrument.

a. Temperature cor-
rectly indicated.

b. Temperature indi-
cation too low.

c. Temperature indi-
cation too high.

FIG. 195.   Appearance of the photometric field in the Pyro optical pyrometer (Fig. 193) when the instrument is adjusted to indicate a temperature higher, lower, or equivalent to the temperature of the radiating source.

It can be shown that the intensity of radiation (I) emerging from the distant side of a partial absorber, such as an optical wedge, is equal to

$$I = I_o e^{-\mu t} \tag{43}$$

where I_o = intensity of radiation made incident on the absorbing medium
μ = coefficient of absorption
t = thickness of absorbing medium

The value of μ is equal to the reciprocal of that thickness of absorbing material that would reduce the intensity of the emergent radiation to 1/2.718 of the intensity of the incident radiation. Thus, in effect, the thickness or density of the optical wedge is so calibrated with the fixed intensity of the filament lamp that the temperature may be indicated directly, the numerical value depending upon the position of the wedge.

The filament lamp of each pyrometer is so calibrated that its intensity of radiation will be correctly related to the optical wedge when a definite amount of current is used. Generally this correct current value is given on the base of the instrument. Provision is made in the form of a rheostat which permits the regulation of this current to the correct value, as indicated by the ammeter located on the side of the instrument.

STANDARDIZATION. The Pyro pyrometer may be standardized by any of the methods described before. However, calibration is facilitated by the use of a master red-tip lamp, which is generally supplied with each instrument.

The master lamp is inserted in the instrument and the current is adjusted by the rheostat to the correct current value for the master lamp. The pyrometer is then sighted on a hot body of constant temperature and the temperature is measured in the usual manner. Without change in any of the adjustments, the service lamp is then inserted and the pyrometer is sighted on the same source. In this instance, the photometric match is made, not by rotating the optical wedge, but by adjusting the current through the lamp by means of the rheostat. Thus, by noting the reading of the ammeter, a new current value of the service lamp is obtained, which is to be used for all subsequent temperature measurements until a further check shows that the standardization has changed.

Pyro Bioptical Pyrometer. It was shown in the past discussion that optical pyrometry is based in principle upon measurements of spectral radiant energy, and that temperatures can be determined by a measurement of the intensity of a narrow wave band of radiation being emitted by a hot body. The temperatures so determined may at times be in error owing to the surface condition of the radiating body. Both the intensity of radiation and the emissivity of a hot body are in part influenced by the condition of the radiating surface, whereas the color or spectral composition of the radiation is effected very little, if at all.

A measurement of the intensity conditions of the separate colors in the radiation will indicate the color temperature of the radiating body, which approaches more closely the actual temperature, with little regard to surface condition, than the temperature indicated by the ordinary optical pyrometer. If the emission differs considerably from that of the black body, the color temperature will indicate an upper limit.

Pyrometers designed primarily to measure color temperatures are known as color pyrometers.   The principle upon which they operate is based either on a direct comparison of the color of a standard lamp with that of the radiant body or, if the color lamp is not used, on matching with a mixed color of two different colors.   The separation of colors from the radiating source whose intensity conditions serve for the measurement of the temperature is essentially brought about either by a prismatic analysis or by the use of color filters.   The separated colors are then remixed by an optical apparatus.

The Pyro bioptical pyrometer shown in Fig. 196, is essentially an optical and color pyrometer combined in the same unit, permitting both the color temperature and the apparent temperature to be measured simultaneously.

As shown in Fig. 197, the radiation from the source a is passed into the instrument through an objective b, eyepiece f, and diaphragm g, to the observer's eye h.   Located near the objective in the optical axis of the pyrometer is an optical bichromatic color wedge c, through which the light rays must pass before reaching the observer's eye.   This wedge consists of two optical filters placed one behind the other.   The first filter, nearest the objective, is of plane form, which transmits only red and green wave lengths of the visible spectrum. Directly behind this first filter is placed a gradient filter, which is not of plane form and which transmits only a part of the colors

Fig. 196.   The Pyro bioptical pyrometer.   (*Courtesy of Pyrometer Instrument Co.*)

allowed to pass through the first filter, the remainder being absorbed to an extent which increases gradually or in stages.   Because of the variation or changes in the degree of absorption of different colors, a mixture of light rays is produced which has at all points a different spectral distribution and, therefore, a different color.   The colors which are transmitted by the first filter (red and green) merge at a particular point of the second filter to form a mixed or neutral color.   The part of the filter at which this mixed or neutral color is formed depends upon the spectral distribution of the radiation emitted by the radiating body, which is a function of its temperature.   The point at which the mixed color appears in the second filter may, therefore, serve as an index of the temperature of the source.

In the actual operation of the pyrometer, the mixing filter is rotated by an adjusting scale ring until the red and green colors transmitted by the first filter are remixed to form such a neutral color that it matches the standard color from the comparison lamp *i*.   As will be noted, a glass cube *e*, which is partly silvered and cemented diagonally, directs the colored light rays from the comparison lamp into the optical path of the principal beam. When a correct color match is obtained in the comparison field, as observed through the eyepiece of the instrument, the color or true temperature of the source may be read directly from a dial on the side of the instrument.

The apparent temperature of the source, *i.e.*, the temperature referred to the black-body temperature as measured with the ordinary optical pyrometer, is determined in the same manner as with the Pyro optical pyrometer previously discussed.   By rotation of a second adjusting scale ring, which is coupled to a neutral optical wedge *d*, the intensity of the radiation from

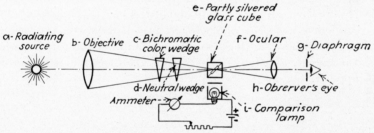

Fig. 197.   Illustrating the principle of the Pyro bioptical pyrometer. (*Courtesy of Pyrometer Instrument Co.*)

the source that is transmitted by the color filters is made to match the intensity of the radiation from the standard lamp.   When a perfect intensity match is obtained, the temperature may be read directly, as in the case of the color temperature, from a second indicating dial on the side of the instrument.

When a perfect visual match is obtained between the focused image of the radiating body and the comparison field with regard to color and intensity of radiation, the entire field of view appears homogeneous and the spectral energy distribution is therefore the same in both fields.

The color and intensity of radiation from the comparison lamp are maintained standard by suitable color filters and by accurate control of a definite amount of current, which is passed through the lamp filament.

Conditions Affecting the Operation of Optical Pyrometers.   Often conditions are encountered where certain factors—the presence of smoke and flame, reflected light, etc.—affect the pyrometer readings to such an extent that they are grossly in error.

Incandescent objects which are in the open usually do not approximate black bodies, and because of this the pyrometer indicates too low a temperature. Under such conditions, more accurate temperature measurements can be made by sighting the pyrometer into a cavity in the object, where more nearly black-body conditions exist. However, certain materials, such as carbon and a few oxides, approximate black-body conditions in the open to so great an extent that direct temperature measurements can be considered fairly reliable.

The presence of flame and smoke contributes to inaccurate temperature measurements. Generally, where smoke is present, the temperature readings are low, due to absorption of part of the radiation, whereas the presence of flame increases the temperature readings, because of reflected light. Where these conditions exist and cannot be eliminated, satisfactory temperature measurements can be obtained by placing the closed end of a refractory tube in the hot medium and sighting the pyrometer into the opened end. The closed end of the tube, if uniformly heated over a sufficient area, will closely approximate black-body conditions.

It is frequently necessary to sight an optical pyrometer into a furnace through a glass window. Glass absorbs a certain amount of the radiation passing through it, and for ordinary clean glass the following correction may be applied:[1]

$$\frac{1}{T} - \frac{1}{T_a} = -0.0000046 \qquad (44)$$

where T = true absolute temperature

T_a = apparent absolute temperature

This formula has been based on an average transmission coefficient of 0.904 for ordinary glass at a wave length of radiation of 0.65 micron.

TOTAL RADIATION PYROMETERS

Radiation pyrometers are temperature-measuring devices whereby the temperature of a radiating body may be determined by measuring the intensity of all wave lengths of radiation being emitted by the hot source. They differ from optical pyrometers in that a combination of the intensities of light and heat radiation is utilized, rather than the intensity of an approximate single wave length of visible radiation.

All total radiation pyrometers operate on the principle of focusing or concentrating radiant energy of all wave lengths upon a sensitive actuating element, causing an emf to be developed therein which is subsequently measured by a suitable emf-measuring instrument. This actuating element is usually a small, sensitive thermocouple appropriately shielded, or in some cases a thermopile, to be described.

[1] Foote, P. D., C. O. Fairchild, and T. R. Harrison: Pyrometric Practice, *Nat. Bur. Standards, Tech. Paper* 170, 1921.

Radiation pyrometers, like optical pyrometers, are calibrated to indicate true temperatures when sighted on black bodies. The principles of black bodies, nonblack bodies, and emissivities which were discussed previously under optical pyrometry are similarly applicable to the principles of radiation pyrometry and need not be repeated here.

Stefan-Boltzmann Radiation Law. The rise in temperature of the thermocouple in a radiation pyrometer is approximately proportional to the rate at which radiant energy is made incident upon it. The Stefan-Boltzmann law, which is the basis of the temperature scale for all radiation pyrometers, shows that this rate of incident energy from a black body is proportional to the fourth power of its absolute temperature. Thus

$$J = K(T^4 - T_1^4) \tag{45}$$

where J = rate at which energy is being emitted by a black body for all wave lengths of radiation

K = proportionality constant

T = absolute temperature of the black body

T_1 = absolute temperature of the surroundings

The value of T_1^4 is generally small as compared to T^4, and for all practical purposes may be neglected and dropped from the equation. Hence

$$J = KT^4 \tag{46}$$

In the case of nonblack bodies, the value of J as compared to its value for a black body will be decreased according to the value of the emissivity of the material. Thus

$$J' = KE_tT^4 = KT_a^4 \tag{47}$$

where E_t = total emissivity of the nonblack body for all wave lengths of radiation (see Tables 57 and 58, pages 492 and 493)

T_a = apparent absolute temperature of the nonblack body as indicated by the pyrometer

Equation (47) may be rearranged and expressed in the following logarithmic form:

$$T_a^4 = E_tT^4$$

or

$$4 \log T_a = \log E_t + 4 \log T \tag{48}$$

Thus, if the total emissivity of a material is known, the apparent temperature as indicated by the pyrometer may be readily corrected to the true temperature, equivalent to the temperature that would be read by the pyrometer if the material were under black-body conditions. Table 60, page 495, shows the relationship between true and apparent temperatures corresponding to various values of total emissivity.

In general, radiation pyrometers depart somewhat from the Stefan-Boltzmann law. This departure may be attributed to the fact that not all of the radiation which is received from the radiating source is effective in heating the thermocouple. There is always a certain amount of radiation which is lost within the instrument by absorption. Another contributing factor toward this departure is that the emf-temperature relationship of the thermocouple is not strictly a straight-line function and, therefore, the emf developed is not proportional to the difference in temperature of the hot and cold junctions.

Thermocouple Element. The thermocouple element is so located within the pyrometer that most of the radiation entering it from the source is focused onto the hot junction of the element. The cold junction is shielded from direct radiation, but nevertheless it is heated to a temperature proportional to the temperature of the hot junction, as no attempt is made to keep the cold-junction temperature constant. When temperature equilibrium of the thermocouple is attained, there is a direct differential between the hot and cold junctions. As long as this temperature difference remains constant and reproducible for a particular hot-junction temperature, then the emf developed by the thermocouple will be approximately proportional to the temperature of the radiating source.

It was shown that the rate at which radiant energy is emitted from a black body is proportional to the fourth power of the absolute temperature of the source. The temperature to which the hot junction of the thermocouple is heated (usually never higher than 100°C) is approximately proportional to the rate at which radiant energy is made incident upon it. Since the emf developed by the thermocouple is nearly proportional to the hot-junction temperature or the difference in temperature between the hot and cold junctions, then there exists a definite relation between the temperature of the radiating source and the emf developed by the thermocouple. This relationship is empirical and may be expressed as

$$e = aT^b \tag{49}$$

where e = emf developed by the thermocouple
T = absolute temperature of the radiating source
a, b = constants determined by two standardization points
This equation is particularly useful in the calibration of radiation pyrometers.

Types of Radiation Pyrometers. All radiation pyrometers consist essentially of two distinct parts: (1) a suitable arrangement whereby the radiation emitted by a hot body is focused upon the hot junction of the thermocouple, and (2) an emf-measuring instrument, such as a millivoltmeter or a potentiometer, to measure the emf developed by the thermocouple. The emf-measuring instrument may be contained within the pyrometer or may be, as a separate unit, connected to the pyrometer by extension lead wires.

The radiation emitted from the radiating source may be focused upon the hot junction of the thermocouple by means of a lens system, usually of quartz, or by means of a concave or a hollow conical mirror, or by a combination of both. The mirror-type instrument generally follows more closely the Stefan-Boltzmann law, because the absorption of radiation by the mirror is considerably less than the absorption by an optical quartz lens. Furthermore, the mirror will focus all wave lengths of radiation at the same point, whereas the lens, because of its dispersive power, will focus each particular wave length at successive points along its principal axis. The net result is that the radiation focused by the lens will not be so concentrated on the hot junction of the thermocouple as will the radiation which is focused by the mirror.

Mirror-type Radiation Pyrometer. The Thwing radiation pyrometer, a cross section of which is shown in Fig. 198, is of the hollow-conical-mirror type. The radiation from the source is collected by a dead-black receiving tube, which is vaned to prevent stray radiation from reaching the mirror.

FIG. 198.   Illustrating the construction and essential parts of the Thwing total-radiation pyrometer.   (*Courtesy of Thwing Instrument Co.*)

The radiation which is passed through the receiving tube is impinged upon a hollow conical mirror, where by multiple reflections the radiation is focused at the apex of the mirror. At this point the hot junction of the thermocouple is located. Any radiation that is passed around and not impinged upon the thermocouple element is reradiated from a small concave mirror located at the back of the thermocouple.

To make a temperature measurement, the pyrometer is merely pointed at the source of radiation and, after a few seconds' time to allow the thermocouple to reach temperature equilibrium, the temperature may be read directly from the emf-measuring instrument.

It can be shown by theoretical considerations that the Thwing pyrometer as well as other total-radiation pyrometers, will indicate correct temperatures, regardless of the distance between the radiating source and the pyrometer, provided that the area of the source is large enough to form the base of a solid cone defined as AOA' in Fig. 199, at a distance OA from the pyrometer. In practical cases, however, a limit of distance is reached because of significant amounts of radiation being absorbed by the atmosphere at greater distances.

It may be demonstrated that the radiant energy received by the pyrometer from a circular area of diameter AA', at a distance AO from the pyrometer, is equal to the energy received at a distance BO from a circular area of diameter BB'.   From the construction of the instrument and the diameter of the tube opening, the above conditions will be realized if the diameter of the source is approximately 1 in. for every 8 in. of distance between the

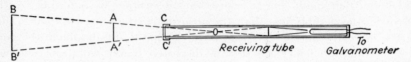

Fig. 199.   Illustrating the principle of the "distance factor" in the Thwing total-radiation pyrometer.

source and the tube end.   When the pyrometer is to be used at high temperatures, the effective aperture of the tube end is decreased by means of a diaphragm, thus permitting a corresponding increase in distance between the source and the pyrometer for the same radiating area.

Lens-type Radiation Pyrometer.   The Pyro radiation pyrometer, shown in Fig. 200, operates on the principle of concentrating the radiation upon the hot junction of a thermocouple by means of a quartz lens.   The instru-

Fig. 200.   The Pyro total-radiation pyrometer.   (*Courtesy of Pyrometer Instrument Co.*)

ment is unique in that the lens, thermocouple, and a millivoltmeter are all enclosed within a single compact housing.   As shown in Fig. 201, the thermocouple is enclosed in an evacuated glass bulb, the hot junction of which is located directly on the principal axis of the optical lens system.   The hot junction of the thermocouple is covered by a blackened receiving disk, to ensure maximum absorption of radiation made incident upon it.

When a temperature measurement is to be made, the instrument is merely sighted onto the radiating source and a button depressed to release the millivoltmeter pointer. When the thermocouple has reached temperature equilibrium, which requires only a few seconds of time, the button is released to lock the pointer of the millivoltmeter in its deflected position so that the emf may be subsequently noted when the instrument is lowered from the eye.

The distance factor associated with the design of the Pyro radiation pyrometer is approximately 15, *i.e.*, for every circular area of radiating surface of 1 in. diameter, the pyrometer may be at a distance of 15 in. from the source. Similar to the Thwing pyrometer, this distance may be increased within limits when the effective aperture of the objective lens is decreased to permit higher temperatures to be measured.

Thermopile-type Radiation Pyrometer. Although the conventional types of total-radiation pyrometers are well suited for use as portable temper-

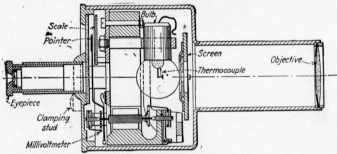

Fig. 201. Illustrating the principle of the Pyro total-radiation pyrometer. (*Courtesy of Pyrometer Instrument Co.*)

ature-measuring devices, the relatively small emf developed by the thermocouple element, even at high temperatures of the radiating source, precludes successful actuation of temperature controllers and recorders. Larger emfs may be developed for this purpose by designing the pyrometer to include a thermopile unit, instead of a single thermocouple element. A thermopile consists of two or more thermocouples of appropriate design that are connected together in series, and so arranged physically that all of the hot junctions are located at about the same point. Radiation pyrometers employing a thermopile as the actuating element function by precisely the same principles associated with other radiation pyrometers, and the emf developed by the unit may be measured by a potentiometer, a millivoltmeter, or an indicating controller or recorder.

The Rayotube radiation pyrometer, parts of which are illustrated in Fig. 202, includes a thermopile as the actuating element and either a spherically shaped mirror or an objective lens to focus the incident radiation from the radiating source onto the locus point of the hot junctions. The thermopile element in both types usually consists of eight very sensitive thermo-couples arranged as the spokes of a wheel with the hot junctions close to-gether, but not touching, at the hub.   The thermocouples are connected in series, and a radiation-receiving disk is located at, but insulated electrically

a. Radiation focused on the thermopile by an adjustable lens.

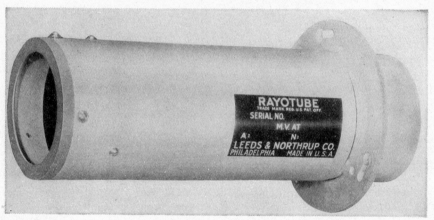

b. Radiation focused on the thermopile by a fixed-focus, spherically shaped mirror.
Fig. 202.   Two types of thermopile assemblies associated with total-radiation pyrometers whose actuating elements are thermopiles.

from, the hot-junction terminal point.   The function of the receiving disk is to ensure maximum absorption of radiant energy.   Both the mirror- and lens-type instruments may be used as portable temperature-measuring units, or they may be installed semipermanently outside of furnaces or other metallurgical units where it is desirable to have a constant check on the temperature of the hot source.

Like in the operation of conventional radiation pyrometers, no attempt is made to keep the cold junctions of the thermopile at a fixed or constant temperature. However, to prevent damage to the sensitive unit due to heat when the pyrometer is semipermanently installed near the hot source, the pyrometer unit is generally surrounded with a water-cooling jacket.

The sensitive unit of the lens-type Rayotube, illustrated in Fig. 202a, is smaller in size than that of the mirror-type instrument, and employs a much smaller target area on the radiating surface. Owing to these advantages, the lens-type instrument is applicable to temperature measurements of a limited part of a hot surface, and it can readily be sighted into a closed-end tube which projects, for example, into a hot furnace or kiln. The instrument is capable of measuring temperatures up to about 1800°C (3272°F), and functions by focusing the incident radiation onto the hot-junction point of the thermopile by means of an adjustable objective lens. In this respect, it is similar to the Pyro radiation pyrometer.

The mirror-type instrument is usually employed for relatively low temperature measurements, having a recommended useful temperature range of about 150° to 550°C (300° to 1020°F). In some cases, however, the temperature range of the instrument can be changed by changing the aperture of the limiting diaphragm mounted on the entrance end of the sighting tube. The reflecting surface of the focusing mirror is concave, and the mirror is located in the instrument behind the thermopile unit at a distance appropriate for focusing the reflected radiation onto the hot-junction point of the thermopile. A sighting aperture through the mirror, and a lens system appropriate for the purpose, is provided to facilitate in sighting the unit on the radiating source.

Standardization. The calibration of total-radiation pyrometers may be carried out by sighting the pyrometer into a furnace simulating black-body conditions and comparing the indicated temperature with that obtained by a calibrated thermocouple. A number of such measurements over the temperature range of the thermocouple will permit a calibration curve of the pyrometer to be plotted, black-body temperatures vs. emf. It is important that the distance factor be carefully considered and that the furnace opening conform to the size of the cone of radiation necessary for correct determinations.

Calibration may be carried out also by comparing the readings of the unknown pyrometer with those obtained by a similar radiation pyrometer, originally calibrated under black-body conditions. The source of radiation need not necessarily be a black body. For all practical purposes, a sheet of nickel electrically heated to elevated temperatures and then allowed to cool slowly in air serves very well as a radiating source.

As was previously shown on page 376, the relation between the emf developed by the thermocouple and the absolute temperature of the radiating source may be empirically expressed as

$$e = aT^b \qquad (50)$$

or

$$\log e = \log a + b \log T \qquad (51)$$

The log form of the equation is the most readily computable form and generally calibration curves are plotted $\log e$ vs. $\log T$. The observed data are plotted on these coordinates and the best straight line drawn between the points to represent the calibration curve.

Conditions Affecting the Operation of Radiation Pyrometers. In general, the conditions of nonblack bodies, smoke, flame, and reflected radiation that affect the operation of optical pyrometers will similarly affect the operation of radiation pyrometers. These conditions have been discussed on page 373 and need not be repeated here. It should be emphasized, how-

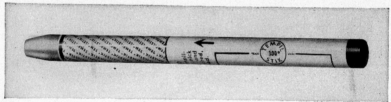

Fig. 203.   Illustrating a Tempilstik which in this case has a melting temperature of 300°F ± 3°.   (*Courtesy of Tempil Corp.*)

ever, that in order to obtain the best possible results with a radiation pyrometer, the optical parts and mirrors must be kept scrupulously clean, so as to minimize absorption effects. Very often, temperature measurements are grossly in error because of dust and dirt adhering to these parts.

TEMPIL TEMPERATURE INDICATORS

Constant melting-point compounds, available under the proprietary names of Tempilstiks, Tempil pellets, and Tempilaq, are temperature indicators in the physical form of sticks or crayons, compressed pellets, and liquid suspensions, respectively. These indicators cover a wide range of temperatures and are particularly useful for determining within close limits the temperature of heated bodies on whose surface these materials may be applied, or under conditions where the material may first be applied and subsequently observed during heating or cooling of the object of interest. Some specific metallurgical operations requiring temperature indications and for which Tempil materials may be conveniently employed are in preheating and postheating of weldments, localized heat-treatment, elevated-temperature forming operations, Martempering, conventional tempering, stress-relief annealing, particularly of massive pieces, hot-hardness testing, etc.

Tempilstiks are essentially extruded crayons of different compositions wherein each exhibits a well-defined melting temperature to within 1 per cent of the stated temperature. Forty-seven different Tempilstiks are available, one of which is illustrated in Fig. 203, which as a series cover a range of temperatures from 52°C (125°F) to 871°C (1600°F) in temperature intervals as shown in Table 61, page 496. The manipulations associated

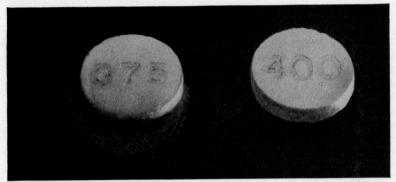

Fig. 204.   Illustrating the use of Tempil pellets.   (*Courtesy of Tempil Corp.*)

with the use of Tempilstiks, as well as the other Tempil materials, are simple and relatively easy to carry out. The surface of a hot body whose temperature is to be determined is merely stroked with an ordered series of Tempilstiks, and when a chalk-like mark is secured with one, and a liquid smear with another next lower in the series, the temperature of the surface is then indicated to be between the stated temperatures of the two Tempilstiks involved.

Tempil pellets are particularly useful for determing the temperature of an object in cases where the observation must be made at some distance. A series of Tempil pellets are placed on the hot body or on an object that is being heated, and the temperature noted as being between the stated temperature of the last pellet to melt and the first in the series to remain solid. This is illustrated in Fig. 204, which shows the surface temperature of the hot body onto which the two pellets were placed to be between 190 and 204°C (375 and 400°F).

Tempilaq is particularly useful as a temperature indicator on glazed or polished surfaces, or under circumstances where Tempilstiks are inconvenient to use, or where the shape of the hot body precludes the use of pellets. Tempilaq is merely applied to the surface of interest by daubing, and after the vehicle has evaporated, the indicating material remains on the surface to indicate the temperature in a manner described for Tempilstiks and Tempil pellets. The temperature intervals of Tempilaq are shown in Table 61.

References

ASM: "Metals Handbook," Cleveland, Ohio, 1948.

Burgess, G. K.: The International Temperature Scale, *Nat. Bur. Standards, Res. Paper* 22, 1928.

Burgess, G. K., and H. LeChâtelier: "The Measurement of High Temperatures," John Wiley & Sons, Inc., New York, 1912.

Callendar, H. L.: On the Practical Measurement of Temperature, *Phil. Trans. (London)*, Vol. 178, 1887.

Ferry, E. S.: "Practical Pyrometry," John Wiley & Sons, Inc., New York, 1917.

Foote, P. D., C. O. Fairchild, and T. R. Harrison: Pyrometric Practice, *Nat. Bur. Standards, Tech. Paper* 170, 1921.

Leeds & Northrup Co.: Catalogues N-33B, N-33D, Philadelphia.

McGraw-Hill Book Company, Inc.: "International Critical Tables," Vol. VI, New York.

Machler, R. C.: Potentiometer Circuit for Portable Optical Pyrometer, *Rev. of Sci. Instruments*, Vol. 10, 1939.

Roeser, W. F.: Thermoelectric Temperature Scales, *Nat. Bur. Standards, Res. Paper* 99, 1929.

"Temperature, Its Measurement and Control in Science and Industry," Reinhold Publishing Corporation, New York, 1941.

The Pyrometer Instrument Co., New York, Catalogues 80, 90, 100.

Van Dusen, M. S.: Platinum Resistance Thermometry at Low Temperatures, *J. Am. Chem. Soc.*, Vol. 47, 1925.

Wood, W. P., and J. M. Cork: "Pyrometry," 2d ed., McGraw-Hill Book Company, Inc., New York, 1941.

CHAPTER 9

THERMAL ANALYSIS
TRANSFORMATIONS OF STATE

It is well known that when a metal undergoes a transformation of state upon heating at constant pressure, *e.g.*, from solid to liquid, or liquid to vapor, the transformation is accompanied by an absorption of heat at the transition temperature. Conversely, in passing from one physical state to another upon cooling, the transformation is accompanied by an evolution of heat. These thermal phenomena are related to a difference in the energy content of the one phase with respect to that of the other being formed, and are manifested accordingly by heat effects, referred to as the latent heats of fusion and of vaporization.

It is a fortunate circumstance in many respects that certain pure metals and alloys undergo phasial transformations in the solid state. Such transformations may be allotropic changes associated with pure metals; or they may be phasial changes in alloys, *e.g.*, decomposition of solid solutions into other phases, formation of solid solutions from two or more existing phases, precipitation of secondary constituents to satisfy equilibrium solubility limits, etc. These solid-state transformations are usually accompanied by heat effects—not always detectable by thermal analysis—and by changes in volume that are best detected by measuring the attendant expansion or contraction of an appropriately shaped specimen. The phenomena described are well illustrated in the changes that occur in pure iron during heating or cooling.

When pure iron is cooled from some elevated temperature, say from about 1000°C (1832°F), where it exists as so-called gamma iron (face-centered cubic arrangement), an allotropic transformation occurs at about 910°C (1670°F). This change involves the formation of alpha iron (body-centered cubic arrangement), with an attendant evolution of heat, referred to as recalescence. The heat effect is of such magnitude that under appropriate conditions of observation, the iron will be seen to glow spontaneously as the transformation proceeds to completion. Because of the difference in atomic packing of the face-centered cubic gamma iron with respect to that of alpha iron, the transformation will also be accompanied by a marked and measurable expansion.

HEATING AND COOLING CURVES

Pure Metals. If a pure metal is uniformly heated from room temperature to the liquid state (assuming no allotropic changes in the solid state) and if the temperature is noted at regular time intervals by means of a thermo-couple or some suitable temperature-measuring device, an irregularity will be noted in the rate of heating which marks the temperature at which melting occurs.[1] This transition point may be shown graphically by plotting

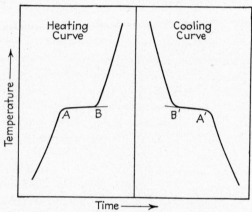

FIG. 205.   Idealized time-temperature heating and cooling curves of a pure metal.

the temperature vs. the time of heating, producing what is known as a simple time-temperature heating curve. In like manner, the plotting of similar data obtained by cooling a metal from the liquid to the solid state gives rise to a time-temperature cooling curve. Such curves for a pure metal are shown in Fig. 205. The abrupt change of slope in the curves indicates the temperature at which a transformation of state occurs.

In the ideal case, the temperature of the metal will remain constant during melting or freezing and, as a consequence, the portions of the curves AB and $A'B'$, respectively, will be perfectly flat and parallel to the time axis. The ideal curve will be further characterized by a discontinuous change of slope, rather than by a gradual change, as shown.

In the practical case, however, only a small part of the relatively horizontal portion of the thermal curves will be flat, and this part of the heating curve will usually possess greater obliquity than the cooling curve. As a consequence, cooling-curve data is considered more reliable in establishing the melting point of a pure metal, although the temperature thus deter-

---

[1] The true freezing or melting point of a pure metal is that temperature at which the solid and liquid phases are in equilibrium under a pressure of 760 mm Hg.

mined is the freezing temperature and is numerically equivalent to that of melting.

The reason for obliquity and its existence to a greater degree in the heating curves is mainly due to conditions relating to equipment and techniques employed in obtaining thermal data. When a crucible of metal is uniformly heated in a furnace, a temperature gradient is established between the walls of the furnace and the crucible. The metal in immediate contact with the crucible is the first to melt, and during the process of melting, the required latent heat of fusion is supplied by the walls of the crucible and the remaining solid metal. The melted, outside layer of metal then tends to remain at fairly constant temperature. The temperature-measuring instrument, located in the center of the charge, measures the temperature of the solid metal and, since heat is being abstracted from this portion of the charge, the rate at which the inside temperature increases is greatly diminished and causes a rounding off of the time-temperature curve.

As the metal progressively melts inwardly, the temperature of the center metal rises slowly. When the metal surrounding the temperature-measuring instrument begins to melt, the temperature remains fairly constant for a short interval of time and indicates the true melting temperature of the metal, as illustrated by point *B*, Fig. 205. While the center of the charge is melting, the temperature of the outside layer of molten metal rises rapidly, owing to the steady supply of heat from the radiating furnace. A large temperature gradient is thus established between the outside and the center of the charge. As a natural consequence, this temperature gradient accelerates the melting process and, if the remaining solid metal is unequally distributed about the temperature-measuring device, tends to increase the temperature indications and cause another rounding off of the time-temperature heating curve.

Upon cooling the molten metal, a temperature gradient is established between the crucible and the walls of the cooling furnace. As a consequence, the metal in immediate contact with the walls of the crucible is the first to solidify, forming an isothermal layer, which decreases the temperature difference between the center and the ouside of the charge. This, in effect, decreases the rate at which the temperature falls at the center of the charge, thereby causing a rounding off of the first part of the cooling curve. During freezing of the center of the charge, the solid outside layer of metal rapidly drops in temperature, thus increasing the temperature gradient between the center and outside portions of the charge. Freezing of the last traces of liquid metal is thereby accelerated, and accounts for further rounding off of the cooling curve.

It is evident, therefore, that the first part of the approximately flat portion of the cooling curve and the latter part of the flat portion of the heating

curve indicate the true freezing and melting points, respectively. Whenever considerable obliquity is present, the true temperature of freezing may
be obtained by extrapolating the straight portion of the cooling curve and
noting at what temperature the constructed straight line deviates from the
original curve. In like manner, the true melting temperature may be
obtained from the heating curve. This method of extrapolation is shown
at points B and B', respectively, in Fig. 205. In general, cooling curves
are more sharply defined than heating curves of the same metal, and are,
therefore, more frequently used to obtain transformation data.

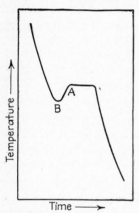

FIG. 206. Time-tempera-
ture cooling curve of a
pure metal exhibiting supercooling.

The cooling curves of certain pure metals, such
as antimony, and eutectic-type alloys are often
poorly defined because of conditions of supercooling or undercooling. Supercooling, in part,
results from experimental conditions which do
not favor the attainment of equilibrium between
the liquid and solid phases at the true freezing
temperature. This circumstance is well illustrated in the cooling curve shown in Fig. 206. As
the molten metal is uniformly cooled, freezing
would normally begin at a temperature equivalent
to point A, but because of conditions favoring
supercooling, the metal actually cools to a temperature B. At this temperature the molten
metal is in a metastable condition. Once freezing is started, however, the rate at which it
proceeds is very high, and, as a consequence,
more heat is evolved during the reaction than is conducted away. The
curve, therefore, rises sharply to the true freezing point A and rounds
off. Subsequent cooling proceeds in the normal way. The amount of
supercooling resulting therefrom is equivalent to the temperature difference
between A and B. Generally, in most metals and alloys the amount of
supercooling rarely exceeds 0.1 to 0.3°C. (0.18 to 0.5°F), but in the case of
antimony, the amount of supercooling may be 30°C (54°F) or more.
Supercooling in metals may be minimized by low rates of cooling, stirring of
the liquid metal during solidification, and the presence of a relatively great
number of nucleuses to act as crystallization centers.

Binary Alloys Forming a Solid Solution. The freezing curve of a binary
alloy whose components are completely soluble in each other in both the
liquid and the solid state,[1] assumes a different form than does the cooling

---

[1] Complete solubility of two or more metals in the solid state is shown by the inability
to distinguish crystals of either of the pure metals when the alloy is examined under
the highest magnification and resolution obtainable. This circumstance is to be
expected, owing to the nature of solid-solution formation. In most alloy systems, solid

curve of a pure metal. It was shown that the transition from the liquid to the solid state of a pure metal, or vice versa, was marked by an irregularity in the cooling and heating curves at nearly constant temperature. The freezing of a binary alloy forming a solid solution takes place over a definite temperature range, the limits of the temperature range depending upon the nature and proportion of the two metals forming the alloy. A typical cooling curve for this type of alloy is shown in Fig. 207. The temperature range of freezing is indicated by the portion of the curve between the two points of inflection, A and B. Point A indicates the beginning of freezing and point B the completion of freezing. At temperatures intermediate between A and B, the alloy is partly solid and partly liquid, the relative amounts of each depending upon the exact temperature.

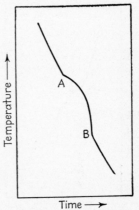

The convexity of the curve between A and B is the resultant of two opposing tendencies. As the alloy cools from A to B, the metal having the highest melting point tends to solidify first, with an accompanying tendency for the curve to flatten out. This flattening tendency is counteracted by the lowering of the average melting point of the remaining liquid alloy as freezing proceeds. This is due to the precipitation, in the form of solid-solution crystals, of a greater proportion of the high-melting-point metal than the low-melting-point metal. The relative proportions of

Fig. 207. Time-temperature cooling curve of a binary alloy forming a solid solution.

the high- and low-melting-point metals are, therefore, altered in the remaining liquid, causing the average melting point of the bath to be lowered.

Binary Alloys Forming a Eutectic. The cooling curve of an alloy composed of two metals completely soluble in each other in the liquid state, but completely insoluble[1] in the solid state, is of a form as shown in Fig. 208. It will be noted that the portion of the curve between the inflection points A and B is similar to that of a solid-solution-forming alloy, and the portion of the curve between B and C is similar to that of a pure metal. In all alloys of this type there is one definite proportion of the two metals that will remain completely liquid at a lower temperature than any other proportions in the alloy series. This lowest-melting-point alloy is known as the eutectic and in the solid state is composed of two phases. The phases

---

solubility occurs by random substitution of solute atoms for solvent atoms on the lattice points of the solvent; in certain other systems, the solute atoms occupy positions within the lattice of the solvent metal.

[1] Theoretically, at least, there is always some mutual solubility in the solid state.

constituting the eutectic are the nearly pure component metals, and the shape and distribution of the phases are characteristics of the alloy system. These latter circumstances can readily be ascertained by appropriate metallographic examination.

During very slow cooling of an alloy of this type, the excess constituent, in the form of very nearly pure crystals of one of the metals, is rejected or precipitated out of the liquid bath, beginning at point A, and is precipitated at an equilibrium rate so that when the temperature reaches point B, the remaining liquid is of eutectic composition. The freezing point of the eutectic is constant and at one particular temperature for any of the alloys composed of the same two metals. Hence, the portion of the curve BC is similar to the corresponding portion of a pure-metal cooling curve.

The net effect of two opposing tendencies, similar to the case of solid-solution-forming alloys, produces a convexity of the curve between points A and B. The curve tends to flatten out due to the heat evolved by solidification of the nearly pure excess constituent, but is opposed by the tendency of the curve to drop rapidly owing to the average melting point of the liquid bath being lowered. The temperature of the bath is lowered because of a change in the composition of the liquid due to rejection of the excess constituent.

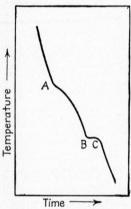

Fig. 208. Time-temperature cooling curve of a binary alloy forming an eutectic and an excess of one of the phases.

Use of Thermal Curves. With regard to pure metals, the principal use of thermal data derived from heating and cooling curves is to establish the temperature at which a change in state occurs (melting or freezing points), and in certain of these metals to determine the temperature at which one or more allotropic changes take place in the solid state. The determinations of phase changes in the solid state by thermal analysis, as compared to other methods to be described, depend mainly upon the magnitude of the heat effects at the allotropic transformation-temperature level. If the heat effects are relatively small and do not effectively alter the slope of the thermal curve, then more appropriate detection methods must be employed.

In the case of alloys, thermal data derived from heating and cooling curves are useful in establishing the temperatures at which melting begins and ends upon heating, or, conversely, the temperatures at which freezing begins and ends upon cooling. In general, the beginning of freezing is best determined by cooling-curve data, whereas the end of freezing or its numerical temperature equivalent—the beginning of melting—is more accurately established from heating-curve data. The latter circumstance is related to the fact that during cooling the precipitated solid phase rarely attains equilibrium with the liquid from which it is forming. Methods other than thermal analysis are often used to determine the temperature at which melting begins, and these other methods, particularly the metallographic procedure, sometimes yield more precise data. Briefly, the metallographic method involves heating a series of small specimens to progressively higher temperatures; quenching each specimen after it has reached equilibrium at the temperature to which it was heated; and finally examining the specimens metallographically to detect evidence of incipient melting.

If the alloy is one that exhibits phase changes in the solid state, the temperature at which these changes occur can often be detected by thermal analysis. The phase changes, as in the case of allotropic changes in pure metals, must be accompanied by large enough heat effects to be appropriately detected on heating or cooling curves. If the heat effects are of relatively small magnitude, the temperature of such transformations are then best determined by other methods to be described.

By appropriate accumulation of thermal data relating to a series of alloys composed of the same two component metals, it is possible to construct an equilibrium diagram of the alloy system involved. Whether or not the equilibrium diagram may be fully or only partially established will depend largely upon the ability of the analysis method to detect changes in the solid state, should such solid-state transformations occur in the system. When used with discretion and intelligent interpretation, an equilibrium diagram will reveal pertinent information about any alloy composition of interest in the system with regard to freezing- and melting-temperature ranges, phasial equilibrium at any temperature of interest, changes in phase concentrations with change in temperature, etc. Because of the importance of equilibrium diagrams when studying the constitution of alloys, the construction of such a diagram of a rather simple alloy system, and one not involving transformations in the solid state, will be discussed briefly.

Consider, for illustrative purposes, a series of alloys between cadmium and bismuth, the pure metals being completely soluble in each other in the liquid state, insoluble in each other in the solid state, and forming a eutectic. If the alloy series are made up in composition variations of 20 per cent from pure cadmium, melting point 321°C (610°F), to pure bismuth, melting point 271°C (520°F), and successive cooling curves obtained for each alloy, the series of cooling curves will be similar to those shown in Fig. 209. It will be noted that all the curves are similar, with the exception of the one obtained from the alloy containing 40 per cent cadmium–60 per cent bismuth. Each curve shows the beginning of freezing of the respective alloy at a temperature corresponding to the respective A point; freezing through the temperature range with the rejection of the excess constituent; and finally forming a eutectic at constant temperature (140°C) marking the completion of freezing. The alloy composed of 40 per cent cadmium–60 per cent bismuth is the eutectic-forming alloy and, therefore, freezes at constant temperature and not through a range of temperatures.

It will be noted that as the percentages of bismuth increase from 0 to 60 per cent, and as the cadmium percentages increase from 0 to 40 per cent, the temperature at which freezing begins is progressively lowered. This is in accordance with Raoult's law,[1] which, by general interpretation, states

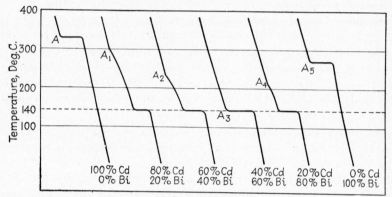

Fig. 209.   Time-temperature cooling curves for a complete series of alloys between cadmium and bismuth.

that the freezing point of a pure substance is lowered by the addition of a second pure substance, provided that they are completely soluble in the liquid state but insoluble in the solid state. The amount by which the freezing point is lowered with respect to the solvent is proportional to the molecular weight and amount of the solute. In the first instance (0 to 60 per cent bismuth), cadmium is the solvent and bismuth the solute, whereas, in the second case (0 to 40 per cent cadmium), bismuth is the solvent and cadmium the solute.

If the eutectic freezing point of each alloy and the temperatures corresponding to the *A* series of points from the cooling curves are plotted on coordinates of temperature as ordinate and per cent composition by weight of the alloy as abscissa, an equilibrium diagram is obtained when the respective plotted points are connected by lines. The completed diagram is shown in Fig. 210.

[1] A more exacting statement of Raoult's law is as follows: At a given temperature, the lowering of the vapor pressure of a solvent by the addition of a nonvolatile solute is equal to the product of the mole fraction of the solute present and the vapor pressure of the pure solvent at this temperature. In general, this law applies with greater accuracy the more dilute the solution.

Since all of the alloys above their respective A temperatures were completely liquid, the curve connecting these plotted points defines the lowest temperature at which any alloy will be completely liquid, and is appropriately designated as the liquidus line. The horizontal line, connecting the temperatures at which solidification is complete for all of the alloys, is known as the solidus line and defines the highest temperature at which any alloy will remain completely solid. The phase fields bounded by the liquidus and solidus lines define the temperature range over which any alloy is partly liquid and partly solid; the solid phase, whether cadmium or bismuth, being subsequently determined by metallographic examination of appropriately selected alloy compositions.

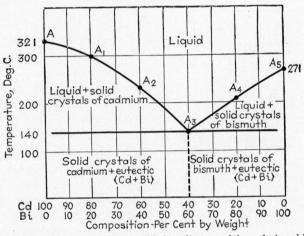

FIG. 210. Schematic presentation of the equilibrium diagram of the cadmium-bismuth system.

In a manner similar to that described, equilibrium diagrams of alloy systems more complicated than the one illustrated may be established, at least in part, from thermal-curve data. The procedure is generally time consuming and, under some circumstances, difficult, and many more specific alloy compositions must be investigated than cited in the above example. As mentioned heretofore, the characteristics of transformations in the solid state may, in some cases, be determined by heating and cooling curves, although more often such transformations are best detected by methods designed particularly for this purpose.

EXPERIMENTAL PROCEDURE

The procedure followed in obtaining time-temperature heating or cooling curves consists of heating or cooling the metal from room temperature to above the melting point, or cooling from above the melting point to below the last transformation temperature, respectively, and noting the temperature of the metal at regular predetermined intervals of time. Temperature readings may be taken at equal time intervals, ranging from 1 to 3 min, depending upon the relative rate of heating or cooling. For very accurate determinations, the time should be noted with a stop watch.

A furnace well suited for securing thermal data on most metals and alloys by heating- and cooling-curve determinations is one that is capable of heating the metal charge to the highest temperature required at about 3°C

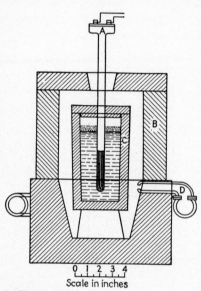

Scale in inches

A. Thermocouple.
B. Furnace.
C. Crucible containing the charge of interest.
D. Gas burner, one of three mounted tangentially to periphery of the furnace hearth.

FIG. 211. A design of gas-fired furnace suitable for thermal analysis of metals and alloys. (*From Nat. Bur. Standards, Tech. Paper* 170, 1921.)

(5.4°F) per minute; and is sufficiently well insulated, or some means of control is provided, so that the specimen charge will cool in the furnace at about the same rate. In the case of low-melting-point alloys, however, lower heating and cooling rates are often more appropriate, whereas in alloys prone to segregation, higher rates are desirable.

A furnace suitable in all respects for this particular purpose is shown in Fig. 174, page 338. Under circumstances where such a design of furnace is not available, other furnaces may be used, provided they are appropriately designed to render uniform and low rates of heating and cooling. During the heating cycle, slight variations in the power supply or heat losses arising from improper insulation may alter considerably the shape of the thermal curves and, as a consequence, mask important transformation points. Gas furnaces, although not so satisfactory in many respects as electrically heated furnaces, may be used if the required temperature range is not too high, and if uniform heating of the metal charge may be obtained. A recommended type of gas furnace is shown in Fig. 211. The furnace is provided with three burners *D*, only one of which is shown, mounted tangentially to the circumference, which provide a spiraling flame. This arrangement of burners, as contrasted to others, assures fairly good temperature uniformity.

In the preparation of the metal charge to be heated for the first time, the metal should be broken into small pieces, placed in a crucible of proper size and composition, and heated approximately 10°C (18°F) beyond the melting point. If the charge under investigation is an alloy, the molten bath should be stirred, to ensure complete and intimate mixture of the components. The surface of the molten metal, except when the metal is nickel, should be covered with a layer of graphite or charcoal to prevent oxidation of the metal. Small amounts of oxide in a pure metal, such as in copper, may appreciably alter the shape of the cooling curve and indicate the occurrence of transformations at temperatures lower than normal. Graphite should be added only after the metal has become molten or after it is stirred. This order of procedure should be followed in order to prevent graphite from becoming entrapped in the molten metal, forming graphite pockets.

The selection of a crucible to hold the charge is of considerable importance. For most metals, crucibles composed of Acheson graphite are found to be satisfactory. At high temperatures the gases formed from its oxidation provide a reducing atmosphere which, in addition to the powdered graphite, protects the surface of the molten metal. Metals—such as iron and nickel—that react with graphite at high temperatures should be melted in crucibles composed of magnesia or alumina, or a mixture of the two.

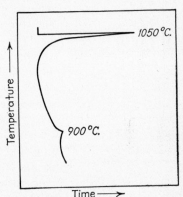

Fig. 212. An inverse-rate cooling curve of a single-phase, copper-tin alloy (*From Nat. Bur. Standards, Tech. Paper* 170, 1921.)

Generally, almost any type of thermocouple may be used for measuring temperatures, provided that the working range of the thermocouple is within the limits of the temperature range to be explored. Where high accuracy is required at elevated temperatures, noble-metal couples are recommended, such as a platinum–platinum-10 per cent rhodium thermocouple.

The most accurate data are obtained only when the metal bath and the thermocouple wires are protected from contamination. The metal bath may be protected as described before, and the thermocouple elements, by means of a suitable protection tube. For low-temperature investigations, up to approximately 500°C (932°F), pyrex-glass tubing sealed at one end serves very well. The use of porcelain or quartz protection tubes is recommended for higher temperatures, and with all metals except aluminum, which readily attacks silica at elevated temperatures. Above 1100°C (2012°F), protection tubes of Usalite or Impervite, having a composition approximating that of sillimanite, are recommended.[1]

Inverse-rate Method. When a transition point is accompanied by only a relatively small heat effect, the change in slope of the direct time-

[1] *Nat. Bur. Standards, Tech. Paper* No. 170, 1921.

temperature cooling curve may be so gradual that it will be difficult to detect the correct transformation temperature. Indications that are more definite may be obtained by means of the inverse-rate curve, which consists essentially of plotting the absolute time interval required for a definite and arbitrary temperature change. This method, in effect, amounts to differentiating the direct time-temperature curve with respect to the temperature. Figure 212 illustrates the inverse-rate cooling curve through the melting range of bronze. It will be noted that the lower transition point at 900°C is easily detected, although it was not apparent by the ordinary time-temperature cooling method.

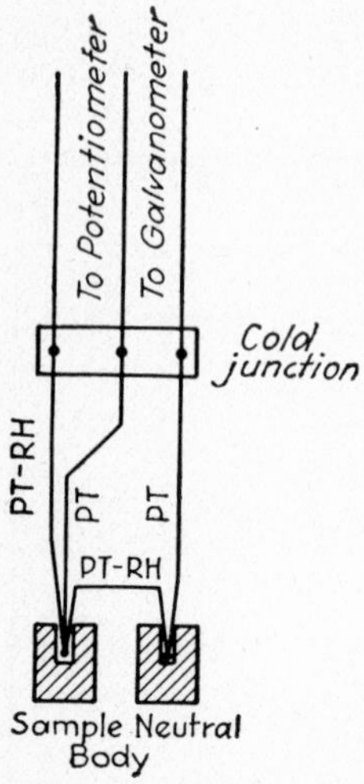

The experimental procedure for obtaining data is the same as in the time-temperature method, except that the time must be more accurately measured. This may be readily accomplished by the use of two stop watches, one of which is stopped and one started at the end of each time interval. The temperature interval may be between 1 and 10°C (1.8 and 18°F), depending upon the accuracy desired.

In constructing the inverse-rate curve, the lower temperature of each successive temperature interval is plotted against the time. Thus, if it requires 40 sec for the temperature to drop through an arbitrary temperature interval of 3°C, from 800 to 797°C, the 40-sec time interval is plotted against 797°C. The subsequent temperature drop from 797 to 794°C may require 50 sec. This time is then plotted against 794°C, etc.

FIG. 213. Illustrating the experimental arrangement of thermocouples, specimen, and neutral body for securing differential and derived differential heating- or cooling-curve data. (*From Nat. Bur. Standards, Tech. Paper* 170, 1921.)

Differential Method. Another sensitive method for detecting minor transformation points in a metal is the differential method, which consists essentially of plotting the temperature of the specimen under investigation against the temperature difference between the specimen and a neutral body. The method is of importance, as the results are not influenced to any great extent by variations in heat losses due to external influences. Two thermocouples are required—one a differential thermocouple with a hot junction in both the specimen and the neutral body; and the other, a thermocouple to indicate the temperature of the specimen. This arrangement is shown in Fig. 213. The neutral body consists of some metal, usually platinum or nickel, that undergoes no transformations in the temperature range through which the specimen is investigated. To ensure that both the specimen and the neutral body are thermally affected simultaneously during the heating and cooling cycles, the two bodies are placed in close contact with one another within the furnace. If the sizes and specific heats of the two bodies are approximately the same, the temperature differ-

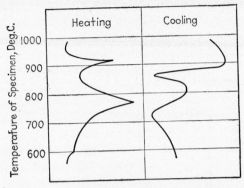

Fig. 214. Differential heating and cooling curves of pure iron.

ence between the two will be small, except at a transformation point of the specimen, and the effects of variations in the rate of heating and cooling will have negligible effects on the results.

When a transformation point of the specimen is reached, the uniform drop in temperature of the specimen is temporarily arrested, whereas the temperature of the neutral body continues to fall with practically no interruption. Minor transformation points are thereby easily detected, being marked by abrupt increases in the temperature differences between the two bodies.

Differential heating and cooling curves of pure iron are shown in Fig. 214. It will be noted that the transformation points on heating occur at a higher temperature than the same transformations which occur on cooling. This is due in part to

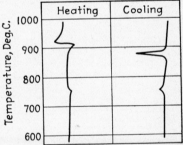

Fig. 215. Derived differential heating and cooling curves of pure iron.

the imposed rates of heating and cooling, and is a manifestation of the tendency and persistence of an equilibrium phase to remain unaltered when, by changes in temperature, the phase is removed from its thermal region of stability. This phenomenon is encountered in many other instances, as exemplified in supercooling and surfusion of a metal or alloy.

Derived Differential Method. The derived differential thermal curve, shown in Fig. 215, is essentially the result of differentiating the differential curve with respect to the temperature and plotting the values so obtained against the temperature. This method requires no deviation from the experimental procedure followed in the differential method, except in the manner in which data are recorded. The derived differential curve is advantageous, in that transformation points are made apparent which would otherwise not be detected. With respect to the method of recording data and constructing the curve, temperature differences between the specimen and the neutral body are noted at equal, successive temperature intervals of the specimen. These temperature differences are then plotted, in the case of cooling curves, against the upper temperature value of the temperature interval. In constructing heating curves, they are plotted against the lower temperature value of the interval. For example, it is found by observation that, in cooling from 800°C, the differences in temperature between the specimen and neutral body are 2, 3, 5, and 8°C, for every 2°C drop in temperature of the specimen. These values are then plotted against 800, 798, 796, and 794°C, respectively.

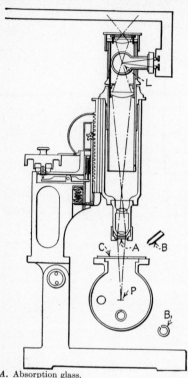

A. Absorption glass.
B. Air-blast pipes for cooling the heating chamber.
C. Cover glass.
L. Comparison lamp.
P. Platinum strip for supporting sample.

FIG. 216. The Burgess micropyrometer. (*From Nat. Bur. Standards, Tech. Paper* 170, 1921.)

The Micropyrometer. It is sometimes necessary to determine the melting point of a microscopic sample of a metal, which by the ordinary methods of thermal analysis would be difficult, if not impossible, to do. Such determinations, however, may be carried out readily by use of the Burgess micropyrometer, shown in Fig. 216. The principle upon which the micropyrometer operates is essentially the same as the principle of the disappearing-filament-type optical pyrometer.[1]

The microscope, containing the disappearing filament bulb, is sighted on the sample, which is supported by a strip of pure platinum. The sample and the platinum strip are enclosed in a water-cooled, gastight chamber, which permits surrounding the sample with an inert or reducing atmosphere when desired. The sample is observed through the top of the chamber, which is constructed of clear plate glass.

[1] See p. 365.

The platinum strip is heated by electrical methods from an external source and, at the instant the sample melts, as determined by observation through the microscope, the temperature of the platinum is measured with the optical pyrometer. The melting point of the sample is readily noted, as it either gathers into a ball or spreads out on the strip.

Melting-point Determinations with Optical Pyrometers. In determining the melting or freezing points of metals at temperatures above 1550°C (2822°F), it is necessary to use temperature-measuring devices other than thermocouples. Disappearing-filament-type optical pyrometers are generally used, as they afford high accuracy at high temperatures and their accuracy is not impaired when they are sighted on very small sources.

An experimental device, primarily for determining high-temperature melting points, is shown in Fig. 217. The outer tube is constructed of sillimanite and constitutes that unit which is inserted into a furnace. A graphite crucible containing the charge is placed at the bottom of the tube. There are two holes through the cover of the crucible—one in the center, through which the progress of melting or freezing can be observed, and another near the outer edge, through which the pyrometer is sighted. The surface of the molten metal, because of surface tension, forms a V-shaped depression with the side walls of the crucible. This depression, when viewed through the outer hole of the crucible cover, serves as a satisfactory black body upon which to sight the pyrometer. When a determination of either the melting or the freezing point of a metal is to be made, the tube and its contents are uniformly heated or cooled in a furnace. When the transformation point occurs, as determined by visual examination through the center hole of the crucible cover, temperature measurements of the metal are made with the optical pyrometer in the usual way.

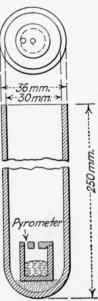

FIG. 217. Furnace apparatus for determining the melting and freezing points of metals and alloys with an optical pyrometer. (*From Nat. Bur. Standards, Tech. Paper* 170, 1921.)

SOLID-STATE TRANSFORMATIONS

The methods of thermal analysis described are primarily techniques for determining transformation temperatures that are associated with reactions accompanied by detectable thermal changes. The preciseness of the analysis depends to a great extent upon the rates of heating and cooling and, because of this, errors of large magnitude are often associated with determining transformation points in solid solubilities whose constituents have slow reaction rates.   The quantitative analysis of reaction rates requires quite precise measurements and when the rates vary over wide limits, it is practically impossible to study the phenomenon by ordinary methods of thermal analysis.

As mentioned heretofore, transformations in the solid state are best determined by either metallographic methods, electrical resistivity measurements, X-ray diffraction analyses, or volume and dimensional changes of the alloy of interest.

Metallographic Method.   The metallographic method of analysis is ideally suited for determining solid state transformations, owing to the visual revelation of the constitutions of the alloy of interest.   The presence of one or more phases is readily detectable; the size, shape, distribution, and homogeneity (in some cases) of the phases are easily determined; and it is often possible to determine if the solid-state reaction has gone to completion.   The method involves appropriate metallographic preparation of specimens that have been heated to various temperatures and subsequently quenched; and heat-treating specimens at the same temperature for increasing lengths of time, subsequently quenching, and observing in the metallographic structure the approach to, and final attainment of, equilibrium at the temperature of interest.   The former procedure is well suited for determining temperatures at which solid solutions decompose, or are formed from two or more phases in the solid state.

Electrical-resistivity Method.   Electrical-resistivity measurements for determining transformations in the solid state, as well as solid-liquid equilibria, are particularly advantageous in that the method is sensitive to the kind and amount of phases present; the measurements can be made at the equilibrium temperature of interest; and the attainment of equilibrium at temperature is easily recognized from the data obtained.   Although equally precise measurements can be made at ordinary temperature on quenched specimens, the interpretation of such measurements may be in error owing to structural changes likely to occur in some alloys when the specimens are removed from the heating furnace, when quenched, and during storage at ordinary temperature.   Electrical resistivity measurements may be carried out with a Wheatstone bridge, or by a similar potentiometric method described on page 358.

X-ray Diffraction Method.   Perhaps the most precise method for determining phasial equilibria in the solid state is by X-ray diffraction, employing powder or back-reflection analytical techniques at ordinary and elevated temperatures.   When appropriately carried out, these techniques are capable of detecting phases that might escape notice by metallographic methods, and offering a convenient means of positive identification of the one or more phases present.   In general, X-ray diffraction techniques for the purpose described are involved and highly specialized, but are adequately discussed elsewhere.[1]

Magnetic Method. Solid-state transformations that are accompanied by magnetic changes in the alloy may be determined by a magnetic method, first used by Sauveur to detect A_2 temperature levels in pure iron and very low-carbon steels.

The specimen of interest, in the form of a round rod about 3 in. in length, is securely held in a vertical position between two rods of ingot iron, forming essentially a composite rod. The length of the ingot iron rods are such that they extend approximately 5 or 6 in. above the top and below the bottom of a vertical electrical furnace. The portion of the bar extending above the top of the furnace is surrounded by an electromagnet. Located at the lower end of the bar is an iron needle, so balanced that one end is readily attracted by the bar when the latter is magnetized. When a magnetic transformation point of the specimen is reached on heating, the magnetic flux ceases to pass through the sample and the needle suddenly drops. In cooling, when the transformation temperature is again passed, the sample regains its magnetism and the needle is again attracted. The temperature of the specimen at which the transformation occurs is measured by a thermocouple, the hot junction of which is located within a hole in the specimen.

Dilatometry. The principles of dilatometry are associated with the phenomenon that practically all solid materials exhibiting phasial transformations in the solid state, *e.g.*, certain metals, alloys, glasses, ceramics, etc., simultaneously exhibit at the transformation temperature a detectable change in volume and in length dimensions of an appropriately shaped specimen. Dilatometric methods are applicable, however, only for determining solid-state transformations that occur well below the melting temperature of the material of interest. At or near the melting point, the method cannot be used with any degree of certainty.

In general, dilatometric measurements are simple in principle to carry out, and the data secured are precise, reliable, and relatively easy to interpret. Secondary reactions that may occur in the material, arising from the presence of impurities, usually introduce no significant errors. The dilatometric characteristics of a metal are influenced to a degree commensurate with the proportion of foreign element present and, therefore, may be compensated for.

The dilatometric characteristics of a metal or an alloy are best observed by the use of a dilatometer, which is essentially a sensitive measuring device that magnifies, through mechanical or optical levers, the length changes of an appropriately shaped specimen with changes in temperature. In a manner to be described, an acceleration or reversal of these dimensional

[1] See references pertaining to X-ray diffraction and crystal structure, p. 325.

changes indicates the occurrence of a phasial transformation, which may be correlated subsequently, with the temperature at which it occurs.

Dilatometric instruments are available commercially in the form of stationary units that indicate dimensional changes of the specimen through a system of mechanical levers and linkages, or through a combined system of mechanical and optical levers. The mechanical-type instrument records expansion and contraction characteristics of the specimen by mechanical arrangements, whereas these characteristics are generally recorded photographically in the optical type. Often, portable dilatometers of the mechanical type are designed so that observations may be made during

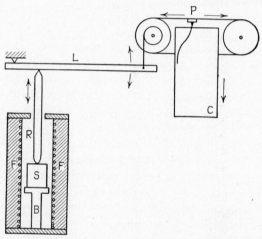

B. Specimen support.
C. Recording chart paper.
F. Vertical-type electric tube furnace.
L. Lever.

P. Stylus for inscribing curve on chart paper.
R. Quartz contact rod which transmits length changes in specimen S to actuating lever L.

Fig. 218. Illustrating schematically the principle of the simple mechanical-type dilatometer.

quenching of the specimen, although one of the stationary units to be described is designed so that it may be used in this manner if desired.

Mechanical-type Dilatometers. The principle of operation of the mechanical-type dilatometer in its simplest form is illustrated in Fig. 218. The specimen S under investigation is supported on a resting platform B composed of an alloy possessing a nearly uniform and linear coefficient of expansion. A quartz rod R is in contact with the upper surface of the specimen and extends through the top of the furnace, where it actuates a lever L. Movement of the lever through a small arc actuates a stylus P in a horizontal direction, which traces a curve, in accordance with its position, on a movable chart C. The chart is unrolled beneath the stylus at a constant rate by means of a mechanical motor. It is evident, therefore, that as the specimen is heated by furnace F, the stylus will be actuated by the lever arrangement and trace out a curve corresponding to the combined length changes of S, specimen support B, and part of rod R.

The Bristol dilatometer, illustrated in Fig. 219, is representative of commercial mechanical-type instruments that are designed to record automatically temperature-dilation and temperature-time changes of the specimen under investigation. The dilation-measuring system consists of two quartz

FIG. 219. The Bristol mechanical-type dilatometer. (*Courtesy of Bristol Co.*)

rods, one resting directly on the specimen of interest, and the other on the platform supporting the specimen within the split-type furnace. Through an ingenious system of multiplying levers and interrelated linkages, the expansion and contraction of the specimen platform and the heat-affected parts of the lever mechanism are compensated for, so that the graphic recording produced represents only the dilatometric characteristics of the specimen. By appropriate insertion of a thermocouple into a drilled hole in the specimen, the temperature of the specimen may be recorded through a potentiometric recording system onto the same chart as the dilatometric-characteristic curve.

The Bristol instrument is designed so that a graphic record of expansion and contraction may be secured when the specimen is quenched. When the specimen has reached the temperature of interest, the furnace is opened and swung out of its normal position, leaving exposed the specimen, supporting platform, and contacting quartz rods. A quenching tank, nor-

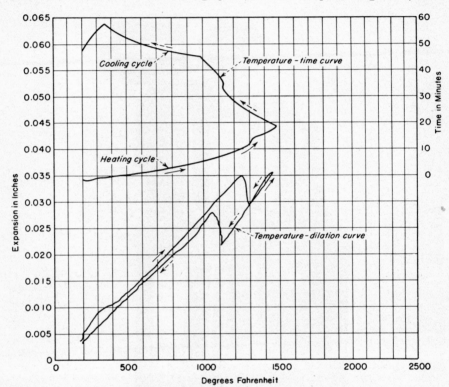

FIG. 220. Graphical record of the expansion and contraction characteristics of an SAE 1112 steel heated to, and cooled from, 1450°F in the Bristol dilatometer. (*Courtesy of Bristol Co.*)

mally positioned below the instrument bench top (see Fig. 219), is then quickly raised by means of a chain lift actuated by a hand crank. This operation envelopes the specimen and parts of the instrument with the quenching medium.

A typical temperature-dilation curve, and a temperature-time curve, automatically recorded during test, are illustrated in Fig. 220. These curves represent the characteristics of an SAE 1112 steel placed within a hot furnace at the beginning of the test; heated to a temperature of 788°C

(1450°F); cooled from this temperature in the furnace to 516°C (960°F); and finally cooled in air to about 88°C (190°F). As indicated by the temperature-dilation curve, the specimen begins to undergo, upon heating, a marked contraction at about 1250°F, which defines the Ac_1 temperature for this particular steel as related to the rate of heating. Upon cooling from 788°C (1450°F)—a temperature between the A_1 and A_3 transformation temperatures for this steel—a measurable expansion begins at about

Fig. 221. The Chevenard differential dilatometer employing photographic methods for recording the expansion and contraction characteristics of the specimen of interest. (*Courtesy of the R. Y. Ferner Co.*)

1100°F, defining the Ar_1 temperature as related to the rate of cooling. These transformation temperatures, upon heating and cooling, are likewise indicated by abrupt changes of slope in the temperature-time curve, illustrated in Fig. 220.

Photographic-type Dilatometers. In general, dilatometers designed on the principle of optical levers and associated photographic recording methods are more sensitive to small dilation changes than are the mechanical-type instruments. The Chevenard dilatometer, illustrated in Fig. 221, is a combination of mechanical levers and an optical arrangement, so as to record on a photographic plate the expansion and contraction characteristics of the specimen of interest.   The high degree of sensitivity attending this instrument is related to the differential principle incorporated in its design.

With reference to Fig. 222, the specimen E_1, in the form of a small rod, and a standard rod of pyros[1] E_2 are contained within two quartz tubes, T_1 and T_2, symmetrically arranged in a tube-type furnace in the order shown.   The changes in length of the specimen and the pyros standard are appropriately transmitted by two quartz rods, t_1 and t_2, through the connecting rods c_1 and c_2, respectively, to points p_1 and p_2 of the optical lever mounted in the photographic chamber.   The Invar tripod L carries a concave mirror M that reflects a pencil of light made incident upon it onto a photographic plate at the end of the photographic chamber.   The points

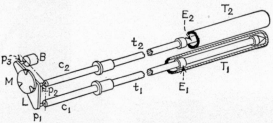

FIG. 222.   Illustrating schematically the differential principle of the Chevenard differential dilatometer.

of the tripod, p_1, p_2, and p_3, respectively, are fitted into small depressions in such a way that all lateral play is eliminated.   The support B, on which point p_3 rests, is stationary.

As the specimen and the standard rod are heated simultaneously, the expansion resulting therefrom causes the mirror to move, and this traces a curve on the photographic plate.   Each point of the curve has for its ordinate the difference of the dilations of the specimen and the standard, and the abscissa is proportional to the dilation of the pyros standard with respect to the quartz tube.

The stationary support B may be replaced by another heating-tube arrangement similar to T_1 containing another sample of the material.   The photographic registration is then the difference in the dilation between the two samples, as a function of a third sample, which is used as a pyrometric standard.   This triple arrangement is very sensitive and the curves pro-

[1] Pyros is a nickel-chromium-tungsten alloy.   Its coefficient of expansion at 0°C (32°F) is about 12.57×10^{-6} and increases at a very nearly linear rate to 21.24×10^{-6} at 1000°C (1832°F).   The alloy remains rigid up to 1000°C (1832°F), oxidizes very little, and its dilatometric properties are not modified by repeated heatings.

duced are nearly horizontal, being in a favorable direction for the detection of slight irregularities.

References

The Bristol Company: Waterbury, Conn., *Bull.* W1803.

Epstein, S.: "The Alloys of Iron and Carbon," Vol. I, "Constitution," The Engineering Foundation, McGraw-Hill Book Company, Inc., New York, 1936.

Evans, V. R.: "Metals and Metallic Compounds," Longmans, Green & Co., Inc., New York.

The R. Y. Ferner Company: "Industrial Thermal Analysers and Differential Dilatometers," Catalogue D-1, Washington, D. C.

Fink, W. L.: Determination of Phase Fields, *Metal Progress*. Vol. 53, 1948.

Foote, P. D., C. D. Fairchild, and T. R. Harrison: Pyrometric Practice, *Nat. Bur. Standards, Tech. Paper* 170, 1921.

Portevin, A., and P. Chevenard: Principles and Chief Applications of Dilatometric Analyses of Materials, *Trans. Am. Foundrymen's Assoc.*, Vol. 34, 1926.

Sauveur, A.: "The Metallography and Heat Treatment of Iron and Steel," University Press, Cambridge, Mass., 1935.

Wood, W. P., and J. M. Cork: "Pyrometry," 2d ed., McGraw-Hill Book Company, Inc., New York, 1941.

APPENDIX

Metal or alloy	Composition of electrolyte	Current density, amp per sq dm	Voltage	Time	Remarks
No. 1. Aluminum and aluminum alloys............	Perchloric acid—2 parts Glacial acetic acid — 7 parts Aluminum—3 to 5 g dissolved in electrolyte by electrolytic polishing of a specimen of aluminum	3–5	50–100	5–15 min	Aluminum cathode; temperature less than 50°C. Polish specimen through No. 600 alundum. In heterogeneous alloys, electrolytic polishing may in some cases remove certain constituents, and in others the constituents may be left in relief
No. 2. Aluminum..	Perchloric acid (sp gr 1.48) —2 parts Acetic anhydride—7 parts Aluminum—4 to 5 g dissolved in electrolyte by electrolytic polishing of a specimen of aluminum	3–5	50–100†	15 min	*Explosive mixture!* Aluminum cathode; temperature less than 50°C
No. 3. Aluminum..	Methyl alcohol (abs.)—2 parts Nitric acid (conc)—1 part	31–93	4–7	....	Stainless-steel cathode; ordinary temperature; distance between anode and cathode ½–1 in.
No. 4. Aluminum bronze	Orthophosphoric acid -- 990 g per liter	1–2			
No. 5. Brass.......	Chromic acid—200 g Water—1000 ml	248 or more	....	....	Platinum-gauze cathode
No. 6. Brass.........	Orthophosphoric acid — 430 g per liter	13–15	1.9		
No. 7. Brass (70% Cu, 30% Zn)	Same as No. 3	....	40–50†	Few sec	Same as No. 3
No. 8. Brass (70% Cu, 30% Zn)	Orthophosphoric acid—5 to 14%; 8% solution preferred, 5.8 deg Baumé, sp gr 1.042	78–310	30	60 sec	Stainless-steel cathode; preferred electrolytic cell arrangement, Fig. 14, p. 30. For electrolytic etching in same electrolyte, see No. 4, Table 13
No. 9. Brass (free-cutting grade)	Same as No. 8	78–310	25	45 sec.	Same as No. 8
No. 10. Brass (66.7% Cu, 33.3% Zn)	Orthophosphoric acid— 990 g per liter	2.5–3			
No. 11. Brass (66.7% Cu, 33.3% Zn)	Same as No. 3	....	40–50†	Few sec	Same as No. 3
No. 12. Brass........	Same as No. 8	78–310	25	60 sec	Same as No. 8
No. 13. Brass.......	Pyrophosphoric acid—530 g per liter	9–11	1.9		
No. 14. Brass (60% Cu, 40% Zn)	Same as No. 3	....	40–50†	Few sec	Same as No. 3

Note: Time of polishing will depend upon concentration of electrolyte, cell arrangement, and initial smoothness of the surface.

* From ASTM Standards, Part I-B, 1946; *Metal Progress*, October, 1940; *Mining and Met.*, July, 1943; and *Metals & Alloys*, February, 1945.

† External applied voltage.

TABLE 1. ELECTROLYTES FOR ELECTROLYTIC POLISHING OF
METALLOGRAPHIC SPECIMENS.*—(Continued)

Metal or Alloy	Composition of electrolyte	Current density, amp per sq dm	Voltage	Time	Remarks
No. 15. Brass (admiralty compositions)	Same as No. 8	78–310	25	60 sec	
No. 16. Bronze (85% Cu, 10% Sn, 3% Zn, 2% Pb)	Orthophosphoric acid— 990 per liter	1–2			
No. 17. Bronze (Phosphor and silicon)	Same as No. 3	....	40–50†	Few sec	
No. 18. Cobalt......	Orthophosphoric acid—sp gr 1.35	....	1–2	....	Cobalt cathode
No. 19. Copper......	Orthophosphoric acid—sp gr 1.3 to 1.4	0.65–0.75	2	5 min or more	Copper cathode. Anode-cathode separation, 2.2 cm
No. 20. Copper......	Pyrophosphoric acid—530 g per liter	8–10	1.6–2.0	10–15 min	Copper cathode
No. 21. Copper......	Same as No. 3	....	40–50†	Few sec	Same as No. 3
No. 22. Copper......	Same as No. 8	78–310	20	45 sec	Same as No. 8
No. 23. Copper + 3.2% cobalt	Same as No. 19	0.07	2	5–10 min	Copper cathode; electrodes horizontal and spaced ½ in. apart
No. 24. Copper + 2.4% iron	Same as No. 19	0.07	2	5–10 min	Same as No. 23
No. 25. Iron and silicon iron	Orthophosphoric acid	0.6	0.75–2.0	....	Iron cathode
No. 26. Iron—Armco and white cast iron	Acetic anhydride—765 ml Perchloric acid (65%)— 185 ml Water (distilled)—50 ml Aluminum—about 0.5% dissolved in electrolyte by electrolytic polishing of a specimen of aluminum Prepare electrolyte 24 hr before using	4–6	50†	4–5 min	*Explosive mixture!* Iron or aluminum cathode; temperature less than 30°C; use moderate agitation. Dissolved aluminum increases viscosity, which permits more vigorous agitation of electrolyte
No. 27. Lead......	Acetic acid—650 to 750 ml Perchloric acid—350 to 250 ml	1–2	....	3–5 min	*Explosive mixture!* Copper cathode; anode horizontal. Use current density of about 20 amp per sq dm for 1–2 min to remove distorted metal
No. 28. Lead-antimony alloys	Same as No. 27	2			
No. 29. Monel metal..	Same as No. 3	....	40–50†	Few sec	Same as No. 3
No. 30. Nichrome....	Same as No. 3	....	40–50†	Few sec	Same as No. 3
No. 31. Nickel.......	Same as No. 3	....	40–50†	Few sec	Same as No. 3
No. 32. Steel—austentic types	Same as No. 26	4–10	50†	4–5 min.	Same as No. 26

Note: Time of polishing will depend upon concentration of electrolyte, cell arrangement, and initial smoothness of the surface.

* From ASTM Standards, Part I-B, 1946; *Metal Progress*, October, 1940; *Mining and Met.*, July, 1943; and *Metals & Alloys*, February, 1945.

† External applied voltage.

TABLE 1. ELECTROLYTES FOR ELECTROLYTIC POLISHING OF
METALLOGRAPHIC SPECIMENS.*—(*Continued*)

Metal or alloy	Composition of electrolyte	Current density, amp per sq dm	Voltage	Time	Remarks
No. 33. Steel—austenitic type........	Acetic anhydride—2 parts Perchloric acid—1 part Aluminum — about 0.5 per cent, as in No. 26	6	50†	4–5 min	Same as No. 26
No. 34. Steel—austenitic type	Same as No. 3	....	40–50†	Few sec	Same as No. 3
No. 35. Steel—all carbon steels in heat-treated condition	Same as No. 26	4–6	50†	4–5 min	Same as No. 26
No. 36. Steel—3% silicon	Same as No. 26	4–6	50†	4–5 min	Same as No. 26
No. 37. Steel—plain carbon and alloy...	Perchloric acid (65%)—7 parts Glacial acetic acid—10 parts	4–6	40–70	1–2 min	Stainless-steel cathode; ordinary temperature. Solution may be used immediately after mixing. No cooling necessary during use
No. 38. Steel—plain carbon and alloy...	Perchloric acid (65%)—1 part Glacial acetic acid—10 parts	3–6	40–70	15–60 sec	Aluminum or iron cathode; temperature 15–40°C. External resistor and specimen to cathode distance adjusted to give specified voltage and current-density ranges
No. 39. Steel—plain carbon and alloy..	Perchloric acid (65%)—1 part Glacial acetic acid—4 parts Aluminum or iron . . . 0.5% dissolved in electrolyte by electrolytic polishing of one of these metals Prepare electrolyte 24 hr before using	....	27–33	1.5–4 min	Aluminum or iron cathode; temperature less than 30°C; distance between specimen and cathode 3.5–2.5 in. with specimen 0.1–1.0 sq in. in area
No. 40. Tin.......	Perchloric acid—194 ml (sp gr 1.61) Acetic anhydride—806 ml	9–15	25–40†	8–10 min	Tin cathode; temperature 15–22°C; electrodes 2 cm' apart. Stir electrolyte if time of polishing exceeds 10 min
No. 41. Tin + 3% antimony	Same as No. 40	9–15	25–40†	8–10 min	Same as No. 40
No. 42. Zinc.......	Chromic acid—200 g Water—1000 ml	248 or more	....	....	Platinum-gauze cathode
No. 43. Zinc........	Potassium hydroxide — 20% aqueous solution	16	6	15 min	Copper cathode; ordinary temperature; electrodes 2.5–15 mm apart; electrolyte agitated by air or nitrogen
No. 44. Zinc........	Same as No. 3	....	40–50†	Few sec	Same as No. 3

Note: Time of polishing will depend upon concentration of electrolyte, cell arrangement, and initial smoothness of the surface.

* From ASTM Standards, Part I-B, 1946; *Metal Progress*, October, 1940; *Mining and Met.*, July, 1943; and *Metals & Alloys*, February, 1945.

† External applied voltage.

TABLE 2. ETCHING REAGENTS FOR MICROSCOPIC EXAMINATION OF STEELS AND IRONS*

Etching reagent	Composition†	Remarks	Uses
I. General reagents for irons and steels (carbon, low-, and medium-alloy steels)			
No. 1. Nitric acid (nital)	HNO_3, colorless...........1–5 ml Ethyl or methyl alcohol (95% or absolute)........100 ml (also amyl alcohol)	Etching rate is increased, selectively decreased, with increasing percentages of HNO_3. Reagent No. 2 (picric acid) usually superior. 4% in amyl alcohol useful for grain boundary and contrast of low-carbon materials. Etching time, a few sec to 1 min	In carbon steels: (1) to darken pearlite and give contrast between pearlite colonies; (2) to reveal ferrite boundaries; (3) to differentiate ferrite from martensite. The 1 per cent solution is also suitable for uses as noted below for reagent No. 2
No. 2. Picric acid (picral)	Picric acid............... 4 g Ethyl or methyl alcohol (95% or absolute).......100 ml (Use absolute alcohol only when acid contains 10% or more of moisture)	More dilute solutions occasionally useful. Does not reveal ferrite grain boundaries as readily as No. 1. Etching time, a few sec to 1 min or more	For all grades of carbon steels: annealed, normalized, quenched, quenched and tempered, spheroidized, austempered. For all low-alloy steels attacked by this reagent
No. 3. Hydrochloric and acids..............	HCl.................... 5 ml Picric acid............... 1 g Ethyl or methyl alcohol (45% or absolute)........100 ml	Best results are obtained when the martensite is tempered for 15 min at 400 to 475°F (205–245C)	For revealing the austenite grain size in quenched, and quenched-and-tempered steels
No. 4. Chromic acid....	CrO_3.................... 10 g H_2O....................100 ml	Used electrolytically, the specimen as anode, the stainless steel or platinum as cathode, ¾–1 in. apart; 6 volts usually used. Etching time, 30 to 90 sec, depending on specimen	For various structures except grain boundaries of ferrite. Attacks cementite very rapidly, austenite less rapidly, ferrite and iron phosphide very slowly if at all
No. 5. Heat tinting.......	Heat only	Clean, dry, polished specimen heated face up on hot plate to 400–700°F (205–370°C). Time and temperature both have decided effects. Bath of sand or molten metal may be used	Pearlite first to pass through a given color, followed by ferrite; cementite less affected, iron phosphide still less. Especially useful for cast irons
No. 6. Heat etching......	Heat only	Specimen is heated 10 to 60 min at 815-1205°C (1500–2200°F) in carefully purified hydrogen, and must have no contact with scale or reducible oxides. After etching, specimen is cooled in mercury to avoid oxidation	For revealing austenitic grain size of polished specimens

* ASTM Standards, Part I-B, 1946.
† The use of concentrated reagents is intended, unless otherwise specified. Only reagents of higher analytical purity should be used.

Table 2. Etching Reagents for Microscopic Examination of Steels and Irons.*—(*Continued*)

Etching reagent	Composition†	Remarks	Uses
II. General reagents for alloy steels (stainless and high-speed steels)			
No. 7. Ferric chloride and hydrochloric acid	FeCl₃.................... 5 g HCl...................... 50 ml H₂O.....................100 ml	....	Structure of austenitic nickel and stainless steels
No. 8. Nitric acid (nital)	HNO₃ (colorless)........5–10 ml Ethyl or methyl alcohol (95% or better).......95–90 ml	....	General structure of high-speed tool steel
No. 9. Chrome regia....	HCl..................... 25 ml CrO₃ solution (10%) in H₂O...................5–50 ml	Activity is controlled by amount of chromic acid	Heat-treated 18% chromium, 8% nickel stainless steels. Useful electrolytically, if diluted with 2 parts alcohol and 2 parts glycerine and applied for 20–60 sec with 6 volts
No. 10. Ferric chloride and nitric acid	Saturated solution of FeCl₃ in HCl, to which a little HNO₃ is added	Use full strength	Structure of stainless steel
No. 11. Mixed acids in glycerol.............	*A.* HNO₃................ 10 ml HCl................20–30 ml Glycerol...........30–20 ml	Warm specimen in water before etching. For best results use method of alternate polishing and etching. If given sufficient time, will etch totally austenitic alloys, but better results are obtained by using reagent No. 11-*C*	Structure of iron-chromium-base alloys, high-speed steels, and austenitic manganese steel. Etches nickel-chromium alloys satisfactorily
	B. HNO₃................ 10 ml HF................. 20 ml Glycerol...........20–40 ml	Amount of glycerol may be varied to suit metal	Structure of high silicon alloys of the Duriron type
	C. HNO₃................ 10 ml HCl.................. 20 ml Glycerol............. 20 ml H₂O₂................. 10 ml	Amount of HCl may be varied if reagents acts too rapidly or slowly. For best results employ method of alternate polishing and etching	To etch iron-chromium-nickel, iron-chromium-manganese, and all other austenitic iron-chromium-base alloys
No. 12. Oxalic acid.....	Oxalic acid............... 10 g H₂O.....................100 ml	Used electrolytically, the specimen as anode, stainless steel or platinum as cathode, about 1 in. apart; 6 volts usually used. Precipitated carbides in stainless steels may be revealed in 10 to 15 sec, the general structure in about 1 min. For study of carbides, 1.5–3 volts may be used, thus increasing the etching time and improving control of etch	For austenitic stainless steels and high-nickel alloys. Carbides and general structure revealed depending on etching time

* ASTM Standards, Part I-B, 1946.

† The use of concentrated reagents is intended, unless otherwise specified. Only reagents of higher analytical purity should be used.

TABLE 2. ETCHING REAGENTS FOR MICROSCOPIC EXAMINATION OF STEELS AND IRONS.*—(*Continued*)

Etching reagent	Composition†	Remarks	Uses
I1.	General reagents for alloy steels (stainless and high-speed steels).—(*Continued*)		
No. 13. Ammonium persulphate............	$(NH_4)_2S_2O_8$............... 10 g H_2O...................... 90 ml	Used electrolytically like the oxalic acid solution, but acts more slowly, requiring longer than 15 sec with 6 volts	For fast-etching stainless steels
No. 14. Cupric chloride and hydrochloric acid	$CuCl_2$.................. 5 g HCl....................100 ml Ethyl alcohol............100 ml H_2O.....................100 ml	Used cold (Kalling's reagent)	For austenitic and ferritic steels, the ferrite being most easily attacked, carbides not attacked, and austenite attacked appreciably only when it has begun to decompose to martensite
No. 15. Mixed acids and cupric chloride.......	HCl..................... 30 ml HNO_3................... 10 ml Saturate with cupric chloride and let stand 20 to 30 min before use	Apply by swabbing	For stainless alloys and others high in nickel or cobalt
No. 16. Nitric and acetic acids.............	HNO_3................... 30 ml Acetic acid............... 20 ml	Apply by swabbing	For stainless alloys and others high in nickel or cobalt
No. 17. Nitric and hydrofluoric acids......	HNO_3.................... 5 ml HF (48%)................ 1 ml H_2O..................... 44 ml	Used cold for about 5 min	For revealing general structure of austenitic stainless steel with avoidance of strain markings
No. 18. Hydrochloric and nitric acids in alcohol..............	HCl.................... 10 ml HNO_3.................... 3 ml Methyl alcohol............100 ml	Etch 2 to 10 min	To reveal the grain size of quenched, or quenched and tempered high-speed steel
No. 19. Hydrochloric acid in alcohol.......	HCl.................... 10 ml Alcohol.................. 90 ml	Used electrolytically for 10–30 sec with 6 volts; must be free from water to prevent tarnishing	For straight chromium and chromium-nickel steels
No. 20. Lactic and hydrochloric acids in alcohol...............	Lactic acid.............. 45 ml HCl..................... 10 ml Alcohol.................. 45 ml	Used electrolytically for 10–30 sec with 6 volts; does not tarnish	For chromium steels (4–30% chromium) or delta ferrite in austenitic stainless steels
No. 21. Ferricyanide solution..............	Potassium ferricyanide..... 30 g KOH..................... 30 g H_2O..................... 60 ml	Must be fresh; use boiling	To distinguish between ferrite and sigma phase in iron-chromium, iron-chromium-nickel, iron-chromium-manganese, and related alloys. Colors: sigma phase, light blue; ferrite, yellow

*ASTM Standards, Part I-B, 1946.

† The use of concentrated reagents is intended, unless otherwise specified. Only reagents of higher analytical purity should be used.

TABLE 2. ETCHING REAGENTS FOR MICROSCOPIC EXAMINATION OF STEELS AND IRONS.*—(*Continued*)

Etching reagent	Composition†	Remarks	Uses
II. General reagents for alloy steels (stainless and high-speed steels).—(*Continued*)			
No. 22. Cupric sulphate	$CuSO_4$ .... 4 g HCl .... 20 ml H_2O .... 20 ml	Marble's reagent	Structure of stainless steels
No. 3. Hydrochloric and picric acids	HCl .... 5 ml Picric acid .... 1 ml Ethyl or methyl alcohol (95% or absolute) .... 100 ml	....	To etch many steels of the iron-chromium, iron-chromium-nickel, and iron-chromium-manganese types
No. 23. Heat tinting	Heat only in air for 10–60 sec at about 595–650C (1100–1200°F)	Carbides remain white, and austenite darkens less rapidly than ferrite, on sections preferably etched first with a chemical reagent	For austenitic stainless steel containing ferrite and carbide
III. Segregation, primary structure, and strain lines			
No. 24. Cupric chloride	A. $CuCl_2$ .... 1 g $MgCl_2$ .... 4 g HCl .... 1 ml H_2O .... 20 ml Alcohol (absolute) ...... 100 ml	Dissolve salts in least possible quantity of hot water. Etch for about 1 min, repeating if necessary. Stead's reagent	For showing segregation of phosphorus or other elements in solid solution; copper tends to deposit first on areas lowest in phosphorus. The structure may be more clearly delineated by light hand-polishing to remove the copper deposit after etching
	B. $CuCl_2$ .... 5 g HCl .... 40 ml H_2O .... 30 ml Ethyl alcohol .... 25 ml	May be used cold; etching time, about 10 sec; Fry's reagent	To reveal strain lines and their microstructure, and precipitation hardening in steel
No. 25. Modified ferric chloride	$FeCl_3$ .... 30 ml $CuCl_2$ .... 1 ml $SnCl_2$ .... 0.5 ml HCl .... 50 ml Ethyl alcohol .... 500 ml H_2O .... 500 ml	Oberhoffer's reagent	For showing phosphorus segregation and dendritic structure
No. 26. Alkaline chromate	CrO_3 .... 16 g H_2O (distilled) .... 145 ml NaOH .... 80 g	Add NaOH slowly, and use when not over one day old, boiling at 118–120°C (244–248°F) for 7–20 min	Shows oxygen segregation by darkening martensite rapidly, ferrite more slowly, and zones of high oxygen content much more slowly

* ASTM Standards, Part I-B, 1946.
† The use of concentrated reagents is intended, unless otherwise specified. Only reagents of higher analytical purity should be used.

TABLE 2. ETCHING REAGENTS FOR MICROSCOPIC EXAMINATION OF STEELS AND IRONS.*—(*Continued*)

Etching reagent	Composition†	Remarks	Uses
	IV. Structure and depth of case of nitrided steels		
No. 27. Cupric sulphate and cupric chloride...	$CuSO_4$...................1.25 g $CuCl_2$...................2.50 g $MgCl_2$...................10 g HCl...................2 ml H_2O...................100 ml Dilute above solution to 1000 ml with 95% ethyl alcohol	Proportions must be accurate. Etch by immersion to avoid confusing edge effects	For showing total depth, structure, and various zones of nitrided chromium-vanadium steels and nitralloy
No. 28. Picric and nitric acids..............	Picric acid (4%) (No. 2) . 10 parts HNO_3 (4%) (No. 1)..... 1 part	Best results are obtained when the specimen is annealed in lead at 1475°F (800°C) before etching	For depth of case and structure of nitralloy
No. 1. Nitric acid (nital)	HNO_3...................2 ml Ethyl or methyl alcohol (95% or absolute).......100 ml	....	For structure and depth of case of nitrided steels
No. 22. Cupric sulphate	$CuSO_4$...................4 g HCl...................20 ml H_2O ....20 ml	Marble's reagent	Total depth of nitrided case
	V. Reagents for carbides, phosphides, nitrides, and tungstides		
No. 29. Sodium picrate (neutral)............	Sodium picrate........... 1 g H_2O...................100 ml (Wash salt well with alcohol to remove excess acid or alkali)	Use boiling; etching time, 20 min	Shows difference between phosphides and cementite; iron phosphide attacked, cementite unattacked
No. 30. Chromic acid and heat tinting......	CrO_3...................8 g H_2O...................100 ml Followed by heat tinting	Etch first in picric acid (No. 2), then for 1 min in chromic acid; heat tint by heating face upon hot plate at about 500°F (260°C) for 1 min	Distinguishes between iron phosphide and cementite in phosphide eutectic of cast iron: iron phosphide is colored darker
No. 31. Sodium picrate, alkaline............	Picric acid.............. 2 g NaOH................... 25 g H_2O...................100 ml	Use boiling 5–10 min, or preferably electrolytically at room temperature. For the latter, specimen is anode, platinum or stainless steel is cathode; with 6 volts about 40 sec is usually sufficient	Colors cementite but not carbides high in chromium. In tungsten steels, iron tungstide (Fe_nW) and iron tungsten carbide (Fe_4W_2C) are colored more rapidly than cementite, but tungsten carbide is unaffected. Attacks sulphides. Delineates grain boundaries in hypereutectoid steels in slowly cooled condition

* ASTM Standards, Part I-B, 1946.

† The use of concentrated reagents is intended, unless otherwise specified. Only reagents of higher analytical purity should be used.

TABLE 2. ETCHING REAGENTS FOR MICROSCOPIC EXAMINATION OF STEELS AND IRONS.*—(*Continued*)

Etching reagent	Composition†	Remarks	Uses
V. Reagents for carbides, phosphides, nitrates, and tungstides			
No. 32. Hydrogen peroxide and sodium hydroxide...........	H_2O_2.................... 10 ml NaOH (10% solution in water)................. 20 ml	Must be fresh; etching time, 10–12 min	Attacks and darkens iron tungstide in carbon-free iron tungsten alloys. When carbon is present this solution darkens the compound (FeW WC?) in proportion to the amount of carbide present; tungsten carbide is darkened
No. 33. Ferricyanide solution.............	A. $K_3Fe(CN)_6$...........1–4 g KOH................. 10 g H_2O...................100 ml	Must be freshly made; etch 15 min in boiling solution. Seven grams of NaOH may be substituted for 10 g of KOH in either A or B	Differentiates between carbides and nitrides. Cementite is blackened, pearlite turned brown, and massive nitrides remain unchanged
	B. $K_3Fe(CN)_6$........... 10 g KOH................. 10 g H_2O ....100 ml	May be used cold, but preferably hot, should be freshly made; etching time, 5–10 min. Murakami's reagent	Darkens carbide containing chromium, carbides, and tungstides in tungsten and high-speed steels. At room temperature colors ternary carbides (Fe_3W_3C or Fe_3W_2C) in a few seconds, iron tungstide (Fe_3W_2) in several minutes, and barely colors cementite
No. 34. Sodium cyanide	NaCN................. 10 g H_2O..................... 90 ml	Used electrolytically, the specimen as anode; cathode, similar material about 1 in. apart; 6 volts (not less than 5). Etching time, 5 min or more	Darkens carbides without attacking austenite or grain boundaries
No. 4. Chromic acid....	CrO_3.................... 10 g H_2O....................100 ml	See No. 4 in Group I	Attacks carbides in stainless steels very rapidly, austenite less rapidly, and ferrite very slowly if at all. For various structures of stainless steels
No. 12. Oxalic acid.....	Oxalic acid............... 10 g H_2O....................100 ml	See No. 12 in Group II. If strongly etched general structure is revealed; therefore for study of carbides reduced voltage is used for etching, giving better control of etch	Reveals carbides in stainless steels

*ASTM Standards, Part I-B, 1946.

† The use of concentrated reagents is intended, unless otherwise specified. Only reagents of higher analytical purity should be used.

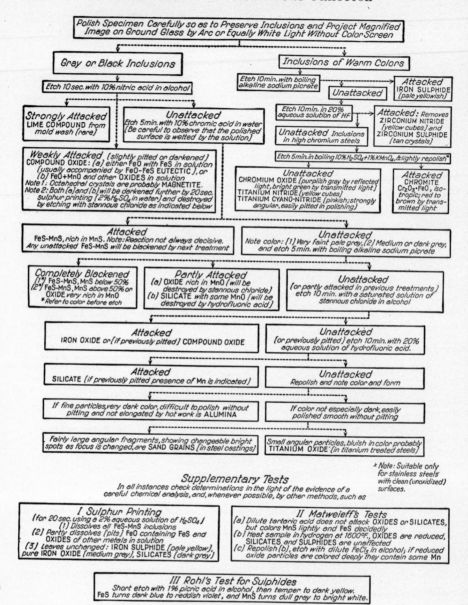

CHART 1. Method for identification of inclusions in iron and steel. (*From Metal Progress, Vol. 45, 1944.*)

TABLE 3. ETCHING REAGENTS FOR COPPER AND ITS ALLOYS*

Etching reagent	Composition	Remarks	Use
No. 1. Ammonium hydroxide–hydrogen peroxide.............	NH_4OH (sp gr 0.88)..... 5 parts H_2O.................. 5 parts H_2O_2 (3 per cent)......2–5 parts	Peroxide content varies directly with copper content of alloy to be etched. Immersion or swabbing for about 1 min. Fresh H_2O_2 is desirable for good results	Generally used for copper and many of its alloys. Film on etched aluminum bronze removed by weak Grard's solution
No. 2. Ammonium hydroxide	Dilute solutions	Immersion	Polish-attack etching of brass and bronze
No. 3. Ammonium hydroxide – ammonium persulphate..........	NH_4OH.................1 part H_2O.....................1 part $(NH_4)_2S_2O_8$ (2.5 per cent solution)...............2 parts	Immersion	Polish-attack of copper and some alloys
No. 4. Ammonium persulphate.............	$(NH_4)_2S_2O_8$.................10 g H_2O.....................90 ml	Use either cold or boiling; immersion	Copper, brass, bronze, nickel silver, aluminum bronze
No. 5. Chromic acid......	Saturated aqueous solution (CrO_3)	Immersion or swabbing	Copper, brass, bronze, nickel silver (plain etch)
No. 6. Chromic acid......	1% aqueous solution.............	Use electrolytically at 6 volts, with aluminum cathode, 3–6 sec	Aluminum bronze and beryllium copper
No. 7. Chromic acid–hydrochloric acid.......	CrO_3 (10 to 15 per cent).....50 ml HCl.................1–2 drops	Add HCl at time of use; immersion	Same as reagent No. 5. Color by electrolytic etching or $FeCl_3$ reagents
No. 8. Chromic acid–nitric acid	HNO_3 (conc)..............50 ml H_2CrO_4....................20 g H_2O.....................30 ml *or* HNO_3 (conc)............... 5 ml H_2CrO_4...................20 g H_2O.....................75 ml	Immersion	Aluminum bronze; film from polishing removed by 10% HF
No. 9. Copper ammonium chloride–ammonium hydroxide	10 per cent aqueous solution of copper ammonium chloride plus NH_4OH to neutral or alkalinity	Immersion; wash specimen thoroughly	Best for darkening large areas of beta in alpha-beta brass. Copper, brass, nickel silver
No. 10. Ferric chloride..	PARTS $FeCl_3$ HCl H_2O 5 50 100 19 6 100 5 10 100 25 25 100 1 20 100 8 25 100 10 1 100 3 10 100	Immersion or swabbing Grard's No. 1 Grard's No. 2 Plus 1 part CrO_3 Plus 1 part $CuCl_2$ and 0.05 part $SnCl_2$	Copper, brass, bronze, aluminum bronze; darkens beta in brass; gives contrast following bichromate and other etches. Etch lightly or by successive light etches to required results

* From ASTM Standards, Part I·B, 1946.

TABLE 3.   ETCHING REAGENTS FOR COPPER AND ITS ALLOYS.*—*(Continued)*

Etching reagent	Composition	Remarks	Use
No. 11. Ferric chloride..	FeCl₃..................... 5 g Ethyl alcohol..............96 ml HCl..................... 2 ml	Immersion or swabbing for 1 sec to several min	Copper, aluminum, magnesium, nickel, and zinc alloys, etc
No. 12. Nitric acid.......	Various concentrations	Immersion or swabbing	Deep etching
No. 13. Potassium bichromate..........	K₂Cr₂O₇.................. 2 g H₂SO₄ (sp gr 1.84)........ 8 ml NaCl (saturated solution).. 4 ml H₂O.....................100 ml	NaCl can be replaced by 1 drop HCl to 25 ml solution added just before using. Immersion	Copper, copper alloys of beryllium, manganese, silicon, nickel silver, and bronze. Followed by FeCl₃ or other contrast etch
No. 14. Electrolytic etch	FeSO₄.................. 30 g NaOH.................. 4 g H₂SO₄.................. 100 ml H₂O..................1900 ml	Use 0.1 amp at 8 to 10 volts. Generally not over 15 sec. Do not swab surface after etching	Darkens beta in brass, gives contrast after H₂O₂ etch. Nickel silver, bronze, and other alloys

* From ASTM Standards, Part I-B, 1946.

TABLE 4. ETCHING REAGENTS FOR TIN AND ITS ALLOYS*

Etching reagent	Composition	Remarks	Use
No. 1. Nital	HNO$_3$ (conc)............2–5 ml C$_2$H$_5$OH..............95–98 ml	Swab or immerse for several minutes	Tin-cadmium or tin-iron alloys
No. 2. Potassium dichromate	Acidified, dilute solution	....	Tin-cadmium alloys
No. 3. Mixed acids in glycerol	A. HNO$_3$ (conc)..........1 part Acetic acid............1 part Glycerol..............8 parts B. HNO$_3$ (conc)..........1 part Acetic acid............3 parts Glycerol..............5 parts	Immerse ½–10 min at 38 to 42°C Same as for *A*, above	Tin-lead alloys Pure tin
No. 4. Hydrochloric acid	A. HCl (conc) B. HCl (conc)..........10–20 ml H$_2$O................90–80 ml C. HCl (conc)............ 10 ml H$_2$O................. 90 ml	Immerse for several seconds Immerse ½–5 min following *A* Electrolytic etch at low current density	To remove surface flow Follows *A* Tin-iron alloys
No. 5. Ferric chloride	FeCl$_3$.................... 10 g HCl (conc)............... 2 ml H$_2$O.................... 95 ml	Immerse ½–5 min at room temperature	Microetching tin-rich babbitt metal
No. 6. Nitric and picric acids	Alcoholic solution	....	Tin-iron alloys
No. 7. Ammonium polysulphide	Concentrated solutions	Immerse 20–30 min at room temperature	Macroetching of tin-rich babbitt metal
No. 8. Nitric acid	HNO$_3$ (conc).............. 5 ml C$_2$H$_5$OH (absolute)........95 ml	....	Pure tin
No 9. Ammonium persulphate	Ammonium persulphate..... 5 ml H$_2$O....................95 ml	....	Tin coatings on steel
No. 10. Acetic acid	Acetic acid................50 ml H$_2$O....................50 ml H$_2$O$_2$....................1 drop	....	Soldered joint
No. 11. Acid ferric chloride	FeCl$_3$.................... 2 g HCl (conc)............... 5 ml H$_2$O................30 ml C$_2$H$_5$OH................60 ml	Alternate polish and etch	For use on block tin
No. 12. Nitric and hydrofluoric acids	HNO$_3$ (conc)............1 drop HF....................2 drops Glycerine................ 25 ml	Etch for 1 min at 70–80°F	To reveal structure of tin coating on tin plate stock

* ASTM Standards, Part I-B, 1946; and ASM: "Metals Handbook," 1948.
Solutions of silver nitrate or 10 per cent nitric acid plus 5 per cent chromic acid are also occasionally used.

TABLE 5. ETCHING REAGENTS FOR ZINC AND ITS ALLOYS*

Etching reagent	Composition	Remarks	Use
No. 1. Palmerton reagent...............	CrO_3 (99.95%) ....200 g Na_2SO_4 (c.p.) .... 15 g H_2O ....1000 ml	Immersion with gentle agitation. Follow with rinse in solution of CrO_3.......... 200 g H_2O ....1000 ml	General. (Reduce Na_2SO_4 to 7.5 g when using solution to develop grain structure in alloys containing copper)
No. 2. Dilute Palmerton reagent.............	CrO_3 (99.95%) .... 50 g Na_2SO_4 (c.p.) .... 4 g H_2O ....1000 ml	Immerse for 2–3 sec, follow with rinse as in No. 1	Structure of die-casting alloys, also for contrast between the same and plated coatings
No. 3. Dilute nitric acid	*A.* HNO_3 (conc) ....1 drop Ethyl alcohol......... 10 ml *or* *B.* HNO_3 (conc) ....3 drops Amyl alcohol.......... 50 ml	Etch by immersion. Reagent *B* should be used within 1 hr after mixing, and the specimen should be rinsed in either absolute ethyl or amyl alcohol	For showing the zinc-iron alloys in galvanized steel or iron. Reagent *B* is preferred for this purpose
No. 4. Chromic anhydride...............	CrO_3 .... 200 g Na_2SO_4 .... 7 g NaF.................... 2 g H_2O ....1000 ml	Immersion for about 2 sec. Alloys containing copper should be rinsed (1 sec) in a solution of CrO_3.......... 50 g Na_2SO_4........ 4 g H_2O ....1000 ml Final rinsing in a solution of CrO_3 (pure)..... 200 g H_2O ....1000 ml	Die-casting alloys. Reveals gamma phase
No. 5. Hydrochloric acid	HCl (conc)............... 5 ml Ethyl alcohol..............95 ml	Immersion for 2–3 sec	Contrast between die casting and plated coatings
No. 6. Hydrochloric acid..	HCl (conc).....................................	....	Macrostructure of pure zinc

* ASTM Standards, Part I-B, 1946; and ASM: "Metals Handbook," 1948.

TABLE 6. ETCHING REAGENTS FOR LEAD AND ITS ALLOYS*

Etching reagent	Composition	Remarks	Use
No. 1. Ammonium molybdate	Molybdic acid (85%)......100 g NH₄OH (sp gr 0.9)........140 ml H₂O ....240 ml Filter and add to HNO₃ (sp gr 1.32)........ 60 ml	Alternately swab specimen and wash in running water	Rapid etch; very suitable for removing thick layers of worked metal
No. 2. Mixed acids	Glacial acetic acid........ 3 parts HNO₃ (conc)............ 4 parts H₂O....................16 parts	Use freshly prepared solution at 40–42°C (104–108°F). Immerse 4–30 min depending on depth of worked metal layer. Clean with wet cotton in running water	Gives excellent detail for photographing
No. 3. Mixed acids in glyserol	Glacial acetic acid........ 1 part HNO₃ (conc)............ 1 part Glycerol................ 4 parts	Use freshly prepared solution at 80°C (176°F)	Used for alternate polishing and etching, and to reveal grain boundaries in pure lead
No. 4. Ammonium molybdate–mixed acids	See reagents Nos. 1 and 2	Swab with reagent No. 1 until structure is clearly visible; immerse in reagent No. 2 for about 6 min; clean by swabbing in running water	Combines advantages of reagents Nos. 1 and 2
No. 5. Acetic acid–hydrogen peroxide	A. Glacial acetic acid......3 parts H₂O₂ (9%)............1 part	Etch 10–30 min, depending on depth of worked metal layer; clean in concentrated HNO₃, if necessary	Recommended for antimony alloys up to 2 per cent antimony
	B. Glacial acetic acid......3 parts H₂O₂ (30%)............1 part	Etch for 6–15 sec	For lead-antimony alloys
	C. Glacial acetic acid......2 parts H₂O₂ (30%)............1 part	Etch 8–15 sec by immersion	For pure lead and lead-calcium alloys
No. 6. Perchloric acid	HClO₄....................60 ml H₂O......................40 ml	Electrolytic etching; specimen cathode, platinum anode	Recommended for antimony alloys over 2 per cent antimony
No. 7. Silver nitrate	AgNO₃................5–10 g H₂O..................95–90 ml	Use by swabbing	For antifriction metals
No. 8. Nitric acid	HNO₃ (conc)............. 50 ml H₂O..................... 50 ml	Etch in boiling solution 5–10 min	Macroscopic etching; welds, laminations, etc.
No. 9. Nitric acid	HNO₃ (conc)...................	Alternate in acid and running water	Pure lead

* ASTM Standards, Part I-B, 1946.

TABLE 7. ETCHING REAGENTS FOR ALUMINUM AND ITS ALLOYS*

Etching reagent	Composition	Remarks	Use
No. 1. Hydrofluoric acid	HF (conc) .... 0.5 ml H_2O ....99.5 ml	Swab with soft cotton for 15 sec	General; microscopic
No. 2. Sodium hydroxide	NaOH.................. 1 g H_2O .... 99 ml *or* NaOH.................. 10 g H_2O .... 90 ml	Swab for 10 sec Immerse for 5 sec at 70°C (160°F); rinse in cold water	General; microscopic Can be used for both micro- and macroetching
No. 3. Sulphuric acid...	H_2SO_4 (conc) .... 20 ml H_2O .... 80 ml	Immerse for 30 sec at 70°C (160°F); quench in cold water	Aluminum-copper-iron-manganese from aluminum-iron-manganese or aluminum-copper-iron
No. 3A. Sulphuric acid..	H_2SO_4 (conc) .... 10 ml H_2O .... 90 ml	Immerse at 60–70°C	FeAl₃
No. 4. Nitric acid......	HNO_3 (conc) .... 25 ml H_2O .... 75 ml	Immerse for 40 sec at 70°C (160°F); quench in cold water	αAluminum-iron-silicon from FeAl₃; microscopic
No. 5. Keller's etch....	HF (conc) ....1.0 ml HCl (conc)...............1.5 ml HNO_3 (conc) ....2.5 ml H_2O ....95.0 ml	Immerse for 10–20 sec; wash in stream of warm water	Microstructure of duralumin-type alloys, 17ST, and alclad
No. 6. Keller's concentrated etch..........	HF (conc) .... 10 ml HCl (conc).............. 15 ml HNO_3 (conc) .... 25 ml H_2O .... 50 ml	Use concentrated for macroetching, dilute 9 to 1 with water for microetching. Store concentrated solution in wax bottle	Excellent micro- and macroetchant for copper-bearing alloys
No. 7. Double etch.....	A. HNO_3 (conc)........... 25 ml 　　H_2O .... 75 ml B. NaF...................0.5 g 　　HNO_3 (conc) ....1.0 ml 　　HCl (conc) ....2.0 ml 　　H_2O ....97.0 ml	Immerse specimen in solution A for 60 sec at 70°C (160°F), quench in cold water. Then immerse in solution B for 15–30 sec, wash in warm running water	This method is used for determining quality of heat-treatment of duralumin alloys and presences of CuAl₂ precipitate at grain boundaries
No. 8. Flick's etch.....	HF (conc) .... 10 ml HCl (conc).............. 15 ml H_2O .... 90 ml	Immerse for 10–20 sec; wash in warm water followed by dip in concentrated HNO_3	Macroscopic etching
No. 9. Tucker's etch...	HF (conc) .... 15 ml HCl (conc).............. 45 ml HNO_3 (conc) .... 15 ml H_2O .... 25 ml	Etch by immersion	Macroscopic
No. 10. Vilella's etch...	HF (conc) ....2 parts HNO_3 (conc)............1 part Glycerin.................3 parts	....	General
No. 11. Bossert's etch..	NaOH.................. 1 g Na_2CO_3.................. 1 g H_2O ....94 ml Solution containing 0.5% each of $ZnCl_2$ and $SnCl_2$... 4 ml	Solutions to be fresh. Immersion until surface becomes black (3–5 min). Black deposit removed by immersion in conc HNO_3	Aluminum-copper-magnesium-manganese alloy. Microscopic examination of cold-worked 24S or annealed 24S and 24OS

* ASTM Standards, Part I-B, 1946; and ASM: *Metal Progress*. Also see pp. 425 and 426.

Note: Successful etching of aluminum and aluminum alloys depends upon the care with which the temperature, etching time, and concentration of the reagent is controlled.

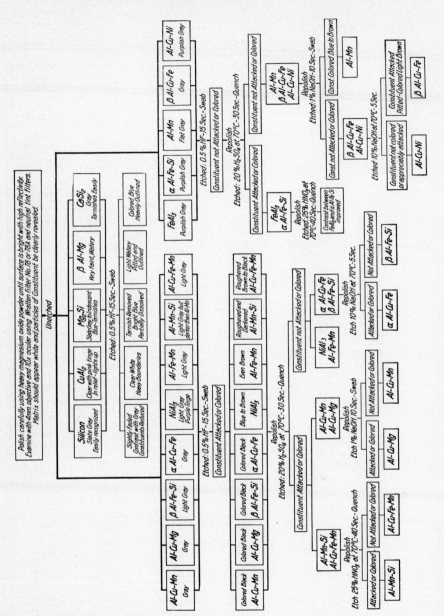

CHART 2. Method for identification of the common metallographic phases in aluminum alloys. (*From Metal Progress, Vol. 23, 1933.*)

Constituents	Reagent					
	0.5% HF Swab for 15 seconds, wash in cold water	1% NaOH Swab for 10 seconds, wash in running water	20% H_2SO_4 at 70°C, immerse specimen for 30 seconds, quench in cold water	25% HNO_3 at 70°C, immerse specimen for 40 seconds, quench in cold water	10% NaOH at 70°C, immerse specimen for 5 seconds, rinse in cold water	0.5% HF, 1.5% HCl, 2.5% HNO_3, immerse specimen for 15 sec, rinse in warm water
Silicon	Outlined. Unattacked. Color lightened.	Outlined. Unattacked. Color slightly lightened.	Unattacked. Color lightened.	Outlined. Unattacked. Color lightened.	Outlined. Unattacked. Color lightened.	Outlined. Unattacked. Color lightened.
Mg_2Si	Colored bright blue.	Outlined. Color unchanged.	Action violent. Some particles dissolved, any left have a blue color.	Colored brown or black.	Outlined. Color lightened.	Outlined. Colored blue to brown.
$CuAl_2$	Outlined. Part of pinkish tinge removed. Constituent light and clear.	Outlined. Part of pinkish tinge removed. Constituent light and clear.	Outlined. Part of pinkish tinge removed. Constituent light and clear.	Colored brown or black.	Pitted. Colored light to dark brown.	Outlined. Constituent light and clear.
β Al-Mg	Outlined. Slightly clearer and more watery. Black pits appear in particles.	Not Outlined. Unattacked. Uncolored.	Attacked vigorously resulting in pitting. Some particles dissolved.	Heavy attack, particles grayish and watery.	Outlined. Unattacked. Uncolored.	Heavily outlined. Attacked by pitting.
$FeAl_3$	Slightly darkened. Brown stains appear on large primary particles.	Outlined. Slightly darkened.	Heavily attacked. Particles often dissolved or deeply pitted. Color darkened.	Contrast with Al-Fe-Si improved.	Outlined. Colored deep brown.	Outlined. Uncolored. Slightly attacked.
α Al-Fe-Si	Outlined. Not colored.	Outlined. Not colored.	Outlined. Blackened and attacked.	Contrast with $FeAl_3$ improved.	Attacked. Blackened.	Heavily outlined. Darkened and roughened.
β Al-Fe-Si	Blackened and attacked	Outlined. Uncolored. Unattacked.	Outlined. Slightly darkened and pitted.	Outlined. Uncolored. Unattacked.	Outlined. Slightly darkened and attacked.	Outlined. Slightly darkened and roughened.
Al-Mn	Outlined. Slightly darkened.	Attacked. Colored brownish or bluish but coloring is uneven.	Outlined. Uncolored. Unattacked.	Not Outlined. Unattacked. Uncolored.	Colored blue or brown.	Outlined. Unattacked. Uncolored.
$NiAl_3$	Outlined. Colored blue and brown.	Outlined. Darkened slightly. Not colored.	Outlined. Darkened slightly. Not colored.	Outlined. Uncolored. Unattacked.	Colored blue to deep brown.	Colored brown to black.
Al-Fe-Mn	Outlined. Colored brown. (Sometimes bluish)	Outlined. Particles pitted. (Often a rough blue color on a few particles.)	Outlined. Unattacked. Uncolored.	Outlined. Unattacked. Uncolored.	Colored deep brown to blue.	Outlined. Attacked. Darkened.
Al-Cu-Ni	Outlined. Unattacked. Darkened.	Outlined. Slightly darkened. Not attacked.	Outlined. Slightly darkened. Not attacked.	Blackened. Some particles dissolved.	Outlined. Unattacked. Uncolored.	Outlined. Some large particles stained unevenly.
α Al-Cu-Fe	Outlined. Blackened.	Outlined. Unattacked. Uncolored.	Outlined. Unattacked. Uncolored.	Outlined. Unattacked. Uncolored.	Outlined. Blackened. Attacked.	Blackened.
β Al-Cu-Fe	Outlined. Unattacked. Uncolored.	Outlined. Slightly darkened.	Outlined and uncolored. Often show black cores which are probably $FeAl_3$	Outlined. Slightly darkened. Unattacked.	Pitted. Colored light brown.	Outlined. Blackened.
Al-Cu-Fe-Mn	Outlined. Colored light brown to black. Usually appears roughened.	Outlined. Uncolored.	Outlined. Blackened.	Outlined. Unattacked. Uncolored.	Outlined. Uncolored.	Outlined. Colored brown to black.
Al-Mn-Si	Outlined. Colored light brown to black. Usually appears roughened.	Outlined. Usually appears rough and attacked. Slightly darkened.	Outlined. Appears rough and attacked. Darkened slightly.	Outlined. Appears rough. Darkened slightly.	Outlined. Attacked. Color not changed.	Outlined. Slightly darkened.
Al-Cu-Mg	Outlined. Blackened.	Outlined. Colored light brown.	Outlined. Attacked. Blackened.	Outlined. Attacked. Blackened.	Outlined. Unattacked. Colored brown.	Colored brown to black.
$CaSi_2$	Colored blue. Heavily outlined.	Outlined. Color unchanged.	Outlined. Colored blue. Roughened.	Outlined. Unattacked. Uncolored.	Outlined. Blackened. Attacked and roughened.	Colored brown to blue. Mottled.
Al-Cu-Mn	Outlined. Attacked. Blackened.	Outlined. Unattacked. Uncolored.	Outlined. Colored light brown.	Outlined. Unattacked. Uncolored.	Outlined. Attacked. Blackened.	Attacked. Blackened.
$CrAl_7$	Outlined. Unattacked. Uncolored.	Outlined. Unattacked. Uncolored.	Outlined. Unattacked. Uncolored.	Outlined. Unattacked. Uncolored.	Outlined. Colored blue to brown unevenly.	Outlined. Unattacked. Uncolored.
Al-O-Fe	Colored light brown. Unattacked.	Not outlined. Unattacked. Uncolored.	Outlined. Unattacked. Uncolored.	Not outlined. Unattacked. Uncolored	Small particles colored brown to black. Large particles stained all colors.	Outlined. Colored light brown. Not attacked.

CHART 3. The etching characteristics of the common metallographic phases in aluminum alloys. (*From Metal Progress, Vol. 23, 1933.*)

TABLE 8. ETCHING REAGENTS FOR MAGNESIUM AND ITS ALLOYS*

Etching reagent	Composition	Remarks	Use
No. 1. Acetic acid........	10% aqueous solution	Swab with cotton for ½–2 min	Macroetching
No. 2. Oxalic acid........	2% aqueous solution	Swab for 2–5 sec	Cast and wrought magnesium and most alloys in cast form
No. 3. Nitric acid........	2% aqueous solution	Swabbing	To develop coring and for etching some casting alloys
No. 4. Tartaric acid......	2% aqueous solution	Etch 10–20 sec	For magnesium-manganese wrought alloys and cast and heat-treated magnesium-aluminum-manganese-zinc alloys
No. 5. Tartaric acid......	10% aqueous solution..........	Immerse specimen with polished surface up	Macroetching for flow lines in forgings, revealing grain size in cast alloys
No. 6. Acetic-picral....	6% picric acid in 95% ethanol...............100 ml Glacial acetic acid....... 10 ml	Immerse polished face up. Make up etchant fresh just before using	Macroetch for grain size in solution-heat-treated castings
No. 7. Phosphopicral...	Ethyl alcohol...........100 ml Picric acid.............. 4 g Orthophosphoric acid (sp gr-1.70)............0.7 ml	Immerse specimen face up until stained	For extreme contrast between compounds and solid solutions
No. 8. Acetic-glycol....	Ethylene glycol......... 60 ml H_2O.................... 19 ml Glacial acetic acid....... 20 ml HNO_3 (conc).......... 1 ml	Swab or immersion for 2–15 sec	All wrought alloys and solution-heat treated cast alloys. Also for Mg-Mn alloys
No. 9. Glycol.........	Ethylene glycol......... 75 ml H_2O.................... 24 ml HNO_3 (conc).......... 1 ml	Immerse specimen face up and agitate during etching. Time, 3–60 sec. Remove specimen and plunge into running water	For almost all magnesium alloys. Excellent for revealing and outlining constituents without pitting. Excellent for alloys in the aged conditions
No. 10. Citric acid.....	Citric acid............... 5 g H_2O....................95 ml	Swab with soft cotton, rinse in hot water	Wrought magnesium-manganese alloys. Reveals grain boundaries
No. 11. Malic acid.....	Malic acid.............. 5 g HNO_3 (conc)............ 2 ml Ethyl alcohol............93 ml	Swab, rinse in hot water, immerse in acetone, and dry quickly	Wrought magnesium-aluminum alloys. Reveals grain boundaries

*ASTM Standards, Part I-B, 1946; and ASM: "Metals Handbook," 1948.

TABLE 9. MICROSCOPIC IDENTIFICATION OF CONSTITUENTS IN SAND-CAST MAGNESIUM ALLOYS*

Element	Detection		Unetched		Etched with glycol etchant; No. 9, Table 8	
	Optimum condition	Approx. min, per cent	Shape	Color	Shape	Color
Al ($Mg_{17}Al_{12}$): 　Massive.....	Etched	2.0	....	....	Filigreed network[b]	White
Precipitated.	Etched	......	....	....	Lamellar or fine particles	Appears dark at low power
Zn ($MgZn_2$): 　Massive.....	Etched	1.5	....	....	Filigreed network[b]	White
Precipitated.	Etched	......	....	....	Fine particles	Appears dark at low power
Zn (Mg-AlZn): 　Massive.....	Etched	2.0	....	....	Massive[b]	White
Precipitated.	Etched	......	....	....	Lamellar or fine particles	Appears dark at low power
Mn..........	Unetched or etched	.... 0.5	[d] Angular	Bluish gray	Angular	Bluish gray
Cd..........	[a]	[a]	[a]	[a]	[a]	[a]
Si (Mg_2Si).....	Unetched or etched	0.03	Plates or script	Light blue	Angular plates or script	Blue
Sn (Mg_2Sn)....	Etched	4.0	Network	Blue	Filigreed or massive network[e]	Brown, dark blue, purple
Cu (Mg_2Cu)...	Etched	0.5	....	....	Globular or network[b,c]	White
Ni (Mg_2Ni)....	Etched	0.5	....	....	Globular or network[b,c]	White

[a] Cadmium is completely soluble in solid magnesium and its alloys. No constituent visible under microscope. Coring generally visible when etched.
[b] Outlined.
[c] Depending on concentration.
[d] Manganese constituent very hard and polishes in relief. Usually appears as pits at low magnification and particles are resolved at a magnification of $250 \times$ or more.
[e] Usually shows coring around constituent.

* ASTM Standards, Part I-B, 1946.

TABLE 10. ETCHING REAGENTS FOR NICKEL AND ITS ALLOYS*

Etching reagent	Composition	Remarks	Use
No. 1. Flat solution....	HNO_3 (conc)........... 50 ml Glacial acetic acid....... 50 ml	Make up fresh daily; use clear white HNO_3 to avoid staining; etch by immersion at room temperature for 5–20 sec	Nickel, Monel metal and other nickel-copper alloys. Dilute with 25–50% acetone for alloys of less than 25% nickel
No. 2. Electrolytic contrast...............	Glacial acetic acid....... 5 ml H_2O.................... 85 ml HNO_3 (conc)........... 10 ml	Electrolytic etch for 20–60 sec using 1.5-volt dry cell and platinum wires	Very satisfactory for grain-size studies. Stains less than flat solution
No. 3. Nitric acid......	HNO_3 (conc)........... 30 ml H_2O.................... 70 ml	Etch by immersion	Macroetch for nickel silver
No. 4. Potassium cyanide...............	KCN.................. 5 g H_2O.................... 95 ml H_2O_2...............few drops	Etch by immersion	For low-zinc nickel silver
No. 5. Ammonium hydroxide–hydrogen peroxide...............	NH_4OH.............. 85 ml H_2O_2.................. 15 ml	Etch by immersion	For high-zinc nickel silver
No. 6. Electrolytic-sulphuric.............	H_2SO_4 (conc)........... 5 ml H_2O.................... 95 ml	Electrolytic etch for 5–15 sec. Use 2 or 3 1.5-volt dry cells and platinum wires	For nickel and Inconel
No. 7. Aqua regia	HNO_3 (conc)........... 5 ml HCl (conc)............. 25 ml H_2O.................. 30 ml	Immersion for 30 sec–2 min	Inconel
No. 8. Nitric-hydrofluoric.................	HNO_3 (conc)........... 20 ml HF (48%).......... 15 drops	Etch by immersion	Inconel

* Mainly from ASTM Standards, Part I-B, 1946.

TABLE 11. ETCHING REAGENTS FOR THE PRECIOUS METALS*

Etching reagent	Composition	Remarks	Use
No. 1. Potassium cyanide–ammonium persulphate	A. KCN (5% solution).. 1 part (NH₄)₂S₂O₈ (5% solution)..............1 part	Etch 1–2 min Addition of 2% KI will produce more rapid attack.	Pure Silver
	B. KCN (10% solution).1 part (NH₄)₂S₂O₈ (10% solution)..............1 part	Etch ½–3 min. Make up fresh each time	Gold, nearly all karat golds
	C. KCN (20% solution)..1 part (NH₄)₂S₂O₈ (20% solution)..............1 part	3% KI may be added	A slow etch for palladium or the complicated dental alloys
No. 2. Chromate	A. HNO₃ (1:1)..........100 ml K₂Cr₂O₇............. 2 g B. H₂CrO₄............. 20 g Na₂SO₄.............1.5 g H₂O................100 ml	Dilute A to 20 vol, add equal amount of B. Apply with camel's-hair brush. Non-adherent film of red silver chromate should form. If film adheres, add more of A; if none forms, add B	Used for silver alloys
No. 3. Chromic acid–sulphuric acid	Aqueous solution, 0.2% each H₂CrO₄, and H₂SO₄	Etch 1 min	Silver alloys
No. 4. Electrolytic etching	A. 1% solution of HF plus small amount of stannous chloride	....	Used for silver-tin alloy with over 73% silver
	B. Dilute solutions of HCl, KCN or KCN+KI	....	Occasionally used for gold
	C. KCN (5%)................	....	Used particularly for silver where silver is in contact with other materials, as in plated materials
No. 5. Dichromate	K₂Cr₂O₇ (saturated solution)................100 ml NaCl (saturated solution). 2 ml	Use 1 part solution to 9 parts water, apply by swabbing	Silver and silver alloys
No. 6. Ferric chloride	2% aqueous solution	Immersion for 5–30 sec	For silver solders to show structural details
No. 7. Iodine	50% solution of USP tincture of iodine in aqueous KI solution	Film of silver iodide may remain on surface of gold-silver alloys; can be removed with KCN solution	Gold alloys
No. 8. Aqua regia	A. Slightly dilute	Etch 5 min in warm solution	Pure platinum
	B. Concentrated	Use hot	Platinum alloys. Also used for gold alloys but will form chloride film when much silver is present. Ammonia or KCN will remove film
	C. In glycerin	Use cold	Palladium alloys

* Mainly from ASTM Standards, Part I-B, 1946.

TABLE 11. ETCHING REAGENTS FOR THE PRECIOUS METALS.*—(*Continued*)

Etching reagent	Composition	Remarks	Use
No. 9. Potassium sulphide	....	Use hot	Gold-nickel alloys
No. 10. Fused salts......	KOH + 10% KNO₃ or KHSO₄	Etch in fused salt	Used for platinum if aqua regia is ineffective
No. 11. Nitric acid.......	Concentrated	Use hot	Palladium
No. 12. Ammonium hydroxide–hydrogen peroxide.............	NH₄OH..............5 parts H₂O₂...............1–3 parts	Immersion	For silver or silver-palladium alloys
No. 13. Iodine...........	1–5% solution of tincture of iodine	Immersion. Iodine stain removed by immersion in a solution of sodium thiosulphate	Silver-zinc alloys

* Mainly from ASTM Standards, Part I-B, 1946.

TABLE 12. ETCHING REAGENTS FOR MISCELLANEOUS METALS AND ALLOYS

Alloy	Composition of reagent	Remarks	Uses
No. 1. Cadmium–bismuth	Iodine............... 1 part KI................... 3 parts H_2O.................10 parts	Etch by immersion	To reveal structure of cadmium–bismuth alloys
No. 2. Beryllium	H_2SO_4 (conc)........... 5 ml H_2O................... 95 ml *or* HF (conc)............. 0.5 ml H_2O.................99.5 ml	Etch by immersion for 1–15 sec	To reveal grain boundaries in beryllium, and general structure
No. 3. Silver–tin alloys	HF (conc)............. 1 ml H_2O................... 99 ml Add a small amount of stannous chloride	Electrolytic	For silver-tin alloys containing over 73% silver
No. 4. Zinc–iron alloys	*A.* HNO_3 (conc)....... 1 drop Ethyl alcohol........ 10 ml *or* *B.* HNO_3 (conc)....... 3 drops Amyl alcohol........ 50 ml	Etch by immersion Etchant *B* should be used within 1 hr after mixing, and the specimen should be rinsed in either absolute ethyl or amyl alcohol	To show the zinc-iron alloys in galvanized steel or iron Reagent *B* is preferred for this purpose
No. 5. Silver solders	2% aqueous solution of $FeCl_3$	Etch by immersion for 5–30 sec	To show structural details in silver solders
No. 6. Indium–zinc alloys	*A.* CrO_3............... 10 g H_2O............... 90 ml *B.* $K_2Cr_2O_7$............. 6 g H_2SO_4 (conc)........ 20 ml NaCl (saturated solution)............... 12 ml H_2O...............300 ml *C.* CrO_3............... 20 g H_2O................ 80 ml Na_2SO_4............. 1 %	Approximately equal parts of solutions *A* and *B*, mixed together just before using	Mixture of solutions *A* and *B* for indium-rich alloys Solution for zinc-rich alloys
No. 7. Tantalum	HF (60% aqueous solution)...................1 part NH_4F (20% aqueous solution)...................1 part	Etch by immersion at 50°C	To reveal general structure
No. 8. Tungsten	Basis solution of K_3Fe (CN)$_6$ Various concentrations *or* H_2O_2...................... NH_4OH...................... Various proportions *or* HNO_3 (conc).................. HF.......................... Various proportions	Etch by immersion	General microscopic etching
No. 9. Cemented tungsten carbide	KOH (10% aqueous solution)...............1–2 parts K_3Fe (CN)$_6$ (10% aqueous solution)...............1 part	Etch by immersion in boiling solution. Time about 1–5 min	For general microscopic etching. Reveals tungsten carbide particles and produces contrast with bonding material, such as cobalt
No. 10. Cemented tungsten carbide	KOH (5% aqueous solution)...................1 part K_3Fe (CN)$_6$ (5% aqueous solution).........3 parts	Etch by immersion in boiling solution. Time about 10–30 sec	Same as No. 9

TABLE 13. ELECTROLYTES FOR ELECTROLYTIC ETCHING*

Metal or alloy	Composition of electrolyte	Cell data	Remarks
No. 1. Antimony alloys (over 2% antimony)..	$HClO_4$.............. 60 ml H_2O................. 40 ml	Use low current densities	Specimen is cathode, platinum anode
No. 2. Brass..........	$FeSO_4$.............. 30 g NaOH.............. 4 g H_2SO_4 (conc)..........100 ml H_2O................1900 ml	Use 0.1 amp at 8–10 volts; time usually not over 15 sec	Darkens beta in brass and gives contrast after H_2O_2–NH_4OH etch. Also for nickel, silver, bronze, and other alloys. Do not swab surface after etching
No. 3. Brass..........	0.10 m. (NH_4) $C_2H_3O_2$.... 10 ml 0.50 m. $Na_2S_2O_3$........ 30 ml 0.14 m. NH_4OH........ 30 ml H_2O.................. 30 ml	2 amp per sq in.	Useful for etching cold-worked brass. Time of etch varies with the composition of the brass and previous treatment of the specimen
No. 4. Brass............	Orthophosphoric acid (5–14%); 8% solution preferred, 5.8 deg Baumé, sp gr 1.042	Voltage range, 1–8 volts, time of etching, 5–7 sec, depending upon voltage and composition of brass	Useful for etching cartridge and free-cutting brass, and admiralty and gilding-metal compositions. *Note:* Same electrolyte as No. 8, Table 1
No. 5. Copper..........	Same as No. 4	Voltage range, 1–4; time of etching, about 10 sec	
No. 6. Copper containing beryllium or aluminum	1% aqueous solution of CrO_3	Use 6 volts and aluminum cathode; time of etching, about 3–6 sec	For general microscopic etching of beryllium copper and aluminum bronze
No. 7. Copper.........	Glacial acetic acid....... 5 ml HNO_3 (conc).......... 10 ml H_2O................. 30 ml	....	Useful for etching copper-nickel alloys, avoiding the appearance of "coring"
No. 8. Gold.............	Dilute solutions of HCl, KCN, or KCN + KI		
No. 9. Iridium..........	Dilute HCl solutions	0.02–0.1 amp per sq cm; time of etching, 30 min to 3 hr	A-c electrolysis. Useful for grain boundary delineation
No. 10. Iron..........	Copper ammonium chloride..................0.5 g H_2O...................100 ml	0.5 amp per sq dm; time of etching, 30 min. Repeat polishing and etching	To develop etching pits
No. 11. Magnesium and alloys..............	NaOH.............. 10 g H_2O.............. 90 ml	4 volts and about 0.5 amp per sq dm; time of etching, 2 to 4 min; copper cathode	Useful for the complex magnesium alloys containing aluminum, zinc, cadmium, and bismuth. Etching must follow immediately the polishing operation
No. 12. Nickel and nickel alloys.........	H_2SO_4 (conc).......... 5 ml H_2O.................. 95 ml	1.5 volts from a dry cell; time of etching, about 5–12 sec; use platinum wires	To reveal grain boundaries in nickel and Inconel

* ASTM Standards, Part I-B, 1946; ASM: "Metals Handbook," 1948; *Metal Progress;* and *Metals & Alloys.*

TABLE 13. ELECTROLYTES FOR ELECTROLYTIC ETCHING.*—(Continued)

Metal or alloy	Composition of electrolyte	Cell data	Remarks
No. 13. Nickel and nickel alloys.........	Glacial acetic acid....... 5 ml HNO_3 (conc).......... 10 ml H_2O.................. 85 ml	1.5 volts from a dry cell; time of etching, 20–60 sec; use platinum wires	Produces good structural contrast. Very satisfactory for grain boundary delineation
No. 14. Platinum......	A. HCl (conc)...............	0.5 amp per sq cm; time of etching, about 5 min	A-c electrolysis, using carbon or another specimen as the other electrode. Useful for grain-boundary delineation
	B. HCl (conc).......... 20 ml H_2O................ 80 ml Solution saturated with NaCl	0.02–0.2 amp per sq cm	Same as above, but produces a more brilliant structure
No. 15. Silver-tin alloys	HF (conc)............... 1 ml H_2O.................... 99 ml Add a small amount of stannous chloride		
No. 16. Stainless steel..	Oxalic acid............. 10 g H_2O....................100 ml	Specimen and platinum cathode spaced about 1 in. apart; 6 volts usually used. Precipitated carbides may be revealed in 10–15 sec, the general structure in about 1 min. For study of carbides, 1.5–3 volts may be used to increase etching time and improving control of etching	Particularly useful for austenitic steels and high-nickel alloys
No. 17. Stainless steel..	$(NH_4)_2S_2O_8$........... 10 g H_2O.................... 90 ml	Use as No. 16. Etching usually requires more than 15 sec	For fast etching of stainless steels
No. 18. Stainless steel..	Methyl alcohol......... 45 ml HCl (conc)............. 10 ml Lactic acid............. 45 ml	Time of etching, 10–30 sec at 6 volts	For chromium steels containing 4–30% chromium, or delta ferrite in austenitic stainless steels
No. 19. Stainless steel..	Alcohol................ 90 ml HCl (conc)............. 10 ml	Time of etching, 10–30 sec at 6 volts. Reagent must be free of water to prevent tarnishing	For straight chromium and chromium–nickel steels
No. 20. Stainless steel..	Ferric oxalate........... 10 g H_2O.................... 90 ml	0.1 amp at 60 volts, time, about 2 min	To reveal free carbides in all types of stainless steel
No. 21. Stainless steel..	Potassium ferricyanide... 10 g H_2O.................... 90 ml	0.5 amp at about 6 volts, time about 2 min	Same as reagent No. 20
No. 22. Stainless steel..	Tartaric acid........... 10 g H_2O.................... 90 ml	0.1 amp and 6 volts, time about 5 min	Same as reagent No. 20

* ASTM Standards, Part I-B, 1946; ASM: "Metals Handbook," 1948; *Metal Progress;* and *Metals & Alloys.*

TABLE 13. ELECTROLYTES FOR ELECTROLYTIC ETCHING.*—(*Continued*)

Metal or alloy	Composition of electrolyte	Cell data	Remarks
No. 23. Stainless steel..	"Chrome regia"........1 part Alcohol...............2 parts Glycerol.............2 parts	Time of etching, 20 sec to 1 min	Austenitic stainless steels
No. 24. Stainless steel..	Perchloric acid (70%).... 20 ml H_2O................... 80 ml	Time of etching, 20 sec to 2 min	Same as reagent No. 23
No. 25. Stainless steel..	Picric acid............. 2 g NaOH................ 25 g H_2O...................100 ml	Use boiling; 0.5–2 amp per sq in., time of etching, 2–0.5 min	Colors cementite but not carbides rich in chromium. Iron tungstide and Fe_4W_2C colored more rapidly than cementite. Delineates grain boundaries in hypereutectoid steels in annealed condition
No. 26. Steel.........	CrO_3................. 10 g H_2O...................100 ml	Specimen and stainless-steel or platinum cathode spaced ¾–1 in. apart at 6 volts; time of etching, 30–90 sec, depending upon specimen	For various structures except grain-boundary delineation. Attacks Fe_3C very rapidly, austenite less rapidly, and ferrite very slowly
No. 27. Steel.........	NaCN................. 10 g H_2O................... 90 ml	Specimen and cathode of similar material spaced about 1 in. apart at 6 volts (not less than 5 volts should be used); time of etching, 5 min or more	Darkens carbides without attacking austenite or grain boundaries
No. 28. Tin–iron alloys	HCl (conc)............. 10 ml H_2O................... 90 ml	Use low current densities	

* ASTM Standards, Part I-B, 1946; ASM: "Metals Handbook," 1948; *Metal Progress;* and *Metals & Alloys.*

TABLE 14. HEAT-TINTING CHARACTERISTICS OF GRAY CAST IRON*

Color of specimen to unaided eye	Microconstituent	Color of microconstituent at magnification of	
		100 ×	500 ×
First perceptible yellow	Steadite	White	White matrix with dark dots
	Graphite	Black and lusterless with purple edges	Many blue dots along the edges
	Ferrite	....	Brown, with blue dots
	Pearlite	Average color is brownish yellow	Ferrite is brown, cementite is white
Brown, with little or no trace of color	Steadite	Pale yellow	Grayish matrix with purple dots
	Graphite	....	Red, purple and white along the edges
	Ferrite	....	Brown to purple, with blue dots
	Pearlite	Average color is brown	
Dark brown, changing to purple	Steadite	The eutectic shows dark spots	Blue dots; the dendrites have purple edges
	Graphite	....	More blue and less purple along edges
	Ferrite	....	Similar to graphite
	Pearlite	Average color is brown with purple shades	Purple and brown, the lamellae cannot be distinguished
Purplish-red with blue spots	Steadite	Purple edges	Yellowish–red matrix
	Graphite	As above	As above
	Ferrite	....	Blue and white
	Pearlite	Average color is red and blue	Red and purple
Blue and white	Steadite	....	Red matrix with white dots
	Graphite	....	White and blue edges. Network structure appears at higher magnification
	Ferrite	....	White and blue
	Pearlite	Average color is blue and white	Blue and purple

* From *Materials & Methods*, April, 1946

TABLE 16. VERY HIGH-CONTRAST DEVELOPER FOR PLATES OF HIGH AVAILABLE
CONTRAST*
(Formula D-19)

Constituents	Advoirdupois	Metric
Water (about 125°F) (52°C).....	64 oz	2.0 liters
Elon (metol)...................	128 gr	8.8 g
Sodium sulphite, desiccated......	12 oz 360 gr	384.0 g
Hydroquinone..................	1 oz 75 gr	35.2 g
Sodium carbonate, desiccated....	6 oz 180 gr	192.0 g
Potassium bromide.............	300 gr	20.0 g
Cold water to make............	1 gal	4.0 liters

Dissolve the chemicals in the order given.
Use without dilution.
Development time at 20°C (68°F) for Kodak Metallographic plates:
Very high contrast (about 3.0 gamma): tray—8 min; tank—10 min.

* Eastman Kodak Co.

TABLE 17. HIGH-CONTRAST DEVELOPER FOR PLATES OF HIGH AVAILABLE CONTRAST*
(Formula DK-50)

Constituents	Avoirdupois	Metric
Water (about 125°F) (52°C).....	64 oz	500 cc
Elon (metol)..................	145 gr	2.5 g
Sodium sulphite, desiccated......	4 oz	30.0 g
Hydroquinone.................	145 gr	2.5 g
Kodalk.......................	1 oz 145 gr	10.0 g
Potassium bromide.............	29 gr	0.5 g
Cold water to make............	1 gal	1.0 liter

Dissolve the chemicals in the order given.
Use without dilution.
Development time at 20°C (68°F) for Kodak Metallographic plates:
High contrast (about 2.0 gamma): tray—6 min; tank—8 min.

* Eastman Kodak Co.

TABLE 18. STOCK DEVELOPER SOLUTION TO BE MODIFIED FOR LOW-, MEDIUM-, AND
HIGH-CONTRAST DEVELOPMENT*
(Formula D-76)

Constituents	Avoirdupois	Metric
Water (about 125°F) (52°C)......	96 oz	750 cc
Elon (metol)..................	116 gr	2.0 g
Sodium sulphite, desiccated.......	13¼ oz	100.0 g
Hydroquinone.................	290 gr	5.0 g
Borax†.......................	116 gr	2.0 g
Cold water to make.............	1 gal	1.0 liter

Dissolve the chemicals in the order given.
To modify, see Tables 19 and 20.

* Eastman Kodak Co.
† Such as 20-Mule Team Borax.

TABLE 19. ALTERNATE HIGH-CONTRAST DEVELOPER FOR PLATES OF HIGH AVAILABLE CONTRAST*
(Formula D-42)

Constituents	Avoirdupois	Metric
Developer D-76 solution.......... (see Table 18)	1 gal	1.0 liter
Kodak	1 oz 145 gr	10.0 g
Kodak Anti-Fog No. 1........... (0.2% stock solution)	10 dr	10.0 cc

Use without dilution.
Development time at 20°C (68°F) for Kodak Metallographic plates:
  High contrast (about 2.0 gamma): tray—4.5 min; tank—5.5 min.

* Eastman Kodak Co.

TABLE 20. LOW- AND MEDIUM-CONTRAST DEVELOPER FOR PLATES OF HIGH AVAILABLE CONTRAST*
(Formula D-41)

Constituents	Avoirdupois	Metric
Developer D-76 solution.......... (see Table 18)	1 gal	1.0 liter
Kodak Anti-Fog No. 1........... (0.2% stock solution)	5 dr	5.0 cc

Use without dilution.
Development time at 20°C (68°F) for Kodak Metallographic plates:
  Low contrast (about 0.8 gamma); tray—3.5 min; tank—4.5 min.
  Medium contrast (about 1.2 gamma): tray—5.5 min; tank—7 min.

* Eastman Kodak Co.

TABLE 21. TANK AND TRAY DEVELOPER FOR GENERAL USE*
(Formula D-61a)

Constituents of stock solution	Avoirdupois	Metric
Water (about 125°F.) (52°C.).......	16 oz.	500.0 cc.
Elon.............................	45 gr.	3.1 g.
Sodium sulphite, desiccated........	3 oz.	90.0 g.
Sodium bisulphite.................	30 gr.	2.1 g.
Hydroquinone.....................	85 gr.	5.9 g.
Sodium carbonate, desiccated.......	165 gr.	11.5 g.
Potassium bromide................	24 gr.	1.7 g.
Cold water to make...............	32 oz.	1.0 l.

Dissolve the chemicals in the order given.

For tank development, use one part stock solution and three parts water. Develop about 14 min at 65°F (18°C).

For tray development, use equal parts of stock solution and water. Develop about 7 min at 65°F (18°C).

* Eastman Kodak Co.

TABLE 22. ACID HARDENING FIXING BATH FOR USE AFTER DEVELOPERS SUCH AS
D-19, DK-50, D-41, AND D-42*
(Formula F-10)

Constituents	Avoirdupois	Metric
Water (about 125°F.) (52°C.)	64 oz.	500.0 cc.
Sodium thiosulphate (hypo)........	2¾ lb.	3000.0 g.
Sodium sulphite, desiccated........	1 oz.	7.5 g.
Kodalk..........................	4 oz.	30.0 g.
Acetic acid (28 per cent pure)†......	9 oz.	70.0 cc.
Potassium alum...................	3 oz.	22.5 g.
Cold water to make...............	1 gal.	1.0 l.

Dissolve the chemicals in the order given, making certain that each chemical is completely dissolved before adding the next.

* Eastman Kodak Co.

† To make 28 per cent acetic acid from glacial acetic acid, dilute 3 parts of glacial acetic acid with 8 parts of water.

TABLE 23.   ACID HARDENING FIXING BATH FOR FILMS, PLATES, AND PAPERS*
(Formula F-5)

Constituents	Avoirdupois	Metric
Water (about 125°F.)(52°C.).......	20.0 oz.	600.0 cc.
Sodium thiosulphate (hypo)........	8.0 oz.	240.0 g.
Sodium sulphite, desiccated........	0.5 oz.	15.0 g.
Acetic acid (28 per cent pure)†....	1.5 oz.	48.0 cc.
Boric acid crystals................	0.25 oz.	7.5 g.
Potassium alum..................	0.5 oz.	15.0 g.
Cold water to make..............	32 oz.	1.0 l.

Dissolve the chemicals in the order given, making certain that each chemical is completely dissolved before adding the next.

* Eastman Kodak Co.
† To make 28 per cent acetic acid from glacial acetic acid, dilute 3 parts of glacial acetic acid with 8 parts of water.

TABLE 24.   PAPER DEVELOPER*
(Formula D-72)

Constituents	Avoirdupois	Metric
Water (about 125°F.)(52°C.).......	16 oz.	500.0 cc.
Elon...........................	45 gr.	3.1 g.
Sodium sulphite, desiccated.......	1½ oz.	45.0 g.
Hydroquinone...................	175 gr.	12.0 g.
Sodium carbonate, desiccated......	2¼ oz.	67.5 g.
Potassium bromide...............	27 gr.	1.9 g.
Water to make..................	32 oz.	1.0 l.

For use with: Azo papers—1 part developer to 2 parts of water:
Velox papers—1 part developer to 1 part of water.
Temperature of development: 65 to 70°F.
Development time: 40 to 90 sec for all papers.
Agitate well during development or move the papers about.
Unlike negative material, variations in the development time within the recommended limits do not affect the contrast of the print.

* Eastman Kodak Co.

TABLE 25.   CLEANING SOLUTION FOR PHOTOGRAPHIC TANKS AND TRAYS*
(Formula TC-1)

Constituents	Avoirdupois	Metric
Water.........................	32.0 oz.	1000.0 cc.
Potassium bichromate...........	3.0 oz.	90.0 g.
Sulphuric acid (concentrated).....	3.0 oz.	96.0 cc.

* Eastman Kodak Co.

TABLE 28.   ETCHING REAGENTS FOR MACROSCOPIC EXAMINATION OF IRON AND STEEL*

Etching reagent	Composition	Remarks	Use
No. 1. Hydrochloric acid	HCl (conc)............. 50 ml H_2O.................. 50 ml	Used hot or boiling for about 10–15 min, depending on the steel	Showing segregation, porosity, cracks, depth of hardened zone in tool steel, soft spots, etc
No. 2. Mixed acid.....	HCl (conc)............. 38 ml H_2SO_4 (conc)........... 12 ml H_2O.................. 50 ml	Recommended by Yatsevitch; to be used as above for from 15–45 min	Same as for reagent No. 1; works well for stainless steel
No. 3. Nitric acid......	HNO_3 (conc).......... 25 ml H_2O.................. 75 ml	Used cold for large surfaces such as split ingots which cannot conveniently be heated	Same as for reagents Nos. 1 and 2
No. 4. Ammonium persulphate.............	$(NH_4)_2S_2O_8$............. 10 g H_2O................... 90 ml	Surface should be rubbed with absorbent cotton during etching	To bring out grain structure in cases of excessive grain growth, recrystallization at welds, etc
No. 5. Ammonium persulphate.............	A. $(NH_4)_2S_2O_8$..........2.5 g H_2O...............100 ml B. Same as A, plus KI...1.5 g C. Same as B, plus $HgCl_2$.1.5 g D. Same as C, plus H_2SO_4. 15 ml	After grinding on No. 320 paper, swab 15 min with solution A, then 10 min with B, then 5 min with C, and finally 5 min with D. Wash with water and dry with alcohol	Shows dendritic macrostructure of cast iron
No. 6. Stead's reagent..	Cupric chloride........2.5 g Magnesium chloride....10.0 g HCl (conc)............. 5 ml Alcohol......... Up to 250 ml	The salts are dissolved in the HCl with the addition of the least possible quantity of hot water	To bring out phosphorus-rich areas and phosphorus banding
No. 7. Fry's reagent....	Cupric chloride......... 90 g HCl (conc).............120 ml H_2O..................100 ml	Most useful for mild steel, particularly Bessemer and other high-nitrogen steel; before etching, sample should be heated to 200–250°C(302–482°F)for5–30 min, depending on the condition of the steel. During etching the surface should be rubbed with a cloth soaked in the etching solution. Wash in alcohol or rinse in HCl (1:1) after etching to prevent deposition of copper	To show up strain lines due to cold work
No. 7A. Fry's reagent..	Cupric chloride......... 45 g HCl (conc).............180 ml H_2O..................100 ml	Same as for reagent No. 7 but modified by Wazau; may give more contrast; specimen can be washed in water without depositing copper	Same as for reagent No. 7

* Mainly from ASTM Standards, Part I-B, 1946.

TABLE 28. ETCHING REAGENTS FOR MACROSCOPIC EXAMINATION OF IRON AND STEEL*
—(*Continued*)

Etching reagent	Composition	Remarks	Use
No. 8. Nital	HNO_3 .... 5 ml Ethyl alcohol .... 95 ml	Etch 5 min, followed by 1 sec in HCl (10%)	To determine cleanness and to increase contrast
No. 9. Canfield's reagent	Nickel nitrate .... 5 g Cupric chloride .... 1.5 g Ferric chloride .... 6 g H_2O .... 12 ml If reagent is to be slow working, add methanol .. 150 ml	Etch by immersion for 90 sec to several min	To show segregation of impurities, especially phosphorous. The pure metal is plated. Color of deposit from purple-red to pale brown. Areas rich in phosphorus remain bright
No. 10. Dickenson's reagent	A. HNO_3 .... 10 ml Ethyl alcohol .... 90 ml B. $FeCl_3$ .... 40 g $CuCl_3$ .... 3 g HCl .... 40 ml H_2O .... 500 ml	Etch specimen first in reagent A, followed by reetching in reagent B	To reveal segregation, etc
No. 11. Humfrey's reagent	Copper ammonium chloride .... 120 g HCl .... 50 ml H_2O .... 1000 ml	Reagent can be used without HCl but deposit will not adhere	To reveal segregation, etc

* Mainly from ASTM Standards, Part I-B, 1946.

TABLE 29.   MACROETCHING OF IRON AND STEEL FOR SPECIFIC PURPOSES*

Purpose	Composition	Remarks †	Uses
Blowholes	10 % sulphuric acid in water	Use cold on large sections for 24 hr. Surface B	Shows blowholes, porosity, pipe, and inclusions
Carburized case	5 % solution of nitric acid in alcohol	Etch at room temperature. Surface B	Shows depth of carburized case
Contrast	1 part iodine; 2 parts potassium iodide; and 10 parts water	Use at room temperature. Surface B	Produces contrast for photographing
Cracks	5 % solution of nitric acid in alcohol	Etch at room temperature. Etches within ½ hr. Surface B	Detects fatigue, service, hardening and grinding cracks
Cracks	Kerosene and whiting test	Wash surface with kerosene. Wipe off. Paint with thin mixture of whiting and water. Let dry. Kerosene in cracks, if present, will discolor whiting. Surface B	Detects fatigue, service, hardening and grinding cracks
Cracks	120 g. copper ammonium chloride; 50 cc. concentrated hydrochloric acid, 1000 cc. water	Etch first with a neutral solution to remove machine marks. Use cold for 20 min. to ½ hr. Surface B	Produces a strong relief effect. Shows up dendrites and may indicate cracks
Decarburization	5 % nitric acid in alcohol	Etch at room temperature. Surface C	The decarburized areas will be light. Area not decarburized dark
Defects	10 % sulphuric acid in water	Use cold on large sections for 24 hr. Surface B	Shows porosity, pipe, blowholes, and inclusions
Defects	2 parts concentrated sulphuric acid; 1 part concentrated hydrochloric acid; 3 parts water	Use near boiling for at least ½ hr. Surface A	Shows general structure and defects
Dendrites	2 parts concentrated sulphuric acid; 1 part concentrated hydrochloric acid; 3 parts water	Use near boiling for at least ½ hr. Surface A	Shows general structure and defects
Dendritic pattern	25 % sulphuric acid in water	Use cold, 8 to 16 hr. Surface B	Shows dendritic pattern and flow lines
Dendritic structure	40 g. ferric chloride; 3 g. cupric chloride; 40 cc. hydrochloric acid; 500 cc. water	Etch first with 10 per cent nitric acid. Surface B	Shows dendritic structure
Etch universal	500 cc. hydrochloric acid; 70 cc. sulphuric acid; 180 cc. water	Use near boiling point 1 to 2 hr. Surface A	Good universal etch
Fiber	10–20 % solution of ammonium persulphate in water	Swab on freshly made solution for ½ min. Surface B	Shows fiber and grain contrast
Flow lines	25 % sulphuric acid in water	Use cold, 8 to 16 hr. Surface B	Shows flow lines and dendritic pattern

* ASM: "Metals Handbook," 1948.
† Surface Designation:
  A. Sawed or machined surface.
  B. Average ground surface.
  C. Polished surface.

TABLE 29. MACROETCHING OF IRON AND STEEL FOR SPECIFIC PURPOSES.*—(Continued)

Purpose	Composition	Remarks †	Uses
Flow lines	50 % hydrochloric acid in water	Use hot. Surface A	Shows flow lines
Grain contrast	10 to 20 % solution ammonium persulphate in water	Swab on freshly made solution for 1 to 2 min. Surface B	Shows grain contrast and fiber
Grain contrast	1.5 g. cupric chloride; 30 cc. hydrochloric acid; 95 cc. water; 30 cc. ethyl alcohol	Polish specimen carefully. Surface C	Shows good grain contrast
Grain size	10 to 20 % solution ammonium persulphate in water	Swab on freshly made solution for 1 to 2 min. Surface B	Shows grain size especially in low-carbon steels, wrought iron, and welded sections
Hardness penetration	5 % nitric acid in alcohol	Etch at room temperature on smooth ground surface. Surface B	Shows depth of hardness penetration of heat treated samples
Hardness penetration	50 % hydrochloric acid in water	Use hot 2 to 5 min. To preserve surface after etching, scrub in running water, dip in weak NH_4OH, dip in solution of soluble cutting oil, dry with a rag and compressed air. Surface B	Shows depth of hardening, especially carbon steels. Produces distinct contrast between martensitic and troostitic zones
Heterogeneity	50 % hydrochloric acid in water	Immerse specimen in hot solution from ½ to 1 hr. Surface A	Shows heterogeneity in general
Heterogeneity	Undiluted hydrochloric acid	Use hot (212°F.). Surface A	Shows heterogeneity in 3 to 3.5 % nickel steels
Impurities in ferrite	1 g. cupric chloride; 0.5 g. stannous chloride; 30 g. ferric chloride; 30 cc. hydrochloric acid; 500 cc. distilled water; 500 cc. ethyl alcohol	Etch only well polished surface which has been thoroughly dried. Surface C	Plates out copper on ferrite containing impurities. Gives a good even etch
Inclusions	10 % sulphuric acid in water	Use cold on large sections for 24 hr. Surface B	Shows inclusions, porosity, pipe, and blowholes
Inclusions	10 to 20 % sulphuric acid in water	Use hot. Surface A	Etches sulphide inclusions
Pipe and porosity	10 % sulphuric acid in water	Use cold on large sections for 24 hr. Surface B	Shows pipe, porosity, blowholes, and inclusions
Rail sections	9 parts hydrochloric acid; 3 parts sulphuric; 1 part water	Etch near boiling point for 2 hr. Surface A	Use for etching rail sections
Segregation	3 % solution picric acid in ethyl alcohol	Etch at room temperature for 4 to 5 hr. Surface B	Shows segregation
Segregation	1 g. picric acid; 1 drop concentrated hydrochloric acid in 25 cc. alcohol	Use hot. Surface B	Shows segregation

* ASM: "Metals Handbook," 1948.
† Surface Designation:
  A. Sawed or machined surface.
  B. Average ground surface.
  C. Polished surface.

TABLE 29.   MACROETCHING OF IRON AND STEEL FOR SPECIFIC PURPOSES.*—(*Continued*)

Purpose	Composition	Remarks †	Uses
Segregation	10 to 15 % solution of nitric acid in water or alcohol	Etch at room temperature. Surface B	Shows heavy segregation
Segregation	5 % solution nitric acid in water or alcohol	Etches within $\frac{1}{2}$ hr. Surface is black when etched. Surface B	Shows segregation in low-carbon, low chromium-nickel steels
Segregation	2 to 10 % solution nitric acid in water	Etch first with weak acid to remove machine marks and then increase concentration for structure. Surface B	Shows segregation in ingot sections and large pieces
Segregation	1 g. cupric chloride; 0.5 g. picric acid; 1.3 to 2.5 cc. hydrochloric acid; 10 cc. water; 100 cc. ethyl alcohol	May be used for electrolytic etching. Surface B	Shows segregation
Segregation, carbide	2 % nitric acid in alcohol. For high-speed steel, use 5 % nitric acid in alcohol	Use on a polished, longitudinal section. For high-speed steel, etch until matrix is darkened. Surface C	To detect carbide segregations, particularly in high-speed steels
Segregation, carbon and phosphorous	1 part copper ammonium chloride, 12 parts water	Immerse finely ground, clean sample in solution for 1 min.; wash with water and rub off copper. Surface C	Shows phosphorous and carbon segregation
Segregation, phosphorous	0.5 % solution picric acid in water	Etch at room temperature until staining occurs. Surface C	Uneven staining represents phosphorous segregation
Segregation, phosphorous	30 g. ferric chloride; 100 g. hydrochloric acid; 1 g. cupric chloride; 0.5 g. stannous chloride; 1000 cc. water	Polish as for microscopic work, use etch cold for 10 sec. to 2 min. Surface C	Shows phosphorous segregation
Segregation, phosphorous	10 g. cupric chloride; 40 g. magnesium chloride; 20 cc. hydrochloric acid; 1000 cc. ethyl alcohol. (Stead's No. 1 reagent)	Dissolve salts in small amount of hot water, then add alcohol. Apply solution to polished surface drop by drop. Surface C.	Area not coated by copper shows phosphorous segregation
Segregation, phosphorous	5 g. cupric chloride; 4 g. magnesium chloride; 1 cc. hydrochloric acid; 20 cc. water; 100 cc. alcohol. (Stead's No. 2)	Etch for 1 min. Surface C	Shows phosphorous segregation
Segregation, phosphorous	4 g. cupric chloride; 20 cc. hydrochloric acid; 40 cc. water; 20 cc. ethyl alcohol	Polish specimen carefully. Surface C	Shows phosphorous segregation

* ASM: "Metals Handbook," 1948.

† Surface Designation:

  A. Sawed or machined surface.

  B. Average ground surface.

  C. Polished surface.

TABLE 29. MACROETCHING OF IRON AND STEEL FOR SPECIFIC PURPOSES.*—(Continued)

Purpose	Composition	Remarks†	Uses
Segregation, phosphorous	1.5 g. cupric chloride; 5 g. nickel nitrate crystals; 6 g. ferric chloride; 12 cc. water	Immerse specimen for 90 sec. or more. Surface C	Shows phosphorous segregation
Segregation, sulphur	Soak ordinary semimatte photographic silver bromide paper in a 2% solution of sulphuric acid in water. Apply emulsion side of paper to smooth ground, clean sample for a min. or two. Rinse paper, fix in hypo, wash thoroughly and dry	The brown pattern formed on the paper indicates the relative distribution of sulphides. The darker and heavier the marks, the more sulphur is indicated to be present. Surface B	Shows the distribution of sulphur
Segregation, sulphide inclusions	10 to 20% sulphuric acid in water	Use hot. Surface A	Etches sulphide inclusions
Soft spots	5% nitric acid in alcohol	Etch at room temperature. Surface B	Nondestructive test for hardened tools
Soft spots	50% hydrochloric acid in water	Use cold on ground surface. Surface B	Shows soft spots
Strains	90 g. cupric chloride; 120 cc. hydrochloric acid; 100 cc. water	Copper does not precipitate with this etch. Heat specimen to 400–475°F. for ½ hr., then etch ground surface for 1–3 min. Rub with powdered cupric chloride, then rinse with alcohol. Surface B	Shows strain lines
Strains	6 g. cupric chloride; 6 g. ferric chloride; 10 cc. hydrochloric acid; 100 cc. ethyl alcohol	Heat specimen to 400°F. and immerse ground surface. Surface B	Shows strain lines
Strauss test	3% cupric sulphate; 10% sulphuric acid; 87% distilled water	Specimen ½ × 1 × 4 in. is sensitized by actual welding, or artificially, by heating to 900–1300°F. (generally 1250°F.). Immersed in the boiling cupric sulphate solution for 72 hr. Specimen is then bent 180° around ½-in. pin. Cracking indicates disintegration has taken place due to migration of carbides to grain boundaries. Surface B. (120° fine emery)	To measure intergranular corrosion of 18–8 corrosion resisting steel

* ASM: "Metals Handbook," 1948.
† Surface Designation:
  A. Sawed or machined surface.
  B. Average ground surface.
  C. Polished surface.

TABLE 29.   MACROETCHING OF IRON AND STEEL FOR SPECIFIC PURPOSES.*—(*Continued*)

Purpose	Composition	Remarks†	Uses
Structure	25 % nitric acid in water	Etch at room temperature. Surface B	Shows general structure
Structure	2 parts concentrated sulphuric acid; 1 part concentrated hydrochloric acid; 3 parts water	Use near boiling point for at least ½ hr. Surface A	Shows general structure and defects
Weld examination	50 % hydrochloric acid in water	When testing large sections, the acid may be added by building a dam with paraffin wax around the part to be tested. Surface A	For testing the soundness of welds

* ASM: "Metals Handbook," 1948.
† Surface Designation.
   A. Sawed or machined surface.
   B. Average ground surface.
   C. Polished surface.

TABLE 30. RECOMMENDED TIMES FOR ETCHING VARIOUS STEELS IN THE STANDARD HYDROCHLORIC ACID MACROETCHING REAGENT*
[Etching temperature—70°C (158°F)]

SAE Carbon Steels

Steels	Time, min
1010	15
1015	15
X1015	15
1020	15
X1020	15
1025	30
X1025	30
1030	30
1035	30
1040	30
X1040	30
1045	45
X1045	45
1050	45
X1050	45
1055	45
X1055	45
1060	45
1065	45
X1065	45
1070	45
1075	45
1080	45
1085	45
1090	45
1095	45

SAE Nickel-Chromium Steels

Steels	Time, min
3115	45
3120	45
3125	45
3130	45
3135	45
3140	45
X3140	45
3145	45
3150	45
3215	45
3220	45
3230	45
3240	45
3245	45
3250	45
3312	60
3325	60
3335	60
3340	60
3415	45
3435	45
3450	45

SAE Tungsten Steels

Steels	Time, min
71360	45
71660	45
7260	45

SAE Silicon-Manganese Steels

Steels	Time, min
9255	30
9260	30

SAE Free-cutting Steels

Steels	Time, min
X1314	30
X1315	30
X1330	30
X1335	30
X1340	30

SAE Manganese Steels

Steels	Time, min
T1330	30
T1335	30
T1340	30
T1345	30
T1350	30

SAE Nickel Steels

Steels	Time, min
2015	30
2115	30
2315	30
2320	30
2330	30
2335	30
2340	30
2345	30
2350	30
2515	30

SAE Molybdenum Steels

Steels	Time, min
4130	45
X4130	45
4135	45
4140	45
4150	45
4340	45
4345	45
4615	45
4620	45
4640	45
4815	45
4820	45

SAE Chromium Steels

Steels	Time, min
5120	30
5140	30
5150	30
52100	45

SAE Chromium-Vanadium Steels

Steels	Time, min
6115	45
6120	45
6125	45
6130	45
6135	45
6140	45
6145	45
6150	45
6195	45

Stainless Iron and Steel

	Time, min
Chromium Type	30
Cr-Ni Type	45

Free-machining Steels

	Time, min
Chromium Type	30
Cr-Ni Type	20

Tool Steels

	Time, min
Carbon and Carbon-Vanadium	45
Manganese Oil Hardening	30
Fast Finishing	45
Tungsten Hot Die Steel	45
Chromium Hot-working Die	45
High-Carbon High-Chromium	45
High Speed	45

* ASM: "Metals Handbook," 1939.

TABLE 31. BRINELL HARDNESS NUMBERS*
Steel Ball, 10 mm in Diameter, Pressures of 500 and 3000 kg.

The values given in this table for hardness numbers are merely solutions of the equation given on page 220. They do not imply that Brinell tests are feasible on materials of a hardness indicated by the highest value in the table.

Diameter of indentation, mm	Brinell hardness number		Diameter of indentation, mm	Brinell hardness number		Diameter of indentation, mm	Brinell hardness number	
	500-kg load	3000-kg load		500-kg load	3000-kg load		500-kg load	3000-kg load
2.00	158	945	2.35	114	682	2.70	85.7	514
2.01	156	936	2.36	113	676	2.71	85.1	510
2.02	154	926	2.37	112	670	2.72	84.4	507
2.03	153	917	2.38	111	665	2.73	83.8	503
2.04	151	908	2.39	110	659	2.74	83.2	499
2.05	150	899	2.40	109	653	2.75	82.6	495
2.06	148	890	2.41	108	648	2.76	81.9	492
2.07	147	882	2.42	·107	643	2.77	81.3	488
2.08	146	873	2.43	106	637	2.78	80.8	485
2.09	144	865	2.44	105	632	2.79	80.2	481
2.10	143	856	2.45	104	627	2.80	79.6	477
2.11	141	848	2.46	104	621	2.81	79.0	474
2.12	140	840	2.47	103	616	2.82	78.4	471
2.13	139	832	2.48	102	611	2.83	77.9	467
2.14	137	824	2.49	101	606	2.84	77.3	464
2.15	136	817	2.50	100	601	2.85	76.8	461
2.16	135	809	2.51	99.4	597	2.86	76.2	457
2.17	134	802	2.52	98.6	592	2.87	75.7	454
2.18	132	794	2.53	97.8	587	2.88	75.1	451
2.19	131	787	2.54	97.1	582	2.89	74.6	448
2.20	130	780	2.55	96.3	578	2.90	74.1	444
2.21	129	772	2.56	95.5	573	2.91	73.6	441
2.22	128	765	2.57	94.8	569	2.92	73.0	438
2.23	126	758	2.58	94.0	564	2.93	72.5	435
2.24	125	752	2.59	93.3	560	2.94	72.0	432
2.25	124	745	2.60	92.6	555	2.95	71.5	429
2.26	123	738	2.61	91.8	551	2.96	71.0	426
2.27	122	732	2.62	91.1	547	2.97	70.5	423
2.28	121	725	2.63	90.4	543	2.98	70.1	420
2.29	120	719	2.64	89.7	538	2.99	69.6	417
2.30	119	712	2.65	89.0	534	3.00	69.1	415
2.31	118	706	2.66	88.4	530	3.01	68.6	412
2.32	117	700	2.67	87.7	526	3.02	68.2	409
2.33	116	694	2.68	87.0	522	3.03	67.7	406
2.34	115	688	2.69	86.4	518	3.04	67.3	404

* From ASTM Standards, Part I-A, 1946.

TABLE 31.   BRINELL HARDNESS NUMBERS.*—(*Continued*)

Diameter of indentation, mm	Brinell hardness number		Diameter of indentation, mm	Brinell hardness number		Diameter of indentation, mm	Brinell hardness number	
	500-kg load	3000-kg load		500-kg load	3000-kg load		500-kg load	3000-kg load
3.05	66.8	401	3.40	53.4	321	3.75	43.6	262
3.06	66.4	398	3.41	53.1	319	3.76	43.4	260
3.07	65.9	395	3.42	52.8	317	3.77	43.1	259
3.08	65.5	393	3.43	52.5	315	3.78	42.9	257
3.09	65.0	390	3.44	52.2	313	3.79	42.7	256
3.10	64.6	388	3.45	51.8	311	3.80	42.4	255
3.11	64.2	385	3.46	51.5	309	3.81	42.2	253
3.12	63.8	383	3.47	51.2	307	3.82	42.0	252
3.13	63.3	380	3.48	50.9	306	3.83	41.7	250
3.14	62.9	378	3.49	50.6	304	3.84	41.5	249
3.15	62.5	375	3.50	50.3	302	3.85	41.3	248
3.16	62.1	373	3.51	50.0	300	3.86	41.1	246
3.17	61.7	370	3.52	49.7	298	3.87	40.9	245
3.18	61.3	368	3.53	49.4	297	3.88	40.6	244
3.19	60.9	366	3.54	49.2	295	3.89	40.4	242
3.20	60.5	363	3.55	48.9	293	3.90	40.2	241
3.21	60.1	361	3.56	48.6	292	3.91	40.0	240
3.22	59.8	359	3.57	48.3	290	3.92	39.8	239
3.23	59.4	356	3.58	48.0	288	3.93	39.6	237
3.24	59.0	354	3.59	47.7	286	3.94	39.4	236
3.25	58.6	352	3.60	47.5	285	3.95	39.1	235
3.26	58.3	350	3.61	47.2	283	3.96	38.9	234
3.27	57.9	347	3.62	46.9	282	3.97	38.7	232
3.28	57.5	345	3.63	46.7	280	3.98	38.5	231
3.29	57.2	343	3.64	46.4	278	3.99	38.3	230
3.30	56.8	341	3.65	46.1	277	4.00	38.1	229
3.31	56.5	339	3.66	45.9	275	4.01	37.9	228
3.32	56.1	337	3.67	45.6	274	4.02	37.7	226
3.33	55.8	335	3.68	45.4	272	4.03	37.5	225
3.34	55.4	333	3.69	45.1	271	4.04	37.3	224
3.35	55.1	331	3.70	44.9	269	4.05	37.1	223
3.36	54.8	329	3.71	44.6	268	4.06	37.0	222
3.37	54.4	326	3.72	44.4	266	4.07	36.8	221
3.38	54.1	325	3.73	44.1	265	4.08	36.6	219
3.39	53.8	323	3.74	43.9	263	4.09	36.4	218

* From ASTM Standards, Part I-A, 1946.

TABLE 31. BRINELL HARDNESS NUMBERS.*—(Continued)

Diameter of indentation, mm	Brinell hardness number		Diameter of indentation, mm	Brinell hardness number		Diameter of indentation, mm	Brinell hardness number	
	500-kg load	3000-kg load		500-kg load	3000-kg load		500-kg load	3000-kg load
4.10	36.2	217	4.45	30.5	183	4.80	25.9	156
4.11	36.0	216	4.46	30.3	182	4.81	25.8	155
4.12	35.8	215	4.47	30.2	181	4.82	25.7	154
4.13	35.7	214	4.48	30.0	180	4.83	25.6	154
4.14	35.5	213	4.49	29.9	179	4.84	25.5	153
4.15	35.3	212	4.50	29.8	179	4.85	25.4	152
4.16	35.1	211	4.51	29.6	178	4.86	25.3	152
4.17	34.9	210	4.52	29.5	177	4.87	25.1	151
4.18	34.8	209	4.53	29.3	176	4.88	25.0	150
4.19	34.6	208	4.54	29.2	175	4.89	24.9	150
4.20	34.4	207	4.55	29.1	174	4.90	24.8	149
4.21	34.2	205	4.56	28.9	174	4.91	24.7	148
4.22	34.1	204	4.57	28.8	173	4.92	24.6	148
4.23	33.9	203	4.58	28.7	172	4.93	24.5	147
4.24	33.7	202	4.59	28.5	171	4.94	24.4	146
4.25	33.6	201	4.60	28.4	170	4.95	24.3	146
4.26	33.4	200	4.61	28.3	170	4.96	24.2	145
4.27	33.2	199	4.62	28.1	169	4.97	24.1	144
4.28	33.1	198	4.63	28.0	168	4.98	24.0	144
4.29	32.9	198	4.64	27.9	167	4.99	23.9	143
4.30	32.8	197	4.65	27.8	167	5.00	23.8	143
4.31	32.6	196	4.66	27.6	166	5.01	23.7	142
4.32	32.4	195	4.67	27.5	165	5.02	23.6	141
4.33	32.3	194	4.68	27.4	164	5.03	23.5	141
4.34	32.1	193	4.69	27.3	164	5.04	23.4	140
4.35	32.0	192	4.70	27.1	163	5.05	23.3	140
4.36	31.8	191	4.71	27.0	162	5.06	23.2	139
4.37	31.7	190	4.72	26.9	161	5.07	23.1	138
4.38	31.5	189	4.73	26.8	161	5.08	23.0	138
4.39	31.4	188	4.74	26.6	160	5.09	22.9	137
4.40	31.2	187	4.75	26.5	159	5.10	22.8	137
4.41	31.1	186	4.76	26.4	158	5.11	22.7	136
4.42	30.9	185	4.77	26.3	158	5.12	22.6	135
4.43	30.8	185	4.78	26.2	157	5.13	22.5	135
4.44	30.6	184	4.79	26.1	156	5.14	22.4	134

* From ASTM Standards, Part I-A, 1946.

TABLE 31. BRINELL HARDNESS NUMBERS.*—(*Continued*)

Diameter of indentation, mm	Brinell hardness number		Diameter of indentation, mm	Brinell hardness number		Diameter of indentation, mm	Brinell hardness number	
	500-kg load	3000-kg load		500-kg load	3000-kg load		500-kg load	3000-kg load
5.15	22.3	134	5.50	19.3	116	5.85	16.8	101
5.16	22.2	133	5.51	19.2	115	5.86	16.8	101
5.17	22.1	133	5.52	19.2	115	5.87	16.7	100
5.18	22.0	132	5.53	19.1	114	5.88	16.7	99.9
5.19	21.9	132	5.54	19.0	114	5.89	16.6	99.5
5.20	21.8	131	5.55	18.9	114	5.90	16.5	99.2
5.21	21.7	130	5.56	18.9	113	5.91	16.5	98.8
5.22	21.6	130	5.57	18.8	113	5.92	16.4	98.4
5.23	21.6	129	5.58	18.7	112	5.93	16.3	98.0
5.24	21.5	129	5.59	18.6	112	5.94	16.3	97.7
5.25	21.4	128	5.60	18.6	111	5.95	16.2	97.3
5.26	21.3	128	5.61	18.5	111	5.96	16.2	96.9
5.27	21.2	127	5.62	18.4	110	5.97	16.1	96.6
5.28	21.1	127	5.63	18.3	110	5.98	16.0	96.2
5.29	21.0	126	5.64	18.3	110	5.99	16.0	95.9
5.30	20.9	126	5.65	18.2	109	6.00	15.9	95.5
5.31	20.9	125	5.66	18.1	109	6.01	15.9	95.1
5.32	20.8	125	5.67	18.1	108	6.02	15.8	94.8
5.33	20.7	124	5.68	18.0	108	6.03	15.7	94.4
5.34	20.6	124	5.69	17.9	107	6.04	15.7	94.1
5.35	20.5	123	5.70	17.8	107	6.05	15.6	93.7
5.36	20.4	123	5.71	17.8	107	6.06	15.6	93.4
5.37	20.3	122	5.72	17.7	106	6.07	15.5	93.0
5.38	20.3	122	5.73	17.6	106	6.08	15.4	92.7
5.39	20.2	121	5.74	17.6	105	6.09	15.4	92.3
5.40	20.1	121	5.75	17.5	105	6.10	15.3	92.0
5.41	20.0	120	5.76	17.4	105	6.11	15.3	91.7
5.42	19.9	120	5.77	17.4	104	6.12	15.2	91.3
5.43	19.9	119	5.78	17.3	104	6.13	15.2	91.0
5.44	19.8	119	5.79	17.2	103	6.14	15.1	90.6
5.45	19.7	118	5.80	17.2	103	6.15	15.1	90.3
5.46	19.6	118	5.81	17.1	103	6.16	15.0	90.0
5.47	19.5	117	5.82	17.0	102	6.17	14.9	89.6
5.48	19.5	117	5.83	17.0	102	6.18	14.9	89.3
5.49	19.4	116	5.84	16.9	101	6.19	14.8	89.0

* From ASTM Standards, Part I-A, 1946.

TABLE 31. BRINELL HARDNESS NUMBERS.*—(*Continued*)

Diam- eter of inden- tation, mm	Brinell hardness number		Diam- eter of inden- tation, mm	Brinell hardness number		Diam- eter of inden- tation, mm	Brinell hardness number	
	500- kg load	3000- kg load		500- kg load	3000- kg load		500- kg load	3000- kg load
6.20	14.8	88.7	6.50	13.3	79.6	6.75	12.1	72.8
6.21	14.7	88.3	6.51	13.2	79.3	6.76	12.1	72.6
6.22	14.7	88.0	6.52	13.2	79.0	6.77	12.1	72.3
6.23	14.6	87.7	6.53	13.1	78.7	6.78	12.0	72.1
6.24	14.6	87.4	6.54	13.1	78.4	6.79	12.0	71.8
6.25	14.5	87.1	6.55	13.0	78.2	6.80	11.9	71.6
6.26	14.5	86.7	6.56	13.0	77.9	6.81	11.9	71.3
6.27	14.4	86.4	6.57	12.9	77.6	6.82	11.8	71.1
6.28	14.4	86.1	6.58	12.9	77.3	8.83	11.8	70.8
6.29	14.3	85.8	6.59	12.8	77.1	6.84	11.8	70.6
6.30	14.2	85.5	6.60	12.8	76.8	6.85	11.7	70.4
6.31	14.2	85.2	6.61	12.8	76.5	6.86	11.7	70.1
6.32	14.1	84.9	6.62	12.7	76.2	6.87	11.6	69.9
6.33	14.1	84.6	6.63	12.7	76.0	6.88	11.6	69.6
6.34	14.0	84.3	6.64	12.6	75.7	6.89	11.6	69.4
6.35	14.0	84.0	6.65	12.6	75.4	6.90	11.5	69.2
6.36	13.9	83.7	6.66	12.5	75.2	6.91	11.5	68.9
6.37	13.9	83.4	6.67	12.5	74.9	6.92	11.4	68.7
6.38	13.8	83.1	6.68	12.4	74.7	6.93	11.4	68.4
6.39	13.8	82.8	6.69	12.4	74.4	6.94	11.4	68.2
6.40	13.7	82.5	6.70	12.4	74.1	6.95	11.3	68.0
6.41	13.7	82.2	6.71	12.3	73.9	6.96	11.3	67.7
6.42	13.6	81.9	6.72	12.3	73.6	6.97	11.3	67.5
6.43	13.6	81.6	6.73	12.2	73.4	6.98	11.2	67.3
6.44	13.5	81.3	6.74	12.2	73.1	6.99	11.2	67.0
6.45	13.5	81.0						
6.46	13.4	80.7						
6.47	13.4	80.4						
6.48	13.4	80.1						
6.49	13.3	79.8						

* From ASTM Standards, Part I-A, 1946.

TABLE 32. DIAMOND PYRAMID HARDNESS NUMBERS.*

Load—10 kg

Diagonal of impression, mm	0.000	0.001	0.002	0.003	0.004	0.005	0.006	0.007	0.008	0.009
0.05	7416	7128	6857	6600	6358	6129	5912	5706	5511	5326
0.06	5150	4983	4823	4671	4526	4388	4256	4130	4010	3894
0.07	3784	3678	3576	3479	3386	3296	3210	3127	3047	2971
0.08	2897	2826	2757	2691	2628	2566	2507	2449	2394	2341
0.09	2289	2239	2190	2144	2098	2054	2012	1970	1930	1892
0.10	1854	1817	1782	1748	1714	1682	1650	1619	1590	1560
0.11	1533	1505	1478	1452	1427	1402	1378	1354	1332	1310
0.12	1288	1267	1246	1226	1206	1187	1168	1150	1132	1115
0.13	1097	1081	1064	1048	1033	1018	1003	988	974	960
0.14	946	933	920	907	894	882	870	858	847	835
0.15	824	813	803	792	782	772	762	752	743	734
0.16	724	715	707	698	609	681	673	665	657	649
0.17	642	634	627	620	613	606	599	592	585	579
0.18	572	566	560	554	548	542	536	530	525	519
0.19	514	508	503	498	493	488	483	478	473	468
0.20	464	459	455	450	446	442	437	433	429	425
0.21	421	417	413	409	405	401	397	394	390	387
0.22	383	380	376	373	370	366	363	360	357	354
0.23	351	348	345	342	339	336	333	330	327	325
0.24	322	319	317	314	312	309	306	304	302	299
0.25	297	294	292	289	287	285	283	281	279	276
0.26	274	272	270	268	266	264	262	260	258	256
0.27	254	253	251	249	247	245	243	242	240	238
0.28	236	235	233	232	230	228	227	225	224	222
0.29	221	219	218	216	215	213	212	210	209	207
0.30	206	205	203	202	201	199	198	197	196	194
0.31	193	192	191	189	188	187	186	185	183	182
0.32	181	180	179	178	177	176	175	173	172	171
0.33	170	169	168	167	166	165	164	163	162	161
0.34	160	160	159	158	157	156	155	154	153	152
0.35	151.4	150.5	149.7	148.8	148.0	147.1	146.3	145.5	144.7	143.9
0.36	143.1	142.3	141.5	140.7	140.0	139.2	138.4	137.7	136.9	136.2
0.37	135.5	134.7	134.0	133.3	132.6	131.9	131.2	130.5	129.8	129.1
0.38	128.4	127.7	127.1	126.4	125.8	125.1	124.5	123.8	123.2	122.6
0.39	121.9	121.3	120.7	120.1	119.5	118.9	118.3	117.7	117.1	116.5

* Courtesy of Wilson Mechanical Instrument Co., Inc.

TABLE 32. DIAMOND PYRAMID HARDNESS NUMBERS.*—(*Continued*)

Diag-onal of im-pres-sion, mm	0.000	0.001	0.002	0.003	0.004	0.005	0.006	0.007	0.008	0.009
0.40	115.9	115.3	114.8	114.2	113.6	113.1	112.5	111.9	111.4	110.9
0.41	110.3	109.8	109.3	108.7	108.2	107.7	107.2	106.6	106.1	105.6
0.42	105.1	104.6	104.1	103.6	103.1	102.7	102.2	101.7	101.2	100.8
0.43	100.3	99.8	99.4	98.9	98.5	98.0	97.6	97.1	96.7	96.2
0.44	95.8	95.3	94.9	94.5	94.1	93.6	93.2	92.8	92.4	92.0
0.45	91.6	91.2	90.8	90.4	90.0	89.6	89.2	88.8	88.4	88.0
0.46	87.6	87.3	86.9	86.5	86.1	85.8	85.4	85.0	84.7	84.3
0.47	84.0	83.6	83.2	82.9	82.5	82.2	81.8	81.5	81.2	80.8
0.48	80.5	80.2	79.8	79.5	79.2	78.8	78.5	78.2	77.9	77.6
0.49	77.2	76.9	76.6	76.3	76.0	75.7	75.4	75.1	74.8	74.5
0.50	74.2	73.9	73.6	73.3	73.0	72.7	72.4	72.1	71.9	71.6
0.51	71.3	71.0	70.7	70.5	70.2	69.9	69.6	69.4	69.1	68.8
0.52	68.6	68.3	68.1	67.8	67.5	67.3	67.0	66.8	66.5	66.3
0.53	66.0	65.8	65.5	65.3	65.0	64.8	64.5	64.3	64.1	63.8
0.54	63.6	63.4	63.1	62.9	62.7	62.4	62.2	62.0	61.7	61.5
0.55	61.3	61.1	60.9	60.6	60.4	60.2	60.0	59.8	59.6	59.3
0.56	59.1	58.9	58.7	58.5	58.3	58.1	57.9	57.7	57.5	57.3
0.57	57.1	56.9	56.7	56.5	56.3	56.1	55.9	55.7	55.5	55.3
0.58	55.1	54.9	54.7	54.6	54.4	54.2	54.0	53.8	53.6	53.4
0.59	53.3	53.1	52.9	52.7	52.6	52.4	52.2	52.0	51.9	51.7
0.60	51.5	51.3	51.2	51.0	50.8	50.7	50.5	50.3	50.2	50.0
0.61	49.8	49.7	49.5	49.4	49.2	49.0	48.9	48.7	48.6	48.4
0.62	48.2	48.1	47.9	47.8	47.6	47.5	47.3	47.2	47.0	46.9
0.63	46.7	46.6	46.4	46.3	46.1	46.0	45.8	45.7	45.6	45.4
0.64	45.3	45.1	45.0	44.8	44.7	44.6	44.4	44.3	44.2	44.0
0.65	43.9	43.8	43.6	43.5	43.4	43.2	43.1	43.0	42.8	42.7
0.66	42.6	42.4	42.3	42.2	42.1	41.9	41.8	41.7	41.6	41.4
0.67	41.3	41.2	41.1	40.9	40.8	40.7	40.6	40.5	40.3	40.2
0.68	40.1	40.0	39.9	39.8	39.6	39.5	39.4	39.3	39.2	39.1
0.69	39.0	38.8	38.7	38.6	38.5	38.4	38.3	38.3	38.1	38.0
0.70	37.8	37.7	37.6	37.5	37.4	37.3	37.2	37.1	37.0	36.9
0.71	36.8	36.7	36.6	36.5	36.4	36.3	36.2	36.1	36.0	35.9
0.72	35.8	35.7	35.6	35.5	35.4	35.3	35.2	35.1	35.0	34.9
0.73	34.8	34.7	34.6	34.5	34.4	34.3	34.2	34.1	34.0	34.0
0.74	33.9	33.8	33.7	33.6	33.5	33.4	33.3	33.2	33.1	33.1

* Courtesy of Wilson Mechanical Instrument Co., Inc.

TABLE 32. DIAMOND PYRAMID HARDNESS NUMBERS.*—(*Continued*)

Diagonal of impression, mm	0.000	0.001	0.002	0.003	0.004	0.005	0.006	0.007	0.008	0.009
0.75	33.0	32.9	32.8	32.7	32.6	32.5	32.4	32.4	32.3	32.2
0.76	32.1	32.0	31.9	31.8	31.8	31.7	31.6	31.5	31.4	31.4
0.77	31.3	31.2	31.1	31.0	30.9	30.9	30.8	30.7	30.7	30.6
0.78	30.5	30.4	30.3	30.3	30.2	30.1	30.0	29.9	29.9	29.8
0.79	29.7	29.6	29.6	29.5	29.4	29.3	29.3	29.2	29.1	29.1
0.80	29.0	28.9	28.8	28.8	28.7	28.7	28.6	28.5	28.4	28.3
0.81	28.3	28.2	28.1	28.0	28.0	27.9	27.8	27.8	27.7	27.7
0.82	27.6	27.5	27.4	27.4	27.3	27.3	27.2	27.1	27.0	27.0
0.83	26.9	26.8	26.8	26.7	26.7	26.6	26.5	26.5	26.4	26.3
0.84	26.3	26.2	26.2	26.1	26.0	26.0	25.9	25.8	25.8	25.7
0.85	25.7	25.6	25.6	25.5	25.4	25.4	25.3	25.3	25.2	25.1
0.86	25.1	25.0	25.0	24.9	24.8	24.8	24.7	24.7	24.6	24.6
0.87	24.5	24.4	24.4	24.3	24.3	24.2	24.2	24.1	24.1	24.0
0.88	24.0	23.9	23.8	23.8	23.7	23.7	23.6	23.6	23.5	23.5
0.89	23.4	23.4	23.3	23.3	23.2	23.2	23.1	23.0	23.0	22.9
0.90	22.9	22.8	22.8	22.7	22.7	22.6	22.6	22.5	22.5	22.4
0.91	22.4	22.3	22.3	22.3	22.2	22.2	22.1	22.1	22.0	22.0
0.92	21.9	21.9	21.8	21.8	21.7	21.7	21.6	21.6	21.5	21.5
0.93	21.4	21.4	21.4	21.3	21.3	21.2	21.2	21.1	21.1	21.0
0.94	21.0	20.9	20.9	20.8	20.8	20.8	20.7	20.7	20.6	20.6
0.95	20.5	20.5	20.5	20.4	20.4	20.3	20.3	20.2	20.2	20.2
0.96	20.1	20.1	20.0	20.0	19.96	19.91	19.87	19.83	19.79	19.75
0.97	19.71	19.67	19.63	19.59	19.55	19.51	19.47	19.43	19.39	19.35
0.98	19.31	19.27	19.23	19.19	19.15	19.11	19.07	19.04	19.00	18.96
0.99	18.92	18.88	18.84	18.81	18.77	18.73	18.69	18.66	18.62	18.58
1.00	18.54	18.51	18.47	18.43	18.39	18.36	18.32	18.29	18.25	18.21
1.01	18.18	18.14	18.11	18.07	18.04	18.00	17.96	17.93	17.89	17.86
1.02	17.83	17.79	17.76	17.72	17.69	17.65	17.62	17.58	17.55	17.51
1.03	17.48	17.45	17.41	17.38	17.34	17.31	17.28	17.24	17.21	17.17
1.04	17.14	17.11	17.08	17.05	17.01	16.98	16.95	16.92	16.88	16.85
1.05	16.82	16.79	16.76	16.72	16.69	16.66	16.63	16.59	16.56	16.53
1.06	16.50	16.47	16.44	16.41	16.38	16.35	16.32	16.29	16.26	16.23
1.07	16.20	16.17	16.14	16.11	16 08	16.05	16.02	15.99	15.96	15.93
1.08	15.90	15.87	15.84	15.81	15.78	15.75	15.72	15.69	15.67	15.64
1.09	15.61	15.58	15.55	15.52	15.49	15.47	15.44	15.41	15.38	15.35

* Courtesy of Wilson Mechanical Instrument Co., Inc.

TABLE 32.   DIAMOND PYRAMID HARDNESS NUMBERS.*—(*Continued*)

Diagonal of impression, mm	0.000	0.001	0.002	0.003	0.004	0.005	0.006	0.007	0.008	0.009
1.10	15.33	15.30	15.27	15.24	15.22	15.19	15.16	15.13	15.11	15.08
1.11	15.05	15.02	14.99	14.97	14.94	14.92	14.89	14.86	14.84	14.81
1.12	14.78	14.76	14.73	14.70	14.68	14.65	14.63	14.60	14.57	14.55
1.13	14.52	14.49	14.47	14.45	14.42	14.39	14.37	14.35	14.32	14.29
1.14	14.27	14.24	14.22	14.19	14.17	14.14	14.12	14.09	14.07	14.05
1.15	14.02	13.99	13.97	13.95	13.93	13.90	13.88	13.85	13.83	13.81
1.16	13.78	13.76	13.73	13.71	13.69	13.66	13.64	13.62	13.59	13.57
1.17	13.54	13.52	13.50	13.48	13.45	13.43	13.41	13.39	13.37	13.34
1.18	13.32	13.29	13.27	13.25	13.23	13.21	13.19	13.16	13.14	13.12
1.19	13.10	13.07	13.05	13.03	13.01	12.99	12.96	12.94	12.92	12.90
1.20	12.88	12.86	12.84	12.81	12.79	12.77	12.75	12.73	12.71	12.69
1.21	12.67	12.64	12.62	12.60	12.58	12.56	12.54	12.52	12.50	12.48
1.22	12.46	12.44	12.42	12.40	12.38	12.36	12.34	12.32	12.30	12.28
1.23	12.26	12.24	12.22	12.19	12.18	12.16	12.14	12.12	12.10	12.08
1.24	12.06	12.04	12.02	12.00	11.98	11.96	11.94	11.92	11.91	11.89
1.25	11.87	11.85	11.83	11.81	11.79	11.77	11.75	11.73	11.71	11.69
1.26	11.68	11.66	11.64	11.62	11.61	11.59	11.57	11.55	11.54	11.52
1.27	11.50	11.48	11.46	11.44	11.42	11.40	11.39	11.37	11.35	11.33
1.28	11.32	11.30	11.28	11.26	11.25	11.23	11.21	11.19	11.18	11.16
1.29	11.14	11.12	11.11	11.09	11.07	11.06	11.04	11.02	11.01	10.99
1.30	10.97	10.95	10.94	10.92	10.91	10.89	10.87	10.85	10.84	10.82
1.31	10.80	10.79	10.77	10.75	10.74	10.72	10.70	10.68	10.66	10.65
1.32	10.64	10.62	10.61	10.59	10.58	10.56	10.55	10.53	10.51	10.49
1.33	10.48	10.46	10.45	10.44	10.42	10.40	10.39	10.37	10.36	10.34
1.34	10.33	10.31	10.29	10.28	10.27	10.25	10.24	10.22	10.21	10.19
1.35	10.18	10.16	10.15	10.13	10.12	10.10	10.09	10.07	10.06	10.04
1.36	10.03	10.01	10.00	9.98	9.97	9.95	9.94	9.92	9.91	9.89
1.37	9.88	9.87	9.85	9.84	9.82	9.81	9.79	9.78	9.77	9.75
1.38	9.74	9.72	9.71	9.70	9.68	9.67	9.65	9.64	9.63	9.61
1.39	9.60	9.58	9.57	9.56	9.54	9.53	9.52	9.50	9.49	9.47
1.40	9.46	9.45	9.43	9.42	9.41	9.39	9.38	9.37	9.35	9.34
1.41	9.33	9.31	9.30	9.29	9.27	9.26	9.25	9.24	9.22	9.21
1.42	9.20	9.18	9.17	9.16	9.15	9.13	9.12	9.11	9.09	9.08
1.43	9.07	9.06	9.04	9.03	9.02	9.01	8.99	8.98	8.97	8.96
1.44	8.94	8.93	8.92	8.91	8.89	8.88	8.87	8.86	8.84	8.83

* Courtesy of Wilson Mechanical Instrument Co., Inc.

TABLE 32. DIAMOND PYRAMID HARDNESS NUMBERS.*—(*Continued*)

Diagonal of impression, mm	0.000	0.001	0.002	0.003	0.004	0.005	0.006	0.007	0.008	0.009
1.45	8.82	8.81	8.80	8.78	8.77	8.76	8.75	8.74	8.72	8.71
1.46	8.70	8.69	8.68	8.66	8.65	8.64	8.63	8.62	8.60	8.59
1.47	8.58	8.57	8.56	8.55	8.54	8.52	8.51	8.50	8.49	8.48
1.48	8.47	8.45	8.44	8.43	8.42	8.41	8.40	8.39	8.38	8.36
1.49	8.35	8.34	8.33	8.32	8.31	8.30	8.29	8.27	8.26	8.25
1.50	8.24	8.23	8.22	8.21	8.20	8.19	8.18	8.17	8.15	8.14
1.51	8.13	8.12	8.11	8.10	8.09	8.08	8.07	8.06	8.05	8.04
1.52	8.03	8.02	8.01	7.99	7.98	7.97	7.96	7.95	7.94	7.93
1.53	7.92	7.91	7.90	7.89	7.88	7.87	7.86	7.85	7.84	7.83
1.54	7.82	7.81	7.80	7.79	7.78	7.77	7.76	7.75	7.74	7.73
1.55	7.72	7.71	7.70	7.69	7.68	7.67	7.66	7.65	7.64	7.63
1.56	7.62	7.61	7.60	7.59	7.58	7.57	7.56	7.55	7.54	7.53
1.57	7.52	7.51	7.50	7.49	7.49	7.48	7.47	7.46	7.45	7.44
1.58	7.43	7.42	7.41	7.40	7.39	7.38	7.37	7.36	7.35	7.34
1.59	7.34	7.33	7.32	7.31	7.30	7.29	7.28	7.27	7.26	7.25
1.60	7.24	7.23	7.23	7.22	7.21	7.20	7.19	7.18	7.17	7.16
1.61	7.15	7.15	7.14	7.13	7.12	7.11	7.10	7.09	7.08	7.07
1.62	7.07	7.06	7.05	7.04	7.03	7.02	7.01	7.01	7.00	6.99
1.63	6.98	6.97	6.96	6.95	6.95	6.94	6.93	6.92	6.91	6.90
1.64	6.90	6.89	6.88	6.87	6.86	6.85	6.84	6.84	6.83	6.82
1.65	6.81	6.80	6.80	6.79	6.78	6.77	6.76	6.75	6.75	6.74
1.66	6.73	6.72	6.71	6.71	6.70	6.69	6.68	6.67	6.66	6.66
1.67	6.65	6.64	6.63	6.63	6.62	6.61	6.60	6.59	6.59	6.58
1.68	6.57	6.56	6.56	6.55	6.54	6.53	6.52	6.52	6.51	6.50
1.69	6.49	6.49	6.48	6.47	6.46	6.45	6.45	6.44	6.43	6.42
1.70	6.42	6.41	6.40	6.39	6.39	6.38	6.37	6.36	6.36	6.35
1.71	6.34	6.33	6.33	6.32	6.31	6.31	6.30	6.29	6.28	6.28
1.72	6.27	6.26	6.25	6.25	6.24	6.23	6.22	6.22	6.21	6.20
1.73	6.20	6.19	6.18	6.17	6.17	6.16	6.15	6.15	6.14	6.13
1.74	6.13	6.12	6.11	6.10	6.10	6.09	6.08	6.08	6.07	6.06
1.75	6.06	6.05	6.04	6.03	6.03	6.02	6.01	6.01	6.00	5.99
1.76	5.99	5.98	5.97	5.97	5.96	5.95	5.95	5.94	5.93	5.93
1.77	5.92	5.91	5.91	5.90	5.89	5.89	5.88	5.87	5.87	5.86
1.78	5.85	5.85	5.84	5.83	5.83	5.82	5.81	5.81	5.80	5.80
1.79	5.79	5.78	5.77	5.77	5.76	5.76	5.75	5.74	5.74	5.73

* Courtesy of Wilson Mechanical Instrument Co., Inc.

TABLE 32. DIAMOND PYRAMID HARDNESS NUMBERS.*—(*Continued*)

Diagonal of impression, mm	0.000	0.001	0.002	0.003	0.004	0.005	0.006	0.007	0.008	0.009
1.80	5.72	5.72	5.71	5.70	5.70	5.69	5.68	5.68	5.67	5.67
1.81	5.66	5.65	5.65	5.64	5.64	5.63	5.62	5.62	5.61	5.60
1.82	5.60	5.59	5.59	5.58	5.57	5.57	5.56	5.56	5.55	5.54
1.83	5.54	5.53	5.53	5.52	5.51	5.51	5.50	5.50	5.49	5.48
1.84	5.48	5.47	5.47	5.46	5.45	5.45	5.44	5.44	5.43	5.42
1.85	5.42	5.41	5.41	5.40	5.40	5.39	5.38	5.38	5.37	5.37
1.86	5.36	5.35	5.35	5.34	5.34	5.33	5.33	5.32	5.31	5.31
1.87	5.30	5.30	5.29	5.29	5.28	5.27	5.27	5.26	5.26	5.25
1.88	5.25	5.24	5.24	5.23	5.22	5.22	5.21	5.21	5.20	5.20
1.89	5.19	5.19	5.18	5.17	5.17	5.16	5.16	5.15	5.15	5.14
1.90	5.14	5.13	5.13	5.12	5.12	5.11	5.10	5.10	5.09	5.09
1.91	5.08	5.08	5.07	5.07	5.06	5.06	5.05	5.05	5.04	5.04
1.92	5.03	5.03	5.02	5.01	5.01	5.00	5.00	4.99	4.99	4.98
1.93	4.98	4.97	4.97	4.96	4.96	4.95	4.95	4.94	4.94	4.93
1.94	4.93	4.92	4.92	4.91	4.91	4.90	4.90	4.89	4.89	4.88
1.95	4.88	4.87	4.87	4.86	4.86	4.85	4.85	4.84	4.84	4.83
1.96	4.83	4.82	4.82	4.81	4.81	4.80	4.80	4.79	4.79	4.78
1.97	4.78	4.77	4.77	4.76	4.76	4.75	4.75	4.74	4.74	4.73
1.98	4.73	4.73	4.72	4.72	4.71	4.71	4.70	4.70	4.69	4.69
1.99	4.68	4.68	4.67	4.67	4.66	4.66	4.65	4.65	4.65	4.64
2.00	4.64	—	—	—	—	—	—	—	—	—

* Courtesy of Wilson Mechanical Instrument Co., Inc.

METALLOGRAPHIC LABORATORY PRACTICE

TABLE 34. KNOOP HARDNESS NUMBERS*

Load—1 kg

Length of indentation, mm	0.000	0.001	0.002	0.003	0.004	0.005	0.006	0.007	0.008	0.009
0.010	142,290	117,595	98,812	84,195	72,597	63,240	55,582	49,235	43,917	39,415
0.020	35,572	32,265	29,399	26,898	24,703	22,766	21,049	19,518	18,149	16,919
0.030	15,810	14,806	13,895	13,066	12,309	11,615	10,979	10,394	9,854	9,355
0.040	8,893	8,465	8,066	7,695	7,350	7,027	6,724	6,441	6,176	5,926
0.050	5,692	5,471	5,262	5,065	4,880	4,704	4,537	4,379	4,230	4,088
0.060	3,952	3,824	3,702	3,585	3,474	3,368	3,267	3,170	3,077	2,989
0.070	2,904	2,823	2,745	2,670	2,598	2,530	2,463	2,400	2,339	2,280
0.080	2,223	2,169	2,116	2,065	2,017	1,969	1,924	1,880	1,837	1,796
0.090	1,757	1,718	1,681	1,645	1,610	1,577	1,544	1,512	1,482	1,452
0.100	1,423	1,395	1,368	1,341	1,316	1,291	1,266	1,243	1,220	1,198
0.110	1,176	1,155	1,134	1,114	1,095	1,076	1,057	1,039	1,022	1,005
0.120	988.1	971.9	956.0	940.5	925.4	910.7	896.3	882.2	868.5	855.1
0.130	842.0	829.1	816.6	804.4	792.4	780.7	769.3	758.1	747.2	736.5
0.140	726.0	715.7	705.7	695.8	686.2	676.8	667.5	658.5	649.6	640.9
0.150	632.4	624.0	615.9	607.8	600.0	592.3	584.7	577.3	570.0	562.8
0.160	555.8	548.9	542.2	535.5	529.0	522.6	516.4	510.2	504.1	498.2
0.170	492.4	486.6	481.0	475.4	470.0	464.6	459.4	454.2	449.1	444.1
0.180	439.2	434.3	429.6	424.9	420.3	415.7	411.3	406.9	402.6	398.3
0.190	394.2	390.0	386.0	382.0	378.1	374.2	370.4	366.6	362.9	359.3
0.200	355.7	352.2	348.7	345.3	341.9	338.6	335.3	332.1	328.9	325.7
0.210	322.7	319.6	316.6	313.6	310.7	307.8	305.0	302.2	299.4	296.7
0.220	294.0	291.3	288.7	286.1	283.6	281.1	278.6	276.1	273.7	271.3
0.230	269.0	266.7	264.4	262.1	259.9	257.7	255.5	253.3	251.2	249.1
0.240	247.0	245.0	243.0	241.0	239.0	237.1	235.1	233.2	231.4	229.5
0.250	227.7	225.9	224.1	222.3	220.5	218.8	217.1	215.4	213.8	212.1
0.260	210.5	208.9	207.3	205.7	204.2	202.6	201.1	199.6	198.1	196.6
0.270	195.2	193.7	192.3	190.9	189.5	188.2	186.8	185.4	184.1	182.8
0.280	181.5	180.2	178.9	177.7	176.4	175.2	174.0	172.7	171.5	170.4
0.290	169.2	168.0	166.9	165.7	164.6	163.5	162.4	161.3	160.2	159.2
0.300	158.1	157.1	156.0	155.0	154.0	153.0	152.0	151.0	150.0	149.0
0.310	148.1	147.1	146.2	145.2	144.3	143.4	142.5	141.6	140.7	139.8
0.320	139.0	138.1	137.2	136.4	135.5	134.7	133.9	133.1	132.3	131.5
0.330	130.7	[129.9	129.1	128.3	127.5	126.8	126.0	125.3	124.5	123.8
0.340	123.1	122.4	121.7	120.9	120.2	119.5	118.9	118.2	117.5	116.8
0.350	116.2	115.5	114.8	114.2	113.5	112.9	112.3	*111.6	111.0	110.4
0.360	109.8	109.2	108.6	108.0	107.4	106.8	106.2	105.6	105.1	104.5
0.370	103.9	103.4	102.8	102.3	101.7	101.2	100.6	100.1	99.58	99.06
0.380	98.54	98.02	97.51	97.00	96.50	96.00	95.50	95.01	94.52	94.03
0.390	93.55	93.07	92.60	92.13	91.66	91.20	90.74	90.28	89.83	89.38
0.400	88.93	88.49	88.05	87.61	87.18	86.75	86.32	85.90	85.48	85.06
0.410	84.65	84.23	83.83	83.42	83.02	82.62	82.22	81.83	81.44	81.05
0.420	80.66	80.28	79.90	79.52	79.15	78.78	78.41	78.04	77.68	77.31
0.430	76.95	76.60	76.24	75.89	75.54	75.20	74.85	74.51	74.17	73.83
0.440	73.50	73.16	72.83	72.50	72.18	71.85	71.53	71.21	70.90	70.58
0.450	70.27	69.96	69.65	69.34	69.03	68.73	68.43	68.13	67.83	67.54

* Courtesy of Wilson Mechanical Instrument Co., Inc.

TABLE 34. KNOOP HARDNESS NUMBERS.*—(*Continued*)

Length of indentation, mm	0.000	0.001	0.002	0.003	0.004	0.005	0.006	0.007	0.008	0.009
0.460	67.24	66.95	66.66	66.38	66.09	65.81	65.52	65.24	64.97	64.69
0.470	64.41	64.14	63.87	63.60	63.33	63.06	62.80	62.54	62.28	62.02
0.480	61.76	61.50	61.25	60.99	60.74	60.49	60.24	60.00	59.75	59.51
0.490	59.26	59.02	58.78	58.54	58.31	58.07	57.84	57.61	57.37	57.14
0.500	56.92	56.69	56.46	56.24	56.02	55.79	55.57	55.36	55.14	54.92
0.510	54.71	54.49	54.28	54.07	53.86	53.65	53.44	53.23	53.03	52.82
0.520	52.62	52.42	52.22	52.02	51.82	51.62	51.43	51.23	51.04	50.85
0.530	50.65	50.46	50.27	50.09	49.90	49.71	49.53	49.34	49.16	48.98
0.540	48.80	48.62	48.44	48.26	48.08	47.90	47.73	47.56	47.38	47.21
0.550	47.04	46.87	46.70	46.53	46.36	46.19	46.03	45.86	45.70	45.54
0.560	45.37	45.21	45.05	44.89	44.73	44.57	44.42	44.26	44.10	43.95
0.570	43.79	43.64	43.49	43.34	43.19	43.04	42.89	42.74	42.59	42.44
0.580	42.30	42.15	42.01	41.86	41.72	41.58	41.44	41.29	41.15	41.01
0.590	40.88	40.74	40.60	40.46	40.33	40.19	40.06	39.92	39.79	39.66
0.600	39.52	39.39	39.36	39.13	39.00	38.87	38.75	38.62	38.49	38.37
0.610	38.24	38.11	37.99	37.87	37.74	37.62	37.50	37.38	37.26	37.14
0.620	37.02	36.90	36.78	36.66	36.54	36.43	36.31	36.19	36.08	35.96
0.630	35.85	35.74	35.62	35.51	35.40	35.29	35.18	35.07	34.96	34.85
0.640	34.74	34.63	34.52	34.42	34.31	34.20	34.10	33.99	33.89	33.78
0.650	33.68	33.57	33.47	33.37	33.27	33.17	33.06	32.96	32.86	32.76
0.660	32.67	32.57	32.47	32.37	32.27	32.18	32.08	31.98	31.89	31.79
0.670	31.70	31.60	31.51	31.42	31.32	31.23	31.14	31.05	30.95	30.86
0.680	30.77	30.68	30.59	30.50	30.41	30.32	30.24	30.15	30.06	29.97
0.690	29.89	29.80	29.71	29.63	29.54	29.46	29.37	29.21	29.21	29.12
0.700	29.04	28.96	28.87	28.79	28.71	28.63	28.55	28.47	28.39	28.31
0.710	28.23	28.15	28.07	27.99	27.91	27.83	27.76	27.68	27.60	27.52
0.720	27.45	27.37	27.30	27.22	27.15	27.07	27.00	26.92	26.85	26.77
0.730	26.70	26.63	26.56	26.48	26.41	26.34	26.27	26.20	26.13	26.05
0.740	25.98	25.91	25.84	25.77	25.71	25.64	25.57	25.50	25.43	25.36
0.750	25.30	25.23	25.16	25.09	25.03	24.96	24.90	24.83	24.76	24.70
0.760	24.63	24.57	24.51	24.44	24.38	24.31	24.25	24.19	24.12	24.06
0.770	24.00	23.94	23.87	23.81	23.75	23.69	23.63	23.57	23.51	23.45
0.780	23.39	23.33	23.27	23.21	23.15	23.09	23.03	22.97	22.92	22.86
0.790	22.80	22.74	22.68	22.63	22.57	22.51	22.46	22.40	22.34	22.29
0.800	22.23	22.18	22.12	22.07	22.01	21.96	21.90	21.85	21.79	21.74
0.810	21.69	21.63	21.58	21.53	21.47	21.42	21.37	21.32	21.27	21.21
0.820	21.16	21.11	21.06	21.01	20.96	20.91	20.86	20.80	20.75	20.70
0.830	20.65	20.60	20.56	20.51	20.46	20.41	20.36	20.31	20.26	20.21
0.840	20.17	20.12	20.07	20.02	19.98	19.93	19.88	19.83	19.79	19.74
0.850	19.69	19.65	19.60	19.56	19.51	19.46	19.42	19.37	19.33	19.28
0.860	19.24	19.19	19.15	19.11	19.06	19.02	18.97	18.93	18.89	18.84
0.870	18.80	18.76	18.71	18.67	18.63	18.58	18.54	18.50	18.46	18.42
0.880	18.37	18.33	18.29	18.25	18.21	18.17	18.13	18.09	18.04	18.00
0.890	17.96	17.92	17.88	17.84	17.80	17.76	17.72	17.68	17.64	17.61
0.900	17.57	17.53	17.49	17.45	17.41	17.37	17.33	17.30	17.26	17.22

* Courtesy of Wilson Mechanical Instrument Co., Inc.

TABLE 34. KNOOP HARDNESS NUMBERS.*—(*Continued*)

Length of indentation, mm	0.000	0.001	0.002	0.003	0.004	0.005	0.006	0.007	0.008	0.009
0.910	17.18	17.14	17.11	17.07	17.03	17.00	16.96	16.92	16.88	16.85
0.920	16.81	16.77	16.74	16.70	16.67	16.63	16.59	16.56	16.52	16.49
0.930	16.45	16.42	16.38	16.35	16.31	16.28	16.24	16.21	16.17	16.14
0.940	16.10	16.07	16.04	16.00	15.97	15.93	15.90	15.87	15.83	15.80
0.950	15.77	15.73	15.70	15.67	15.63	15.60	15.57	15.54	15.50	15.47
0.960	15.44	15.41	15.38	15.34	15.31	15.28	15.25	15.22	15.19	15.15
0.970	15.12	15.09	15.06	15.03	15.00	14.97	14.94	14.91	14.88	14.85
0.980	14.82	14.79	14.76	14.73	14.70	14.67	14.64	14.61	14.58	14.55
0.990	14.52	14.49	14.46	14.43	14.40	14.37	14.34	14.31	14.29	14.26
1.000	14.23	14.20	14.17	14.14	14.12	14.09	14.06	14.03	14.00	13.98
1.010	13.95	13.92	13.89	13.87	13.84	13.81	13.78	13.76	13.73	13.70
1.020	13.68	13.65	13.62	13.60	13.57	13.54	13.52	13.49	13.46	13.44
1.030	13.41	13.39	13.36	13.33	13.31	13.28	13.21	13.23	13.21	13.18
1.040	13.16	13.13	13.11	13.08	13.05	13.03	13.00	12.98	12.96	12.93
1.050	12.91	12.88	12.86	12.83	12.81	12.78	12.76	12.74	12.71	12.69
1.060	12.66	12.64	12.62	12.59	12.57	12.55	12.52	12.50	12.47	12.45
1.070	12.43	12.40	12.38	12.36	12.34	12.31	12.29	12.27	12.24	12.22
1.080	12.20	12.18	12.15	12.13	12.11	12.09	12.06	12.04	12.02	12.00
1.090	11.98	11.95	11.93	11.91	11.89	11.87	11.85	11.82	11.80	11.78
1.100	11.76	11.74	11.72	11.70	11.67	11.65	11.63	11.61	11.59	11.57
1.110	11.55	11.53	11.51	11.49	11.47	11.45	11.42	11.40	11.38	11.36
1.120	11.34	11.32	11.30	11.28	11.26	11.24	11.22	11.20	11.18	11.16
1.130	11.14	11.12	11.10	11.08	11.06	11.05	11.03	11.01	10.99	10.97
1.140	10.95	10.93	10.91	10.89	10.87	10.85	10.83	10.82	10.80	10.78
1.150	10.76	10.74	10.72	10.70	10.68	10.67	10.65	10.63	10.61	10.59
1.160	10.57	10.56	10.54	10.52	10.50	10.48	10.47	10.45	10.43	10.41
1.170	10.39	10.38	10.36	10.34	10.32	10.31	10.29	10.27	10.25	10.24
1.180	10.22	10.20	10.18	10.17	10.15	10.13	10.12	10.10	10.08	10.06
1.190	10.05	10.03	10.01	9.998	9.981	9.964	9.947	9.931	9.914	9.898
1.200	9.881	9.865	9.848	9.832	9.816	9.799	9.783	9.767	9.751	9.735
1.210	9.719	9.703	9.687	9.671	9.655	9.639	9.623	9.607	9.591	9.576
1.220	9.560	9.544	9.529	9.513	9.498	9.482	9.467	9.451	9.436	9.420
1.230	9.405	9.390	9.375	9.359	9.344	9.329	9.314	9.299	9.284	9.269
1.240	9.254	9.239	9.224	9.209	9.195	9.180	9.165	9.150	9.136	9.121
1.250	9.107	9.092	9.077	9.063	9.049	9.034	9.020	9.005	8.991	8.977
1.260	8.963	8.948	8.934	8.920	8.906	8.892	8.878	8.864	8.850	8.836
1.270	8.822	8.808	8.794	8.780	8.767	8.753	8.739	8.726	8.712	8.698
1.280	8.685	8.671	8.658	8.644	8.631	8.617	8.604	8.590	8.577	8.564
1.290	8.551	8.537	8.524	8.511	8.498	8.485	8.472	8.458	8.445	8.432
1.300	8.420	8.407	8.394	8.381	8.368	8.355	8.342	8.330	8.317	8.304
1.310	8.291	8.279	8.266	8.254	8.241	8.229	8.216	8.204	8.191	8.179
1.320	8.166	8.154	8.142	8.129	8.117	8.105	8.093	8.080	8.068	8.056
1.330	8.044	8.032	8.020	8.008	7.996	7.984	7.972	7.960	7.948	7.936
1.340	7.924	7.913	7.901	7.889	7.877	7.866	7.854	7.842	7.831	7.819
1.350	7.807	7.796	7.784	7.773	7.761	7.750	7.738	7.727	7.716	7.704

* Courtesy of Wilson Mechanical Instrument Co., Inc.

TABLE 34. KNOOP HARDNESS NUMBERS.*—(*Continued*)

Length of indentation, mm	0.000	0.001	0.002	0.003	0.004	0.005	0.006	0.007	0.008	0.009
1.360	7.693	7.682	7.670	7.659	7.648	7.637	7.626	7.614	7.603	7.592
1.370	7.581	7.570	7.559	7.548	7.537	7.526	7.515	7.504	7.493	7.482
1.380	7.472	7.461	7.450	7.439	7.428	7.418	7.407	7.396	7.386	7.375
1.390	7.365	7.354	7.343	7.333	7.322	7.312	7.301	7.291	7.280	7.270
1.400	7.260	7.249	7.239	7.229	7.218	7.208	7.198	7.188	7.177	7.167
1.410	7.157	7.147	7.137	7.127	7.117	7.107	7.097	7.087	7.077	7.067
1.420	7.057	7.047	7.037	7.027	7.017	7.007	6.997	6.988	6.978	6.968
1.430	6.958	6.949	6.939	6.929	6.920	6.910	6.900	6.891	6.881	6.872
1.440	6.862	6.852	6.843	6.833	6.824	6.815	6.805	6.796	6.786	6.777
1.450	6.768	6.758	6.749	6.740	6.730	6.721	6.712	6.703	6.694	6.684
1.460	6.675	6.666	6.657	6.648	6.639	6.630	6.621	6.612	6.603	6.594
1.470	6.585	6.576	6.567	6.558	6.549	6.540	6.531	6.522	6.514	6.505
1.480	6.496	6.487	6.479	6.470	6.461	6.452	6.444	6.435	6.426	6.418
1.490	6.409	6.401	6.392	6.383	6.375	6.366	6.358	6.349	6.341	6.332
1.500	6.324									

* Courtesy of Wilson Mechanical Instrument Co., Inc.

TABLE 35.   HARDNESS VALUES OF METALS ON THE SHORE SCLEROSCOPE SCALE*

Name of metal	Annealed or cast	Cold-worked	Rapidly cooled
Lead	2–4	3–7	
Aluminum	3½–5	6–12	
Gold, 24 to 14 carat	5–25	24–70	
Silver	6½–14	20–37	
Copper	6–8	14–20	
Zinc	8–10	18–20	
Babbitt metal	4–9		
Tin	8–9	12–14	
Bismuth	8–9		
Brass	7–35	20–45	
Platinum	10–15	17–30	
Bronze, phosphorus	12–21	25–40	
Bronze, manganese	16–21	25–40	
Iron, wrought, pure	16–18	25–30	
Nickel (cast)	13–16		
Nickel, wrought	17–19	35–40	
Mild steel, 0.05 to 0.15 carbon	18–25	30–40	
Iron, gray (sand cast)	25–45		
Tungsten (not ductile)	60–70	.....	60–70
Iron, gray, chilled	.....	.....	50–90
Steel, tool, 1 % carbon	30–35	40–50	90–105
Steel, tool, 1.65 % carbon	38–45	.....	90–105
Steel, vanadium	30–50	50–60	50–105
Steel, chrome nickel	35–50	40–60	60–105
Steel, nickel	25–30	35–45	50–90
Steel, high speed	30–45	40–60	70–100

To be used as a guide in determining the approximate setting of the needle indicator and magnifier when testing these metals.

* The Shore Instrument and Manufacturing Co.

TABLE 36. MICROCHARACTER HARDNESS NUMBERS*
Standard 3-g weight†

λ	K	λ	K	λ	K	λ	K
100.0	1.0	14.5	47.6	9.0	123	3.7	730
76.0	1.7	14.0	51.0	8.8	129	3.6	772
64.0	2.4	13.5	54.8	8.6	135	3.5	816
56.0	3.2	13.0	59.2	8.4	142	3.4	865
50.0	4.0	12.8	61.0	8.2	149	3.3	918
47.5	4.4	12.6	63.0	8.0	156	3.2	977
45.0	4.9	12.4	65.0	7.8	164	3.1	1041
42.5	5.5	12.2	67.2	7.6	173	3.0	1111
40.0	6.3	12.0	69.5	7.4	183	2.9	1189
37.5	7.1	11.8	71.8	7.2	193	2.8	1276
35.0	8.2	11.6	74.3	7.0	204	2.7	1372
32.5	9.5	11.4	76.9	6.8	216	2.6	1479
30.0	11.1	11.2	79.7	6.6	230	2.5	1600
28.0	12.8	11.0	82.6	6.4	244	2.4	1736
26.0	14.8	10.8	85.8	6.2	260	2.3	1891
24.0	17.4	10.6	89.0	6.0	278	2.2	2066
22.0	20.7	10.4	92.5	5.8	297	2.1	2268
20.0	25.0	10.2	96.2	5.6	319	2.0	2500
19.5	26.3	10.0	100	5.4	343	1.9	2770
19.0	27.7	9.9	102	5.2	370	1.8	3087
18.5	29.2	9.8	104	5.0	400	1.7	3460
18.0	30.9	9.7	106	4.8	434	1.6	3906
17.5	32.7	9.6	109	4.6	473	1.5	4444
17.0	34.6	9.5	111	4.4	517	1.4	5102
16.5	36.7	9.4	113	4.2	567	1.3	5917
16.0	39.1	9.3	116	4.0	625	1.2	6944
15.5	41.6	9.2	118	3.9	657	1.1	8264
15.0	44.4	9.1	121	3.8	692	1.0	10000

$K = \lambda^{-2}10^4$ = microhardness number
λ = width of microcut, microns

* ASM, "Metals Handbook," 1939.
† For a 9-g weight use $\frac{1}{2}\lambda$.

TABLE 37. APPROXIMATE HARDNESS CONVERSION NUMBERS FOR STEEL, BASED ON DPH* (VICKERS)

Diamond pyramid hardness number, Vickers, 50-kg load	Brinell hardness number 10-mm ball 3000-kg load			Rockwell hardness number				Rockwell superficial hardness number superficial Brale penetrator			Shore scleroscope hardness number	Diamond pyramid hardness number, Vickers, 50-kg load
	Standard ball	Hultgren ball	Tungsten carbide ball	A scale 60-kg load Brale penetrator	B scale 100-kg load 1/16-in. dia. ball	C scale 150-kg load Brale penetrator	D scale 100-kg load Brale penetrator	15-N scale 15-kg load	30-N scale 30-kg load	45-N scale 45-kg load		
940	....	....	....	85.6	.....	68.0	76.9	93.2	84.4	75.4	97	940
920	....	....	....	85.3	.....	67.5	76.5	93.0	84.0	74.8	96	920
900	....	....	....	85.0	.....	67.0	76.1	92.9	83.6	74.2	95	900
880	....	....	767	84.7	.....	66.4	75.7	92.7	83.1	73.6	93	880
860	....	....	757	84.4	.....	65.9	75.3	92.5	82.7	73.1	92	860
840	....	....	745	84.1	.....	65.3	74.8	92.3	82.2	72.2	91	840
820	....	....	733	83.8	.....	64.7	74.3	92.1	81.7	71.8	90	820
800	....	....	722	83.4	.....	64.0	73.8	91.8	81.1	71.0	88	800
780	....	....	710	83.0	.....	63.3	73.3	91.5	80.4	70.2	87	780
760	....	....	698	82.6	.....	62.5	72.6	91.2	79.7	69.4	86	760
740	....	....	684	82.2	.....	61.8	72.1	91.0	79.1	68.6	84	740
720	....	....	670	81.8	.....	61.0	71.5	90.7	78.4	67.7	83	720
700	....	615	656	81.3	.....	60.1	70.8	90.3	77.6	66.7	81	700
690	....	610	647	81.1	.....	59.7	70.5	90.1	77.2	66.2	..	690
680	....	603	638	80.8	.....	59.2	70.1	89.8	76.8	65.7	80	680
670	....	597	630	80.6	.....	58.8	69.8	89.7	76.4	65.3	..	670
660	....	590	620	80.3	.....	58.3	69.4	89.5	75.9	64.7	79	660
650	....	585	611	80.0	.....	57.8	69.0	89.2	75.5	64.1	..	650
640	....	578	601	79.8	.....	57.3	68.7	89.0	75.1	63.5	77	640
630	....	571	591	79.5	.....	56.8	68.3	88.8	74.6	63.0	..	630
620	....	564	582	79.2	.....	56.3	67.9	88.5	74.2	62.4	75	620
610	....	557	573	78.9	.....	55.7	67.5	88.2	73.6	61.7	..	610
600	....	550	564	78.6	.....	55.2	67.0	88.0	73.2	61.2	74	600
590	....	542	554	78.4	.....	54.7	66.7	87.8	72.7	60.5	..	590
580	....	535	545	78.0	.....	54.1	66.2	87.5	72.1	59.9	72	580
570	....	527	535	77.8	.....	53.6	65.8	87.2	71.7	59.3	..	570
560	....	519	525	77.4	.....	53.0	65.4	86.9	71.2	58.6	71	560
550	505	512	517	77.0	.....	52.3	64.8	86.6	70.5	57.8	..	550
540	496	503	507	76.7	.....	51.7	64.4	86.3	70.0	57.0	69	540
530	488	495	497	76.4	.....	51.1	63.9	86.0	69.5	56.2	..	530
520	480	487	488	76.1	.....	50.5	63.5	85.7	69.0	55.6	67	520
510	473	479	479	75.7	.....	49.8	62.9	85.4	68.3	54.7	..	510
500	465	471	471	75.3	.....	49.1	62.2	85.0	67.7	53.9	66	500
490	456	460	460	74.9	.....	48.4	61.6	84.7	67.1	53.1	..	490
480	448	452	452	74.5	.....	47.7	61.3	84.3	66.4	52.2	64	480
470	441	442	442	74.1	.....	46.9	60.7	83.9	65.7	51.3	..	470
460	433	433	433	73.6	.....	46.1	60.1	83.6	64.9	50.4	62	460
450	425	425	425	73.3	.....	45.3	59.4	83.2	64.3	49.4	..	450
440	415	415	415	72.8	.....	44.5	58.8	82.8	63.5	48.4	59	440
430	405	405	405	72.3	.....	43.6	58.2	82.3	62.7	47.4	..	430
420	397	397	397	71.8	.....	42.7	57.5	81.8	61.9	46.4	57	420

Note. The values in this table shown in **bold-faced type** correspond to the values shown in the corresponding joint SAE–ASM–ASTM Committee on Hardness Conversions as printed in ASTM Spec. E48–43T.

* Courtesy of Society of Automotive Engineers. From SAE Handbook, 1948.

TABLE 37. APPROXIMATE HARDNESS CONVERSION NUMBERS FOR STEEL, BASED ON DPH.* (VICKERS)—*(Continued)*

Diamond pyramid hardness number, Vickers, 50-kg load	Brinell hardness number 10-mm ball 3000-kg load			Rockwell hardness number				Rockwell superficial hardness number superficial Brale penetrator			Shore scleroscope hardness number	Diamond pyramid hardness number, Vickers, 50-kg load
	Standard ball	Hultgren ball	Tungsten carbide ball	A scale 60-kg load Brale penetrator	B scale 100-kg load 1/16-in. dia. ball	C scale 150-kg load Brale penetrator	D scale 100-kg load Brale penetrator	15-N scale 15-kg load	30-N scale 30-kg load	45-N scale 45-kg load		
410	388	388	388	71.4	.....	41.8	56.8	81.4	61.1	45.3	..	410
400	379	379	379	70.8	.....	40.8	56.0	81.0	60.2	44.1	55	400
390	369	369	369	70.3	.....	39.8	55.2	80.3	59.3	42.9	..	390
380	360	360	360	69.8	(110.0)	38.8	54.4	79.8	58.4	41.7	52	380
370	350	350	350	69.2	.....	37.7	53.6	79.2	57.4	40.4	..	370
360	341	341	341	68.7	(109.0)	36.6	52.8	78.6	56.4	39.1	50	360
350	331	331	331	68.1	.....	35.5	51.9	78.0	55.4	37.8	..	350
340	322	322	322	67.6	(108.0)	34.4	51.1	77.4	54.4	36.5	47	340
330	313	313	313	67.0	.....	33.3	50.2	76.8	53.6	35.2	..	330
320	303	303	303	66.4	(107.0)	32.2	49.4	76.2	52.3	33.9	45	320
310	294	294	294	65.8	.....	31.0	48.4	75.6	51.3	32.5	..	310
300	284	284	284	65.2	(105.5)	29.8	47.5	74.9	50.2	31.1	42	300
295	280	280	280	64.8	.....	29.2	47.1	74.6	49.7	30.4	..	295
290	275	275	275	64.5	(104.5)	28.5	46.5	74.2	49.0	29.5	41	290
285	270	270	270	64.2	.....	27.8	46.0	73.8	48.4	28.7	..	285
280	265	265	265	63.8	(103.5)	27.1	45.3	73.4	47.8	27.9	40	280
275	261	261	261	63.5	.....	26.4	44.9	73.0	47.2	27.1	..	275
270	256	256	256	63.1	(102.0)	25.6	44.3	72.6	46.4	26.2	38	270
265	252	252	252	62.7	.....	24.8	43.7	72.1	45.7	25.2	..	265
260	247	247	247	62.4	(101.0)	24.0	43.1	71.6	45.0	24.3	37	260
255	243	243	243	62.0	.....	23.1	42.2	71.1	44.2	23.2	..	255
250	238	238	238	61.6	99.5	22.2	41.7	70.6	43.4	22.2	36	250
245	233	233	233	61.2	.....	21.3	41.1	70.1	42.5	21.1	..	245
240	228	228	228	60.7	98.1	20.3	40.3	69.6	41.7	19.9	34	240
230	219	219	219	....	96.7	(18.0)	....	....	....	....	33	230
220	209	209	209	....	95.0	(15.7)	....	....	....	....	32	220
210	200	200	200	....	93.4	(13.4)	....	....	....	....	30	210
200	190	190	190	....	91.5	(11.0)	....	....	....	....	29	200
190	181	181	181	....	89.5	(8.5)	....	....	....	....	28	190
180	171	171	171	....	87.1	(6.0)	....	....	....	....	26	180
170	162	162	162	....	85.0	(3.0)	....	....	....	....	25	170
160	152	152	152	....	81.7	(0.0)	....	....	....	....	24	160
150	143	143	143	....	78.7	....	....	....	....	....	22	150
140	133	133	133	....	75.0	....	....	....	....	....	21	140
130	124	124	124	....	71.2	....	....	....	....	....	20	130
120	114	114	114	....	66.7	....	....	....	....	....	..	120
110	105	105	105	....	62.3	....	....	....	....	....	..	110
100	95	95	95	....	56.2	....	....	....	....	....	..	100
95	90	90	90	....	52.0	....	....	....	....	....	..	95
90	86	86	86	....	48.0	....	....	....	....	....	..	90
85	81	81	81	....	41.0	....	....	....	....	....	..	85

Note. The values in this table shown in **bold-faced type** correspond to the values shown in the corresponding joint SAE–ASM–ASTM Committee on Hardness Conversions as printed in ASTM Spec. E48–43T.

Values in ( ) are beyond normal range; given for information only.

* Courtesy of Society of Automotive Engineers. From SAE Handbook, 1948.

TABLE 38. APPROXIMATE HARDNESS CONVERSION NUMBERS FOR STEEL, BASED ON BRINELL*

Brinell indentation diameter, mm	Brinell hardness number 10-mm ball 3000-kg load			Diamond pyramid hardness number, Vickers	Rockwell hardness number				Rockwell superficial hardness number superficial Brale penetrator			Shore scleroscope hardness number	Brinell indentation diameter, mm
	Standard ball	Hultgren ball	Tungsten carbide ball		A scale 60-kg load Brale penetrator	B scale 100-kg load 1/16 in. dia.-ball	C scale 150-kg load Brale penetrator	D scale 100-kg load Brale penetrator	15-N scale 15-kg load	30-N scale 30-kg load	45-N scale 45-kg load		
....	....	....	....	940	85.6	....	68.0	76.9	93.2	84.4	75.4	97	....
....	....	....	....	920	85.3	....	67.5	76.5	93.0	84.0	74.8	96	....
....	....	....	....	900	85.0	....	67.0	76.1	92.9	83.6	74.2	95	....
....	....	....	767	880	84.7	....	66.4	75.7	92.7	83.1	73.6	93	....
....	....	....	757	860	84.4	....	65.9	75.3	92.5	82.7	73.1	92	....
2.25	....	....	745	840	84.1	....	65.3	74.8	92.3	82.2	72.2	91	2.25
....	....	....	733	820	83.8	....	64.7	74.3	92.1	81.7	71.8	90	....
....	....	....	722	800	83.4	....	64.0	73.8	91.8	81.1	71.0	88	....
2.30	....	....	712	....	....	....	....	....	....	....	....	..	2.30
....	....	....	710	780	83.0	....	63.3	73.3	91.5	80.4	70.2	87	....
....	....	....	698	760	82.6	....	62.5	72.6	91.2	79.7	69.4	86	....
....	....	....	684	740	82.2	....	61.8	72.1	91.0	79.1	68.6	..	....
2.35	....	....	**682**	**737**	**82.2**	....	**61.7**	**72.0**	**91.0**	**79.0**	**68.5**	84	**2.35**
....	....	....	670	720	81.8	....	61.0	71.5	90.7	78.4	67.7	83	....
....	....	....	656	700	81.3	....	60.1	70.8	90.3	77.6	66.7	..	....
2.40	....	....	**653**	**697**	**81.2**	....	**60.0**	**70.7**	**90.2**	**77.5**	**66.5**	81	**2.40**
....	....	....	647	690	81.1	....	59.7	70.5	90.1	77.2	66.2	..	....
....	....	....	638	680	80.8	....	59.2	70.1	89.8	76.8	65.7	80	....
....	....	....	630	670	80.6	....	58.8	69.8	89.7	76.4	65.3	..	....
2.45	....	....	**627**	**667**	**80.5**	....	**58.7**	**69.7**	**89.6**	**76.3**	**65.1**	79	**2.45**
2.50	....	**601**	....	677	80.7	....	59.1	70.0	89.8	76.8	65.7	..	2.50
	....	....	**601**	640	79.8	....	57.3	68.7	89.0	75.1	63.5	77	
2.55	....	**578**	....	640	79.8	....	57.3	68.7	89.0	75.1	63.5	..	2.55
	....	....	**578**	615	79.1	....	56.0	67.7	88.4	73.9	62.1	75	
2.60	....	**555**	....	607	78.8	....	55.6	67.4	88.1	73.5	61.6	..	2.60
	....	....	**555**	591	78.4	....	54.7	66.7	87.8	72.7	60.6	73	
2.65	....	**534**	....	579	78.0	....	54.0	66.1	87.5	72.0	59.8	..	2.65
	....	....	**534**	569	77.8	....	53.5	65.8	87.2	71.6	59.2	71	
2.70	....	**514**	....	553	77.1	....	52.5	65.0	86.7	70.7	58.0	..	2.70
	....	....	**514**	547	76.9	....	52.1	64.7	86.5	70.3	57.6	70	
2.75	**495**	....	....	539	76.7	....	51.6	64.3	86.3	69.9	56.9	..	2.75
	....	**495**	....	530	76.4	....	51.1	63.9	86.0	69.5	56.2	..	
	....	....	**495**	528	76.3	....	51.0	63.8	85.9	69.4	56.1	68	
2.80	**477**	....	....	516	75.9	....	50.3	63.2	85.6	68.7	55.2	..	2.80
	....	**477**	....	508	75.6	....	49.6	62.7	85.3	68.2	54.5	..	
	....	....	**477**	508	75.6	....	49.6	62.7	85.3	68.2	54.5	66	
2.85	**461**	....	....	495	75.1	....	48.8	61.9	84.9	67.4	53.5	..	2.85
	....	**461**	....	491	74.9	....	48.5	61.7	84.7	67.2	53.2	..	
	....	....	**461**	491	74.9	....	48.5	61.7	84.7	67.2	53.2	65	

[1] Brinell numbers are based on the diameter of impressed indentation. If the ball distorts (flattens) during test, Brinell numbers will vary in accordance with the degree of such distortion when related to hardnesses determined with a Vickers diamond pyramid, Rockwell Brale, or other penetrator which does not sensibly distort. At high hardnesses, therefore, the relationship between Brinell and Vickers or Rockwell scales is affected by the type of ball used. Steel balls (Standard or Hultgren) tend to flatten slightly more than carbide balls, resulting in larger indentation and lower Brinell number than shown by a carbide ball. Thus, on a specimen of 640 Vickers, a Hultgren ball will leave an impression 2.55 mm (578 BHN), and the carbide ball an impression 2.50 mm (601 BHN). Conversely, identical impression diameters for both types of ball will correspond to different Vickers or Rockwell values. Thus, if both impressions show 2.55 mm (578 BHN), material tested with a Hultgren ball has a Vickers hardness 640, while material tested with a carbide ball has a Vickers hardness 615.

Note. The values in this table shown in **bold-faced type** correspond to the values shown in the corresponding joint SAE–ASM–ASTM Committee on Hardness Conversions as printed in ASTM Spec. E48–43T.

* Courtesy of Society of Automotive Engineers. From SAE Handbook, 1948.

TABLE 38. APPROXIMATE HARDNESS CONVERSION NUMBERS FOR STEEL, BASED ON BRINELL.*—(*Continued*)

Brinell indentation diameter, mm	Brinell hardness number 10-mm ball 300-kg load			Diamond pyramid hardness number, Vickers	Rockwell hardness number				Rockwell superficial hardness number superficial Brale penetrator			Shore scleroscope hardness number	Brinell indentation diameter, mm
	Standard ball	Hultgren ball	Tungsten carbide ball		A scale 60-kg load Brale penetrator	B scale 100-kg load 1/16 in. dia.-ball	C scale 150-kg load Brale penetrator	D scale 100-kg load Brale penetrator	15-N scale 15-kg load	30-N scale 30-kg load	45-N scale 45-kg load		
2.90	**444**	....	....	**474**	**74.3**	....	**47.2**	**61.0**	**84.1**	**66.0**	**51.7**	..	2.90
	....	**444**	....	**472**	**74.2**	....	**47.1**	**60.8**	**84.0**	**65.8**	**51.5**	..	
	....	....	**444**	**472**	**74.2**	....	**47.1**	**60.8**	**84.0**	**65.8**	**51.5**	63	
2.95	**429**	**429**	**429**	**455**	**73.4**	....	**45.7**	**59.7**	**83.4**	**64.6**	**49.9**	61	2.95
3.00	**415**	**415**	**415**	**440**	**72.8**	....	**44.5**	**58.8**	**82.8**	**63.5**	**48.4**	59	3.00
3.05	**401**	**401**	**401**	**425**	**72.0**	....	**43.1**	**57.8**	**82.0**	**62.3**	**46.9**	58	3.05
3.10	**388**	**388**	**388**	**410**	**71.4**	....	**41.8**	**56.8**	**81.4**	**61.1**	**45.3**	56	3.10
3.15	**375**	**375**	**375**	**396**	**70.6**	....	**40.4**	**55.7**	**80.6**	**59.9**	**43.6**	54	3.15
3.20	**363**	**363**	**363**	**383**	**70.0**	....	**39.1**	**54.6**	**80.0**	**58.7**	**42.0**	52	3.20
3.25	**352**	**352**	**352**	**372**	**69.3**	(110.0)	**37.9**	**53.8**	**79.3**	**57.6**	**40.5**	51	3.25
3.30	**341**	**341**	**341**	**360**	**68.7**	(109.0)	**36.6**	**52.8**	**78.6**	**56.4**	**39.1**	50	3.30
3.35	**331**	**331**	**331**	**350**	**68.1**	(108.5)	**35.5**	**51.9**	**78.0**	**55.4**	**37.8**	48	3.35
3.40	**321**	**321**	**321**	**339**	**67.5**	(108.0)	**34.3**	**51.0**	**77.3**	**54.3**	**36.4**	47	3.40
3.45	**311**	**311**	**311**	**328**	**66.9**	(107.5)	**33.1**	**50.0**	**76.7**	**53.3**	**34.4**	46	3.45
3.50	**302**	**302**	**302**	**319**	**66.3**	(107.0)	**32.1**	**49.3**	**76.1**	**52.2**	**33.8**	45	3.50
3.55	**293**	**293**	**293**	**309**	**65.7**	(106.0)	**30.9**	**48.3**	**75.5**	**51.2**	**32.4**	43	3.55
3.60	**285**	**285**	**285**	**301**	**65.3**	(105.5)	**29.9**	**47.6**	**75.0**	**50.3**	**31.2**	..	3.60
3.65	**277**	**277**	**277**	**292**	**64.6**	(104.5)	**28.8**	**46.7**	**74.4**	**49.3**	**29.9**	41	3.65
3.70	**269**	**269**	**269**	**284**	**64.1**	(104.0)	**27.6**	**45.9**	**73.7**	**48.3**	**28.5**	40	3.70
3.75	**262**	**262**	**262**	**276**	**63.6**	(103.0)	**26.6**	**45.0**	**73.1**	**47.3**	**27.3**	39	3.75
3.80	**255**	**255**	**255**	**269**	**63.0**	(102.0)	**25.4**	**44.2**	**72.5**	**46.2**	**26.0**	38	3.80
3.85	**248**	**248**	**248**	**261**	**62.5**	(101.0)	**24.2**	**43.2**	**71.7**	**45.1**	**24.5**	37	3.85
3.90	**241**	**241**	**241**	**253**	**61.8**	100.0	**22.8**	**42.0**	**70.9**	**43.9**	**22.8**	36	3.90
3.95	**235**	**235**	**235**	**247**	**61.4**	99.0	**21.7**	**41.4**	**70.3**	**42.9**	**21.5**	35	3.95
4.00	**229**	**229**	**229**	**241**	**60.8**	98.2	**20.5**	**40.5**	**69.7**	**41.9**	**20.1**	34	4.00
4.05	223	223	223	234	....	97.3	(18.8)	....	....	....	....	..	4.05
4.10	217	217	217	228	....	96.4	(17.5)	....	....	....	....	33	4.10
4.15	212	212	212	222	....	95.5	(16.0)	....	....	....	....	..	4.15
4.20	207	207	207	218	....	94.6	(15.2)	....	....	....	....	32	4.20
4.25	201	201	201	212	....	93.8	(13.8)	....	....	....	....	31	4.25
4.30	197	197	197	207	....	92.8	(12.7)	....	....	....	....	30	4.30
4.35	192	192	192	202	....	91.9	(11.5)	....	....	....	....	29	4.35
4.40	187	187	187	196	....	90.7	(10.0)	....	....	....	....	..	4.40
4.45	183	183	183	192	....	90.0	(9.0)	....	....	....	....	28	4.45
4.50	179	179	179	188	....	89.0	(8.0)	....	....	....	....	27	4.50
4.55	174	174	174	182	....	87.8	(6.4)	....	....	....	....	..	4.55
4.60	170	170	170	178	....	86.8	(5.4)	....	....	....	....	26	4.60
4.65	167	167	167	175	....	86.0	(4.4)	....	....	....	....	..	4.65
4.70	163	163	163	171	....	85.0	(3.3)	....	....	....	....	25	4.70
4.80	156	156	156	163	....	82.9	(0.9)	....	....	....	....	..	4.80
4.90	149	149	149	156	....	80.8	....	....	....	....	....	23	4.90
5.00	143	143	143	150	....	78.7	....	....	....	....	....	22	5.00
5.10	137	137	137	143	....	76.4	....	....	....	....	....	21	5.10
5.20	131	131	131	137	....	74.0	....	....	....	....	....	..	5.20
5.30	126	126	126	132	....	72.0	....	....	....	....	....	20	5.30
5.40	121	121	121	127	....	69.8	....	....	....	....	....	19	5.40
5.50	116	116	116	122	....	67.6	....	....	....	....	....	18	5.50
5.60	111	111	111	117	....	65.7	....	....	....	....	....	15	5.60

Note. The values in this table shown in **bold-faced type** correspond to the values shown in the corresponding joint SAE–ASM–ASTM Committee on Hardness Conversions as printed in ASTM Spec. E48–43T.

Values in ( ) are beyond normal range; given for information only.

* Courtesy of Society of Automotive Engineers. From SAE Handbook, 1948.

TABLE 39. APPROXIMATE HARDNESS CONVERSION NUMBERS FOR STEEL, BASED ON ROCKWELL C*

Rockwell C scale hardness number	Diamond pyramid hardness number Vickers	Brinell hardness number 10-mm ball 3000-kg load			Rockwell hardness number			Rockwell superficial hardness number superficial Brale penetrator			Shore scleroscope hardness number	Rockwell C scale hardness number
		Standard ball	Hultgren ball	Tungsten carbide ball	A scale 60-kg load Brale penetrator	B scale 100-kg load 1/6-in.-dia.-ball	D scale 100-kg Brale penetrator	15-N scale 15-kg load	30-N scale 30-kg load	45-N scale 45-kg load		
68	940	....	....	....	85.6	....	76.9	93.2	94.4	75.4	97	68
67	900	....	....	....	85.0	....	76.1	92.9	83.6	74.2	95	67
66	865	....	....	....	84.5	....	75.4	92.5	82.8	73.3	92	66
65	832	....	....	739	83.9	....	74.5	92.2	81.9	72.0	91	65
64	800	....	....	722	83.4	....	73.8	91.8	81.1	71.0	88	64
63	772	....	....	705	82.8	....	73.0	91.4	80.1	69.9	87	63
62	746	....	....	688	82.3	....	72.2	91.1	79.3	68.8	85	62
61	720	....	....	670	81.8	....	71.5	90.7	78.4	67.7	83	61
60	697	....	613	654	81.2	....	70.7	90.2	77.5	66.6	81	60
59	674	....	599	634	80.7	....	69.9	89.8	76.6	65.5	80	59
58	653	....	587	615	80.1	....	69.2	89.3	75.7	64.3	78	58
57	633	....	575	595	79.6	....	68.5	88.9	74.8	63.2	76	57
56	613	....	561	577	79.0	....	67.7	88.3	73.9	62.0	75	56
55	595	....	546	560	78.5	....	66.9	87.9	73.0	60.9	74	55
54	577	....	534	543	78.0	....	66.1	87.4	72.0	59.8	72	54
53	560	....	519	525	77.4	....	65.4	86.9	71.2	58.6	71	53
52	544	500	508	512	76.8	....	64.6	86.4	70.2	57.4	69	52
51	528	487	494	496	76.3	....	63.8	85.9	69.4	56.1	68	51
50	513	475	481	481	75.9	....	63.1	85.5	68.5	55.0	67	50
49	498	464	469	469	75.2	....	62.1	85.0	67.6	53.8	66	49
48	484	451	455	455	74.7	....	61.4	84.5	66.7	52.5	64	48
47	471	442	443	443	74.1	....	60.8	83.9	65.8	51.4	63	47
46	458	432	432	432	73.6	....	60.0	83.5	64.8	50.3	62	46
45	446	421	421	421	73.1	....	59.2	83.0	64.0	49.0	60	45
44	434	409	409	409	72.5	....	58.5	82.5	63.1	47.8	58	44
43	423	400	400	400	72.0	....	57.7	82.0	62.2	46.7	57	43
42	412	390	390	390	71.5	....	56.9	81.5	61.3	45.5	56	42
41	402	381	381	381	70.9	....	56.2	80.9	60.4	44.3	55	41
40	392	371	371	371	70.4	....	55.4	80.4	59.5	43.1	54	40
39	382	362	362	362	69.9	....	54.6	79.9	58.6	41.9	52	39
38	372	353	353	353	69.4	....	53.8	79.4	57.7	40.8	51	38
37	363	344	344	344	68.9	....	53.1	78.8	56.8	39.6	50	37
36	354	336	336	336	68.4	(109.0)	52.3	78.3	55.9	38.4	49	36
35	345	327	327	327	67.9	(108.5)	51.5	77.7	55.0	37.2	48	35
34	336	319	319	319	67.4	(108.0)	50.8	77.2	54.2	36.1	47	34
33	327	311	311	311	66.8	(107.5)	50.0	76.6	53.3	34.9	46	33
32	318	301	301	301	66.3	(107.0)	49.2	76.1	52.1	33.7	44	32
31	310	294	294	294	65.8	(106.0)	48.4	75.6	51.3	32.5	43	31
30	302	286	286	286	65.3	(105.5)	47.7	75.0	50.4	31.3	42	30
29	294	279	279	279	64.7	(104.5)	47.0	74.5	49.5	30.1	41	29
28	286	271	271	271	64.3	(104.0)	46.1	73.9	48.6	28.9	41	28
27	279	264	264	264	63.8	(103.0)	45.2	73.3	47.7	27.8	40	27
26	272	258	258	258	63.3	(102.5)	44.6	72.8	46.8	26.7	38	26
25	266	253	253	253	62.8	(101.5)	43.8	72.2	45.9	25.5	38	25
24	260	247	247	247	62.4	(101.0)	43.1	71.6	45.0	24.3	37	24
23	254	243	243	243	62.0	100.0	42.1	71.0	44.0	23.1	36	23
22	248	237	237	237	61.5	99.0	41.6	70.5	43.2	22.0	35	22
21	243	231	231	231	61.0	98.5	40.9	69.9	42.3	20.7	35	21
20	238	226	226	226	60.5	97.8	40.1	69.4	41.5	19.6	34	20
(18)	230	219	219	219	....	96.7	....	....	....	....	33	(18)
(16)	222	212	212	212	....	95.5	....	....	....	....	32	(16)
(14)	213	203	203	203	....	93.9	....	....	....	....	31	(14)
(12)	204	194	194	194	....	92.3	....	....	....	....	29	(12)
(10)	196	187	187	187	....	90.7	....	....	....	....	28	(10)
(8)	188	179	179	179	....	89.5	....	....	....	....	27	(8)
(6)	180	171	171	171	....	87.1	....	....	....	....	26	(6)
(4)	173	165	165	165	....	85.5	....	....	....	....	25	(4)
(2)	166	158	158	158	....	83.5	....	....	....	....	24	(2)
(0)	160	152	152	152	....	81.7	....	....	....	....	24	(0)

Note. The values in this table shown in **bold-faced type** correspond to the values shown in the corresponding joint SAE–ASM–ASTM Committee on Hardness Conversions as printed in ASTM Spec. E48–43T.

Values in ( ) are beyond normal range; given for information only.

* Courtesy of Society of Automotive Engineers. From SAE Handbook, 1948.

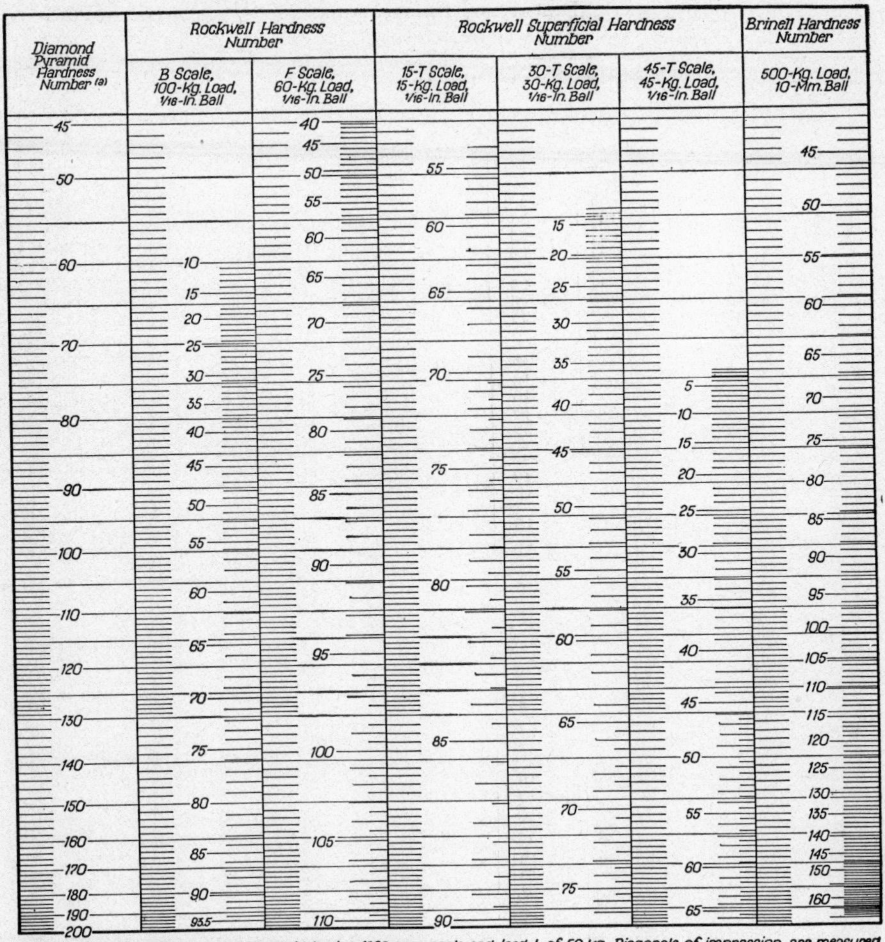

Diamond Pyramid Hardness Number [a]	Rockwell Hardness Number		Rockwell Superficial Hardness Number			Brinell Hardness Number
	B Scale, 100-Kg. Load, 1/16-In. Ball	F Scale, 60-Kg. Load, 1/16-In. Ball	15-T Scale, 15-Kg. Load, 1/16-In. Ball	30-T Scale, 30-Kg. Load, 1/16-In. Ball	45-T Scale, 45-Kg. Load, 1/16-In. Ball	500-Kg. Load, 10-Mm. Ball

[a] Test is made with square base pyramid having 136° apex angle and load L of 50 kg. Diagonals of impression are measured in mm. and averaged as d. Hardness number = $\frac{2L \sin 68°}{d^2}$

Comparative values hold only when standard test procedures are followed, and only when made on flat specimens of cartridge brass thick enough to avoid the anvil effect (roughly 10 times the depth of the indentation).

CHART 4. Conversion of hardness numbers as related to cartridge brass (70 per cent copper, 30 per cent zinc). (*From Metal Progress, Vol. 44,* 1943.)

TABLE 42. APPROXIMATE DEPTH OF PENETRATION OF SUPERSONIC WAVES IN VARIOUS MATERIALS*

Material	Range, ft, unless otherwise marked				Remarks
	Testing frequency, megacycles				
	0.5	1	2.25	5	
Steel:					Generally, depth of penetration into steel depends on the amount of working and processing of the material. Finer grain structure usually permits greater penetration
Ingots..............	4–6	2–4	1–2	1	
Billets, blooms.......	5–12	6–8	3–4	1–2	
Rolled..............	22–25	22–25	10–25	5–8	
Cold drawn........	22–25	22–25	16–20	7–10	
Forged:					
Tool steels.......	22–25	22–25	22–25	22–25	
Spring steels......	22–25	22–25	22–25	22–25	
Stainless steels.....	22–25	22–25	22–25	22–25	
Steel castings:					Cast steels in the lower carbon range permit greater supersonic penetration
Carbon cast steels:					
Low carbon.......	15–20	15–20	10–15	7–10	
Med. carbon......	14–18	14–18	8–12	5–8	
High carbon.......	13–16	13–16	7–10	4–6	
Low alloy castings...	17–20	17–20	12–15	7–10	
High alloy castings...	17–20	17–20	12–15	7–10	
Cast iron:					"Scattering" of supersonic vibrations results where excessive porosity exists
Gray iron...........	1–2	6–12 in.	2–4 in.	—	
Malleable..........	7–10	7–10	3–5	6–12 in.	
Wrought iron..........	7–10	7–10	3–5	6–12 in.	
Aluminum:					Coarse structure or porosity will interfere with penetration. Finer grain structure permits greater penetration
Cast..............	12–15	12–15	8–10	8–10	
Extruded...........	22–25	22–25	22–25	22–25	
Worked.............	22–25	22–25	22–25	22–25	
Magnesium:					Physical properties similar to aluminum
Cast..............	12–15	12–15	8–10	8–10	
Extruded...........	22–25	22–25	22–25	22–25	
Worked.............	22–25	22–25	22–25	22–25	
Copper:					Coarse grain structure disperses vibrations
Cast..............	—	—	—	—	
Worked.............	0–6 in.	0–12 in.	—	—	
Brass and bronze:					Heat-treatment will substantially extend limits of penetration
Cast, fine grain.....	0–1½	0–1½	—	—	
Cast, coarse grain....	—	—	—	—	
Worked.............	1–5	1–3	—	—	
Nickel:					Greater porosity is usually present in cast materials
Cast..............	1–3	1–3	1–2	6–12 in.	
Worked.............	1–15	8–10	5–8	3–5	
Monel—Worked.......	10–15	8–10	5–8	3–5	
Gold—Worked	0–1	0–1	—	—	Testing experience very limited
Silver—Worked	5–10	2–5	1–2	0–1	Application usually confined to small pieces
Platinum.............	7–10	7–10	3–5	3–5	Figures are relative penetration
Tungsten.............	7–10	7–10	3–5	3–5	Sizes tested were not in this range
Sintered carbides.......	0–12 in.	0–6 in.	0–3 in.	0–3 in.	Most of testing done at high frequencies because of nature of flaws to be found
Molybdenum:					
Sintered...........	5–10	5–10	1–5	0–2	
Worked.............	21–25	21–25	21–25	15–18	
Wood:					Direction of grain and amount of moisture in the wood affects transmission into medium
Hard..............	0–8 in.	0–4 in.	—	—	
Soft...............	—	—	—	—	
Masonite...........	0–8 in.	0–4 in.	—	—	
Plastics:					Depends on amount of filler in material
Vinylite.............	6–12 in.	6–12 in.	6 in.–6 ft	6 in.–6 ft	
Catalin...........	6–12 in.	6–12 in.	6 in.–6 ft	6 in.–6 ft	
Bakelite...........	6–12 in.	6–12 in.	6 in.–6 ft	6 in.–6 ft	
Lucite, or Plexiglas...	6–12 in.	6–12 in.	6 in.–6 ft	6 in.–6 ft	
Oil.................	24–25	24–25	24–25	24–25	Indication of penetration limited only by capabilities of instrument
Water...............	24–25	24–25	24–25	24–25	

* Courtesy of Sperry Products, Inc.

TABLE 43.   CHARACTERISTICS OF THE COMMON BASE-METAL AND NOBLE-METAL THERMOCOUPLES

Thermocouple	Chemical composition*		Temperature range		Maximum temperature†	
	Positive element	Negative element	°C	°F	°C	°F
Chromel-alumel..	Chromel 90% Ni 10% Cr	Alumel 94% Ni 2% Al 3% Mn 1% Si	−200 to 1200	−328 to 2192	1350	2462
Iron-constantan	Pure iron	Constantan 60% Cu 40% Ni	−200 to 750	−328 to 1382	1000	1832
Copper-constantan	Pure copper	Constantan 60% Cu 40% Ni	−200 to 300	−328 to 572	600	1112
Platinum–platinum-rhodium	Platinum-rhodium 90% Pt, 10% Rh 87% Pt, 13% Rh	Platinum 100% Pt 100% Pt	0 to 1450	32 to 2642	1700	3092

* Approximate—may vary with different manufacturers.
† For only *short* periods of exposure.

TABLE 44. THERMOCOUPLE-PROTECTION TUBES*

Type of atmosphere	Maximum temperature		Material of tube
	°F	°C	
Natural or oxidizing (ordinary heat-treating)	2200	1205	27% chromium iron
	2400	1315	Pure nickel or porcelain
	2900	1593	Porcelain
Neutral or reducing (carbon monoxide, such as gas curtain hardening)	2200	1205	27% chromium iron
	2400	1315	Pure nickel or porcelain
	2700	1482	Porcelain
Hydrogen, ammonia (such as bright hardening or brazing)	1000	538	Wrought iron or seamless steel
	2200	1205	Inconel, pure nickel, or porcelain
	2700	1482	Porcelain
Nitriding..................	2200	1205	Inconel or 62% nickel–13% chromium
Slagging or corrosive action (neutral or oxidizing, such as forging furnace or kilns)	2700	1482	Fire clay with inner porcelain or P.B. sillimanite
Slagging or corrosive action (neutral or reducing, such as open-hearth furnace)	2500	1371	Silicon carbide with inner porcelain
Sulphur content.............	2000	1093	27% chromium iron
Sulphur dioxide.............	2200	1205	Inconel
Hydrogen sulphide..........	1000	538	Inconel
Lead vapors (rare-metal refining)	2200	1205	Inconel
Steam.....................	1500	816	8% nickel–18% chromium
Steam.....................	2000	1093	Inconel
Carburizing................	2000	1093	27% chromium iron

Note. In general, heavier tubes should be used for excessive corrosion or abrasion.
* From *Materials & Methods*, April, 1947.

TABLE 44. THERMOCOUPLE-PROTECTION TUBES.*—(*Continued*)

Kind of media	Maximum temperature		Material of tube
	°F	°C	
Molten sulphur.............	500	260	Monel metal
Cyanide or salt bath.........	1750	954	Drawn chromium iron (specially treated)
Salt bath..................	2400	1316	Same material as pot
Carburizing salts...........	2000	1093	Same material as pot
Dowtherm.................	....	....	Pure nickel or porcelain
Dilute sulphuric acid........	180	82	Lead or Monel metal
Hydrochloric acid (20%).....	110	43	Monel metal
Hydrochloric acid (10%).....	140	60	Monel metal
Oil baths..................	....	....	Seamless steel or wrought iron
Glass.....................	2800	1538	Porcelain
Tin.......................	....	....	Extra-heavy wrought iron or seamless steel
Lead......................	1000	538	Extra-heavy wrought iron or seamless steel
Lead......................	2000	1093	Inconel
Zinc (galvanizing)...........	1000	538	Extra-heavy wrought iron (Put an open-end pipe sleeve around the protection tube. This should be changed when corroded)
Brass (intermittent only).....	2100	1149	27% chromium iron
Brass or copper alloys (intermittent only)	2300	1260	27% chromium iron
Aluminum.................	1500	816	Cast iron—paint with whiting daily
Magnesium (Dowmetal).....	....	....	Extra-heavy seamless steel or 27% chromium iron
Die-cast metals.............	1200	649	Cast iron or extra-heavy wrought iron—paint with whiting daily

Note. In general, heavier tubes should be used for excessive corrosion or abrasion.

* From *Materials & Methods*, April, 1947.

TABLE 45.   MELTING POINTS OF COMMON METALS*

Metal	Atomic symbol	Melting point	
		°C.	°F.
Aluminum †...............	Al	660.0	1219.8
Antimony...............	Sb	630.5	1166.9
Beryllium...............	Be	1280 ± 40	2336.0 ± 104
Bismuth...............	Bi	271.3	520.3
Boron...............	B	2300 ± 300	4172 ± 572
Cadmium...............	Cd	320.9	609.6
Calcium...............	Ca	850 ± 20	1562 ± 68
Chromium...............	Cr	1800 ± 50	3272 ± 122
Cobalt...............	Co	1490 ± 20	2714 ± 68
Columbium...............	Cb	2000 ± 50	3632 ± 122
Copper †...............	Cu	1083.0	1981.4
Gold...............	Au	1063.0	1945.4
Iron...............	Fe	1535.0	2795.0
Lead †...............	Pb	327.4	621.3
Lithium...............	Li	186.0	366.8
Magnesium...............	Mg	650.0	1202.0
Molybdenum...............	Mo	2625 ± 50	4757 ± 122
Nickel...............	Ni	1453.0	2647.4
Platinum...............	Pt	1773.5	3224.3
Potassium...............	K	62.4	144.3
Silver...............	Ag	960.5	1760.9
Sodium...............	Na	97.6	207.7
Tin †...............	Sn	231.9	449.4
Titanium...............	Ti	1820 ± 100	3308 ± 212
Tungsten...............	W	3410 ± 20	6170 ± 68
Vanadium...............	V	1735 ± 50	3155 ± 122
Zinc†...............	Zn	419.4	787.1

* Metals in **bold-faced type** are suitable for thermocouple calibration.

† Melting-point standards may be purchased at the National Bureau of Standards with certificates giving the freezing point of each particular lot of metal.

TABLE 46. CHROMEL-ALUMEL THERMOCOUPLE*
To convert millivolts to degrees centigrade; cold junction at 0°C.
Type MA

Milli-volts	0	0.1	0.2	0.3	0.4	0.5	0.6	0.7	0.8	0.9	1.0	Milli-volts	°C. per 0.01 mv.
						Degrees Centigrade							
0	0	2	5	8	10	13	15	18	20	23	25	0	0.25
1	25	28	30	33	35	37	40	42	45	47	50	1	0.24
2	50	52	54	57	59	62	64	67	69	71	74	2	0.24
3	74	76	79	81	83	86	88	91	93	95	98	3	0.24
4	98	100	103	105	107	110	112	115	117	119	122	4	0.24
5	122	124	127	129	132	134	137	139	142	144	147	5	0.25
6	147	149	152	154	157	159	162	164	167	169	172	6	0.25
7	172	174	177	179	182	184	187	189	192	194	197	7	0.25
8	197	199	202	204	207	209	212	214	217	219	222	8	0.25
9	222	224	227	229	232	234	236	239	241	244	246	9	0.24
10	246	249	251	254	256	259	261	263	266	268	271	10	0.24
11	271	273	276	278	280	283	285	288	290	293	295	11	0.24
12	295	297	300	302	305	307	309	312	314	317	319	12	0.24
13	319	322	324	326	329	331	333	336	338	341	343	13	0.24
14	343	345	348	350	353	355	357	360	362	365	367	14	0.24
15	367	369	372	374	376	379	381	384	386	388	391	15	0.23
16	391	393	395	398	400	403	405	407	410	412	414	16	0.24
17	414	417	419	421	424	426	429	431	433	436	438	17	0.24
18	438	440	443	445	448	450	452	455	457	459	462	18	0.23
19	462	464	466	469	471	473	476	478	480	483	485	19	0.23
20	485	487	490	492	494	497	499	501	504	506	508	20	0.23
21	508	511	513	516	518	520	523	525	527	530	532	21	0.24
22	532	534	537	539	541	544	546	548	551	553	555	22	0.23
23	555	558	560	562	565	567	569	572	574	577	579	23	0.23
24	579	581	584	586	588	591	593	595	598	600	602	24	0.23
25	602	605	607	609	612	614	616	619	621	623	626	25	0.23
26	626	628	630	633	635	638	640	642	645	647	649	26	0.24
27	649	652	654	657	659	661	664	666	668	671	673	27	0.23
28	673	675	678	680	683	685	687	690	692	694	697	28	0.24
29	697	699	701	704	706	709	711	713	716	718	721	29	0.24
30	721	723	725	728	730	732	735	737	740	742	744	30	0.24
31	744	747	749	752	754	756	759	761	764	766	768	31	0.24
32	768	771	773	776	778	780	783	785	788	790	793	32	0.24
33	793	795	797	800	802	805	807	810	812	815	817	33	0.24
34	817	819	822	824	827	829	832	834	837	839	841	34	0.24
35	841	844	846	849	851	854	856	859	861	864	866	35	0.25
36	866	869	871	874	876	879	881	884	886	888	891	36	0.25
37	891	893	896	898	901	903	906	908	911	913	916	37	0.25
38	916	918	921	923	926	928	931	933	936	938	941	38	0.25
39	941	944	946	949	951	954	956	959	961	964	967	39	0.26
40	967	969	972	974	977	979	982	984	987	989	992	40	0.26
41	992	994	997	999	1002	1005	1007	1010	1013	1015	1018	41	0.26
42	1018	1020	1023	1026	1028	1031	1033	1036	1038	1041	1044	42	0.26
43	1044	1046	1049	1052	1054	1057	1059	1062	1065	1067	1070	43	0.26
44	1070	1073	1075	1078	1080	1083	1086	1088	1091	1094	1096	44	0.26
45	1096	1099	1102	1104	1107	1110	1112	1115	1118	1120	1123	45	0.27
46	1123	1126	1128	1131	1134	1136	1139	1142	1144	1147	1150	46	0.27
47	1150	1152	1155	1158	1160	1163	1166	1169	1171	1174	1177	47	0.28
48	1177	1180	1182	1185	1188	1190	1193	1196	1199	1201	1204	48	0.27
49	1204	1207	1210	1212	1215	1218	1221	1224	1226	1229	1232	49	0.28
50	1232	1235	1237	1240	1243	1246	1249	1252	1254	1257	1260	50	0.28
51	1260	1263	1266	1268	1271	1274	1277	1280	1283	1285	1288	51	0.28
52	1288	1291	1294	1297	1300	1303	1305	1308	1311	1314	1317	52	0.29
53	1317	1320	1323	1326	1329	1332	1334	1337	1340	1343	1346	53	0.29
54	1346	1349	1352	1355	1358	1361	1364	1367	1370	1373	1376	54	0.29

To interpolate between two printed values, add the increase in degrees per 0.01 mv (shown in the right-hand column) for each 0.01 mv above the lower printed value.

Example: 35.33 mv = 35.30 mv + 0.03 mv

$$0.03 \text{ mv} = \frac{0.03}{0.01} \times 0.25° = 0.75°\text{C}$$

35.30 mv = 849°C   35.33 mv = 849° + 0.75° = 849.75°C   (850°C)

* The Brown Instrument Co.

TABLE 47.   CHROMEL–ALUMEL THERMOCOUPLE*
To convert millivolts to degrees Fahrenheit; cold junction at 32°F.
Type MA

Milli-volts	0	0.1	0.2	0.3	0.4	0.5	0.6	0.7	0.8	0.9	1.0	Milli-volts	°F. per 0.01 mv.
						Degrees Fahrenheit							
0	32	36	41	46	50	55	59	64	68	73	77	0	0.45
1	77	82	86	91	95	99	104	108	113	117	121	1	0.44
2	121	126	130	135	139	143	148	152	156	161	165	2	0.44
3	165	169	173	178	182	186	191	195	199	204	208	3	0.43
4	208	212	217	221	225	230	234	238	243	247	251	4	0.43
5	251	256	261	265	269	274	278	283	287	291	296	5	0.45
6	296	301	305	310	314	319	323	328	332	337	341	6	0.45
7	341	346	350	355	359	364	368	373	377	381	386	7	0.45
8	386	391	395	400	404	409	413	418	422	426	431	8	0.45
9	431	436	440	445	449	453	457	462	466	471	475	9	0.44
10	475	480	484	489	493	498	502	506	511	515	519	10	0.44
11	519	524	528	532	537	541	546	550	555	559	563	11	0.44
12	563	568	572	576	581	585	589	594	598	602	607	12	0.44
13	607	611	615	620	624	628	632	636	641	645	650	13	0.43
14	650	654	658	663	667	671	675	680	684	688	693	14	0.43
15	693	697	701	705	710	714	719	723	727	731	735	15	0.42
16	735	740	744	748	752	756	761	765	769	774	778	16	0.43
17	778	782	786	791	795	799	804	808	812	816	821	17	0.43
18	821	825	829	833	838	842	846	850	855	859	863	18	0.42
19	863	867	871	875	880	884	888	893	897	901	905	19	0.42
20	905	909	913	918	922	926	930	935	939	943	947	20	0.42
21	947	951	956	960	964	968	973	977	981	985	990	21	0.43
22	990	994	998	1002	1006	1011	1015	1019	1023	1028	1032	22	0.42
23	1032	1036	1040	1044	1049	1053	1057	1061	1065	1070	1074	23	0.42
24	1074	1078	1083	1087	1091	1095	1099	1103	1108	1112	1116	24	0.42
25	1116	1120	1125	1129	1133	1137	1141	1146	1150	1154	1158	25	0.42
26	1158	1163	1167	1171	1175	1180	1184	1188	1193	1197	1201	26	0.43
27	1201	1205	1209	1214	1218	1222	1226	1231	1235	1239	1243	27	0.42
28	1243	1248	1252	1256	1260	1265	1269	1273	1278	1282	1286	28	0.43
29	1286	1290	1295	1299	1303	1307	1312	1316	1320	1325	1329	29	0.43
30	1329	1333	1338	1342	1346	1350	1355	1359	1363	1368	1372	30	0.43
31	1372	1376	1380	1385	1389	1393	1398	1402	1406	1411	1415	31	0.43
32	1415	1420	1424	1428	1433	1437	1441	1446	1450	1454	1459	32	0.44
33	1459	1463	1467	1472	1476	1481	1485	1490	1494	1498	1503	33	0.44
34	1503	1507	1511	1516	1520	1525	1529	1533	1538	1542	1546	34	0.43
35	1546	1551	1556	1560	1564	1569	1573	1578	1582	1586	1591	35	0.45
36	1591	1596	1600	1605	1609	1613	1618	1622	1627	1631	1636	36	0.45
37	1636	1640	1645	1649	1654	1658	1663	1667	1672	1676	1681	37	0.45
38	1681	1685	1690	1694	1699	1703	1708	1712	1717	1721	1726	38	0.45
39	1726	1731	1735	1740	1744	1749	1753	1758	1762	1767	1772	39	0.46
40	1772	1776	1782	1786	1790	1794	1799	1804	1809	1813	1818	40	0.46
41	1818	1822	1827	1831	1836	1841	1846	1850	1855	1860	1864	41	0.46
42	1864	1869	1873	1878	1883	1887	1892	1897	1901	1906	1911	42	0.47
43	1911	1916	1920	1925	1930	1934	1939	1944	1949	1953	1958	43	0.47
44	1958	1963	1967	1972	1977	1981	1986	1991	1996	2001	2005	44	0.47
45	2005	2010	2015	2020	2025	2030	2034	2039	2044	2049	2053	45	0.48
46	2053	2058	2063	2068	2073	2077	2082	2087	2092	2097	2101	46	0.48
47	2101	2106	2111	2116	2121	2126	2131	2136	2141	2146	2151	47	0.50
48	2151	2155	2160	2165	2170	2175	2180	2185	2190	2195	2200	48	0.49
49	2200	2205	2210	2215	2220	2225	2230	2235	2240	2245	2250	49	0.50
50	2250	2255	2260	2265	2270	2275	2280	2285	2290	2295	2300	50	0.50
51	2300	2305	2310	2315	2321	2326	2331	2336	2341	2346	2351	51	0.51
52	2351	2356	2362	2367	2372	2377	2382	2387	2393	2398	2403	52	0.52
53	2403	2408	2413	2418	2424	2429	2434	2439	2444	2450	2455	53	0.52
54	2455	2461	2466	2471	2476	2482	2487	2492	2497	2503	2508	54	0.53

To interpolate between two printed values, add the increase in degrees per 0.01 mv (shown in the right-hand column) for each 0.01 mv above the lower printed value.
Example: 35.34 mv = 35.30 mv + 0.04 mv

$$0.04 \text{ mv} = \frac{0.04}{0.01} \times 0.45° = 1.80°F$$

35.30 mv = 1560°F   35.34 mv = 1560° + 1.80° = 1561.80°F

* The Brown Instrument Co.

TABLE 48. IRON–CONSTANTAN THERMOCOUPLE*
To convert millivolts to degrees centigrade; cold junction at 0°C.
Type I. C.

Milli-volts	0	0.1	0.2	0.3	0.4	0.5	0.6	0.7	0.8	0.9	1.0	Milli-volts	°C. per 0.01 mv.
						Degrees Centigrade							
0	0	2	4	6	8	10	12	13	15	17	19	0	0.19
1	19	21	23	25	27	29	31	33	35	36	38	1	0.19
2	38	40	42	44	46	48	50	52	54	55	57	2	0.18
3	57	59	61	63	65	67	69	70	72	74	76	3	0.19
4	76	78	80	82	84	86	87	89	91	93	95	4	0.19
5	95	97	98	100	102	104	106	108	110	111	113	5	0.18
6	113	115	117	119	121	122	124	126	128	130	132	6	0.18
7	132	133	135	137	139	141	142	144	146	148	150	7	0.18
8	150	152	154	155	157	159	161	163	164	166	168	8	0.18
9	168	170	172	173	175	177	179	181	183	185	186	9	0.18
10	186	188	190	192	193	195	197	199	201	202	204	10	0.18
11	204	206	208	210	211	213	215	217	219	220	222	11	0.18
12	222	224	226	228	229	231	233	235	237	238	240	12	0.18
13	240	242	244	246	247	249	251	253	255	257	258	13	0.18
14	258	260	262	264	266	267	269	271	273	275	276	14	0.18
15	276	278	280	282	284	285	287	289	291	293	294	15	0.18
16	294	296	298	300	302	303	305	307	309	311	312	16	0.18
17	312	314	316	318	320	321	323	325	327	329	331	17	0.18
18	331	332	334	336	338	340	341	343	345	347	349	18	0.18
19	349	350	352	354	356	358	360	361	363	365	367	19	0.18
20	367	368	370	372	374	376	377	379	381	383	385	20	0.18
21	385	387	388	390	392	394	395	397	399	401	403	21	0.18
22	403	404	406	408	410	412	413	415	417	419	421	22	0.18
23	421	422	424	426	428	430	431	433	435	437	439	23	0.18
24	439	440	442	444	446	448	449	451	453	455	457	24	0.18
25	457	458	460	462	464	466	467	469	471	473	475	25	0.18
26	475	476	478	480	482	483	485	487	489	491	492	26	0.18
27	492	494	496	498	499	501	503	505	507	508	510	27	0.18
28	510	512	514	515	517	519	521	523	524	526	528	28	0.17
29	528	530	531	533	535	537	538	540	542	544	545	29	0.18
30	545	547	549	551	552	554	556	558	559	561	563	30	0.17
31	563	565	566	568	570	572	573	575	577	579	580	31	0.17
32	580	582	584	586	587	589	591	592	594	596	597	32	0.17
33	597	599	601	602	604	606	608	609	611	613	614	33	0.17
34	614	616	618	619	621	623	624	626	628	629	631	34	0.17
35	631	633	635	636	638	640	641	643	645	646	648	35	0.17
36	648	650	651	653	655	656	658	660	661	663	665	36	0.16
37	665	666	668	669	671	673	674	676	678	679	681	37	0.16
38	681	683	684	686	687	689	691	692	694	695	697	38	0.16
39	697	699	700	702	704	705	707	708	710	712	713	39	0.16

* The Brown Instrument Co.

TABLE 48. IRON–CONSTANTAN THERMOCOUPLE.*—(*Continued*)

Milli-volts	0	0.1	0.2	0.3	0.4	0.5	0.6	0.7	0.8	0.9	1.0	Milli-volts	°C. per 0.01 mv.
						Degrees Centigrade							
40	713	715	716	718	720	721	723	724	726	728	729	40	0.16
41	729	731	732	734	736	737	739	740	742	744	745	41	0.16
42	745	747	748	750	752	753	755	756	758	760	761	42	0.16
43	761	763	764	766	767	769	771	772	774	775	777	43	0.16
44	777	778	780	782	783	785	786	788	789	791	793	44	0.16
45	793	794	796	797	799	800	802	804	806	807	808	45	0.16
46	808	810	812	813	815	816	818	819	821	823	824	46	0.16
47	824	826	827	829	830	832	834	835	837	838	840	47	0.16
48	840	841	843	845	846	848	849	851	853	854	856	48	0.16
49	856	857	859	860	862	863	865	867	868	870	871	49	0.16
50	871	873	874	876	878	879	881	882	884	886	887	50	0.16
51	887	889	890	892	893	895	897	898	900	901	903	51	0.16
52	903	904	906	908	909	911	912	914	915	917	919	52	0.16
53	919	920	922	923	925	926	928	930	931	933	934	53	0.16
54	934	936	937	939	941	942	944	945	947	949	950	54	0.16
55	950	952	953	955	956	958	960	961	963	964	966	55	0.16
56	966	967	969	971	972	974	975	977	979	980	982	56	0.16
57	982	983	985	986	988	990	992	993	994	996	997	57	0.16
58	997	999	1001	1002	1004	1005	1007	1008	1010	1012	1013	58	0.16
59	1013	1015	1016	1018	1019	1021	1023	1024	1026	1027	1029	59	0.16
60	1029	1030	1032	1034	1035	1037	1038	1040	1041	1043	1045	60	0.16
61	1045	1046	1048	1049	1051	1052	1054	1056	1057	1059	1060	61	0.16
62	1060	1062	1063	1065	1067	1068	1070	1071	1073	1074	1076	62	0.16
63	1076	1078	1079	1081	1082	1084	1085	1087	1088	1090	1092	63	0.16
64	1092	1093	1095	1096	1098	1099	1101	1103	1004	1106	1108	64	0.16

To interpolate between two printed values, add the increase in degrees per 0.01 mv (shown in the right-hand column) for each 0.01 mv above the lower printed value.

Example: 31.45 mv = 31.40 mv + 0.05 mv

$$0.05 \text{ mv} = \frac{0.05}{0.01} \times 0.17° = 0.85°C$$

31.40 mv = 570°C   31.45 mv = 570° + 0.85° = 570.85°C   (571°C)

* The Brown Instrument Co.

TABLE 49. IRON-CONSTANTAN THERMOCOUPLE*
To convert millivolts to degrees Fahrenheit; cold junction at 32°F.
Type I. C.

Milli-volts	0	0.1	0.2	0.3	0.4	0.5	0.6	0.7	0.8	0.9	1.0	Milli-volts	°F. per 0.01 mv.
					Degrees Fahrenheit								
0	32	36	39	43	46	49	53	56	60	63	67	0	0.35
1	67	70	74	.77	81	84	88	91	95	98	102	1	0.35
2	102	105	108	111	115	118	122	125	129	132	135	2	0.33
3	135	138	142	145	149	152	156	159	162	165	169	3	0.34
4	169	172	176	179	183	186	190	193	196	199	203	4	0.34
5	203	206	210	213	217	220	223	226	230	233	236	5	0.33
6	236	239	243	246	250	253	256	259	263	266	269	6	0.33
7	269	272	276	279	283	286	289	292	296	299	302	7	0.33
8	302	305	309	312	315	318	322	325	328	331	335	8	0.33
9	335	338	341	344	348	351	354	357	361	364	368	9	0.33
10	368	371	374	377	380	383	387	390	393	396	400	10	0.32
11	400	403	406	409	413	416	419	422	426	429	432	11	0.32
12	432	435	439	442	445	448	452	455	458	461	465	12	0.32
13	465	468	471	474	478	481	484	487	491	494	497	13	0.32
14	497	500	504	507	510	513	517	520	523	526	530	14	0.33
15	530	533	536	539	543	546	549	552	556	559	562	15	0.32
16	562	565	569	572	576	579	582	585	589	592	595	16	0.33
17	595	598	602	605	609	612	615	618	622	625	628	17	0.33
18	628	631	635	638	641	644	648	651	654	657	661	18	0.33
19	661	664	667	670	673	676	680	683	686	689	693	19	0.32
20	693	696	699	702	706	709	712	715	719	722	725	20	0.32
21	725	728	732	735	738	741	745	748	751	754	758	21	0.33
22	758	761	764	767	771	774	777	780	784	787	790	22	0.32
23	790	793	797	800	803	806	810	813	816	819	823	23	0.33
24	823	826	829	832	835	838	842	845	848	851	855	24	0.32
25	855	858	861	864	868	871	874	877	881	884	887	25	0.32
26	887	890	894	897	900	903	907	910	913	916	920	26	0.33
27	920	923	926	929	932	935	939	942	945	948	952	27	0.32
28	952	955	958	961	964	967	970	973	977	980	983	28	0.31
29	983	986	989	992	995	998	1002	1005	1008	1011	1015	29	0.32
30	1015	1018	1021	1024	1027	1030	1033	1036	1040	1043	1046	30	0.31
31	1046	1049	1052	1055	1058	1061	1064	1067	1071	1074	1077	31	0.31
32	1077	1080	1084	1087	1090	1093	1096	1099	1102	1105	1108	32	0.31
33	1108	1111	1114	1117	1120	1123	1126	1129	1133	1136	1139	33	0.31
34	1139	1142	1145	1148	1151	1154	1157	1160	1163	1166	1169	34	0.30
35	1169	1172	1175	1178	1181	1184	1187	1190	1193	1196	1199	35	0.30
36	1199	1202	1205	1208	1211	1214	1217	1219	1222	1225	1228	36	0.29
37	1228	1231	1234	1237	1240	1243	1246	1249	1252	1254	1257	37	0.29
38	1257	1260	1263	1266	1269	1272	1275	1278	1281	1283	1286	38	0.29
39	1286	1289	1292	1295	1298	1301	1304	1307	1310	1312	1315	39	0.29

* The Brown Instrument Co.

TABLE 49. IRON–CONSTANTAN THERMOCOUPLE.*—(Continued)

Milli-volts	0	0.1	0.2	0.3	0.4	0.5	0.6	0.7	0.8	0.9	1.0	Milli-volts	°F. per 0.01 mv.
					Degrees Fahrenheit								
40	1315	1318	1321	1324	1327	1330	1333	1336	1339	1341	1344	40	0.29
41	1344	1347	1350	1353	1356	1359	1362	1365	1368	1370	1373	41	0.29
42	1373	1376	1379	1382	1385	1388	1391	1394	1397	1399	1402	42	0.29
43	1402	1405	1408	1411	1414	1417	1420	1422	1425	1428	1431	43	0.29
44	1431	1433	1436	1439	1442	1445	1448	1450	1453	1456	1459	44	0.28
45	1459	1461	1464	1467	1470	1473	1476	1479	1482	1484	1487	45	0.28
46	1487	1490	1493	1496	1499	1502	1505	1507	1510	1513	1516	46	0.29
47	1516	1518	1521	1524	1527	1530	1533	1535	1538	1541	1544	47	0.28
48	1544	1546	1549	1552	1555	1558	1561	1563	1566	1569	1572	48	0.28
49	1572	1574	1577	1580	1583	1586	1589	1592	1595	1597	1600	49	0.28
50	1600	1603	1606	1609	1612	1615	1618	1620	1623	1626	1629	50	0.29
51	1629	1631	1634	1637	1640	1643	1646	1649	1652	1654	1657	51	0.28
52	1657	1660	1663	1666	1669	1672	1675	1677	1680	1683	1686	52	0.29
53	1686	1688	1691	1694	1697	1700	1703	1706	1709	1711	1714	53	0.28
54	1714	1717	1720	1723	1726	1729	1732	1734	1737	1740	1743	54	0.29
55	1743	1745	1748	1751	1754	1757	1760	1762	1765	1768	1771	55	0.28
56	1771	1773	1776	1779	1782	1785	1788	1791	1794	1796	1799	56	0.28
57	1799	1802	1805	1808	1811	1814	1817	1819	1822	1825	1828	57	0.29
58	1828	1830	1833	1836	1839	1842	1845	1848	1851	1853	1856	58	0.28
59	1856	1859	1862	1865	1868	1871	1874	1876	1879	1882	1885	59	0.29
60	1885	1887	1890	1893	1896	1899	1902	1905	1908	1910	1913	60	0.28
61	1913	1916	1919	1922	1925	1928	1931	1933	1936	1939	1942	61	0.29
62	1942	1944	1947	1950	1953	1956	1959	1961	1964	1967	1970	62	0.28
63	1970	1972	1975	1978	1981	1984	1987	1989	1992	1995	1998	63	0.28
64	1998	2000	2003	2006	2009	2012	2015	2018	2021	2023	2026	64	0.28
65	2026	2029	2032	2035	2037	2040	2043	2046	2049	2052	2054	65	0.28
66	2054	2057	2060	2063	2066	2069	2072	2075	2077	2080	2083	66	0.29
67	2083	2086	2089	2092	2095	2097	2100	2103	2105	2108	2111	67	0.28
68	2111	2114	2117	2120	2123	2125	2128	2131	2134	2137	2140	68	0.29
69	2140	2143	2146	2149	2151	2154	2157	2160	2163	2165	2168	69	0.28
70	2168	2171	2174	2177	2179	2182	2185	2188	2191	2194	2197	70	0.29

To interpolate between two printed values, add the increase in degrees per 0.01 mv (shown in the right-hand column) for each 0.01 mv above the lower printed value.

Example: 40.66 mv = 40.60 mv + 0.06 mv

$$0.06 \text{ mv} = \frac{0.06}{0.01} \times 0.29° = 1.74°\text{F}$$

40.60 mv = 1333°F   40.66 mv = 1333° + 1.74° = 1334.74°F

* The Brown Instrument Co.

TABLE 50. PLATINUM–PLATINUM-13 PER CENT RHODIUM THERMOCOUPLE*

To convert millivolts to degrees centigrade; cold junction at 0°C. Type Q. R.

Millivolts	0	0.05	0.10	0.15	0.20	0.25	0.30	0.35	0.40	0.45	0.50	0.55	0.60	0.65	0.70	0.75	0.80	0.85	0.90	0.95	1.00	Millivolts	°C. per 0.005 mv.
								Degrees Centigrade														
0	0	9	18	26	34	42	50	58	65	73	80	87	94	101	107	114	120	127	133	139	145	0	0.73
1	145	152	158	164	169	175	181	187	193	198	204	210	215	221	226	232	237	243	248	253	259	1	0.57
2	259	264	269	275	280	285	290	295	301	306	311	316	321	326	331	336	341	346	351	356	361	2	0.52
3	361	366	371	376	381	386	390	395	400	405	410	415	419	424	429	434	439	443	448	453	458	3	0.48
4	458	462	467	472	476	481	486	490	495	500	504	509	513	518	523	527	532	536	541	545	550	4	0.47
5	550	554	559	563	568	572	577	581	586	590	595	599	603	608	612	617	621	625	630	634	638	5	0.44
6	638	643	647	651	656	660	664	669	673	677	681	686	690	694	698	703	707	711	715	719	724	6	0.43
7	724	728	732	736	740	744	749	753	757	761	765	769	773	777	781	786	790	794	798	802	806	7	0.42
8	806	810	814	818	822	826	830	834	838	842	846	850	854	858	862	866	870	874	878	882	886	8	0.40
9	886	890	894	898	902	906	910	914	918	921	925	929	933	937	941	945	949	953	956	960	964	9	0.39
10	964	968	972	976	979	983	987	991	995	998	1002	1006	1010	1014	1017	1021	1025	1029	1032	1036	1040	10	0.38
11	1040	1044	1047	1051	1055	1059	1062	1066	1070	1073	1077	1081	1084	1088	1092	1096	1099	1103	1107	1110	1114	11	0.37
12	1114	1118	1121	1125	1129	1132	1136	1139	1143	1147	1150	1154	1158	1161	1165	1169	1172	1176	1180	1183	1187	12	0.37
13	1187	1191	1194	1198	1201	1205	1209	1212	1216	1220	1223	1227	1230	1234	1238	1241	1245	1249	1252	1256	1259	13	0.37
14	1259	1263	1267	1270	1274	1277	1281	1285	1288	1292	1295	1299	1303	1306	1310	1314	1317	1321	1324	1328	1332	14	0.36
15	1332	1335	1339	1343	1346	1350	1353	1357	1361	1364	1368	1372	1375	1379	1383	1386	1390	1393	1397	1401	1404	15	0.36
16	1404	1408	1412	1415	1419	1422	1426	1430	1433	1437	1441	1444	1448	1452	1455	1459	1462	1466	1470	1473	1477	16	0.36
17	1477	1481	1484	1488	1492	1495	1499	1502	1506	1510	1513	1517	1521	1524	1528	1532	1535	1539	1543	1546	1550	17	0.37
18	1550	1554	1557	1561	1565	1568	1572	1576	1579	1583	1587	1590	1594	1598	1602	1605	1609	1613	1616	1620	1624	18	0.37
19	1624	1627	1631	1635	1638	1642	1646	1650	1653	1657	1661	1664	1668	1672	1675	1679	1683	1687	1690	1694	1698	19	0.37

To interpolate between two printed values, add the increase per 0.005 mv (shown in the right-hand column) for each 0.005 mv above the lower printed value.

Example: 11.47 mv = 11.45 mv + 0.02 mv

$$0.02 \text{ mv} = \frac{0.020}{0.005} \times 0.37° = 1.48°C$$

11.45 mv = 1073°C   11.47 mv = 1073° + 1.48° = 1074.48°C (1075°C)

* The Brown Instrument Co.

TABLE 51. PLATINUM-PLATINUM-13 PER CENT RHODIUM THERMOCOUPLE*

To convert millivolts to degrees Fahrenheit; cold junction at 32°F.   Type Q. R.

Degrees Fahrenheit

Millivolts	0	0.05	0.10	0.15	0.20	0.25	0.30	0.35	0.40	0.45	0.50	0.55	0.60	0.65	0.70	0.75	0.80	0.85	0.90	0.95	1.00
0	32	48	64	79	94	109	123	136	150	163	176	189	201	213	225	237	249	260	271	282	294
1	294	305	316	326	337	348	358	369	379	389	399	409	420	430	440	449	459	469	479	488	498
2	498	507	517	526	536	545	555	564	573	582	592	601	610	619	628	637	646	655	664	673	682
3	682	691	700	709	717	726	735	744	752	761	770	779	787	796	804	813	821	830	838	847	855
4	855	864	872	881	889	898	906	915	923	931	940	948	956	965	973	981	989	997	1005	1014	1022
5	1022	1030	1038	1046	1054	1062	1070	1078	1086	1094	1102	1110	1118	1126	1134	1142	1150	1158	1165	1173	1181
6	1181	1189	1197	1204	1212	1220	1228	1235	1243	1251	1258	1266	1274	1281	1289	1297	1304	1312	1319	1327	1334
7	1334	1342	1349	1357	1365	1372	1380	1387	1394	1402	1409	1417	1424	1431	1439	1446	1454	1461	1468	1476	1483
8	1483	1490	1497	1505	1512	1519	1527	1534	1541	1548	1555	1563	1570	1577	1584	1591	1599	1606	1613	1620	1627
9	1627	1634	1641	1648	1655	1662	1669	1677	1684	1691	1698	1705	1712	1719	1726	1733	1740	1747	1754	1761	1767
10	1767	1774	1781	1788	1795	1802	1809	1816	1822	1829	1836	1843	1850	1857	1863	1870	1877	1884	1891	1897	1904
11	1904	1911	1917	1924	1931	1938	1944	1951	1957	1964	1971	1977	1984	1991	1997	2004	2011	2017	2024	2030	2037
12	2037	2044	2050	2057	2063	2070	2077	2083	2090	2096	2103	2109	2116	2123	2129	2136	2142	2149	2155	2162	2168
13	2168	2175	2181	2188	2195	2201	2208	2214	2221	2227	2234	2240	2247	2253	2260	2266	2273	2279	2286	2293	2299
14	2299	2305	2312	2318	2325	2331	2338	2344	2351	2357	2364	2370	2377	2383	2390	2397	2403	2410	2416	2423	2429
15	2429	2436	2442	2449	2455	2462	2468	2475	2481	2488	2494	2501	2508	2514	2521	2527	2534	2540	2547	2553	2560
16	2560	2566	2573	2579	2586	2592	2599	2605	2612	2618	2625	2632	2638	2645	2651	2658	2664	2671	2677	2684	2690
17	2690	2697	2704	2710	2717	2723	2730	2736	2743	2750	2756	2763	2769	2776	2782	2789	2796	2802	2809	2815	2822
18	2822	2829	2835	2842	2849	2855	2862	2868	2875	2882	2888	2895	2902	2908	2915	2921	2928	2935	2941	2948	2955
19	2955	2961	2968	2974	2981	2988	2994	3001	3008	3014	3021	3028	3034	3041	3048	3054	3061	3068	3074	3081	3088

Millivolts	°F. per 0.005 mv.
0	1.31
1	1.02
2	0.93
3	0.86
4	0.84
5	0.80
6	0.77
7	0.75
8	0.72
9	0.70
10	0.69
11	0.67
12	0.66
13	0.66
14	0.65
15	0.65
16	0.65
17	0.66
18	0.66
19	0.67

To interpolate between two printed values, add the increase in degrees per 0.005 mv (shown in the right-hand column) for each 0.005 mv above the lower printed value.

Example: 15.37 mv = 15.35 mv + 0.02 mv

$$0.02 \text{ mv} = \frac{0.02}{0.005} \times 0.65° = 2.60°\text{F}$$

$$15.35 \text{ mv} = 2475°\text{F} \qquad 15.37 \text{ mv} = 2475° + 2.60° = 2477.6°\text{F}$$

* The Brown Instrument Co.

TABLE 52. PLATINUM–PLATINUM-10 PER CENT RHODIUM THERMOCOUPLE*
To convert millivolts to degrees centigrade; cold junction at 0°C. Type HER.

Millivolts	0	0.05	0.10	0.15	0.20	0.25	0.30	0.35	0.40	0.45	0.50	0.55	0.60	0.65	0.70	0.75	0.80	0.85	0.90	0.95	1.00
									Degrees Centigrade												
0	0	9	18	26	34	42	50	58	65	73	80	87	94	101	108	114	121	128	134	141	147
1	147	153	159	166	172	178	184	190	196	202	208	213	219	225	231	237	242	248	254	259	265
2	265	271	276	282	287	293	298	304	309	315	320	326	331	336	342	347	352	358	363	368	374
3	374	379	384	389	395	400	405	410	416	421	426	431	436	442	447	452	457	462	467	473	478
4	478	483	488	493	498	503	508	513	518	523	528	533	538	543	548	553	558	563	568	573	578
5	578	583	588	593	598	603	608	612	617	622	627	632	637	642	646	651	656	661	666	670	675
6	675	680	685	690	694	699	704	709	713	718	723	727	732	737	741	746	751	755	760	765	769
7	769	774	779	783	788	793	797	802	806	811	816	820	825	829	834	838	843	847	852	857	861
8	861	866	870	875	879	884	888	893	897	901	906	910	915	919	924	928	933	937	941	946	950
9	950	955	959	963	968	972	977	981	985	990	994	998	1003	1007	1011	1016	1020	1024	1029	1033	1037
10	1037	1042	1046	1050	1054	1059	1063	1067	1071	1076	1080	1084	1088	1093	1097	1101	1105	1110	1114	1118	1122
11	1122	1127	1131	1135	1139	1143	1148	1152	1156	1160	1164	1169	1173	1177	1181	1185	1190	1194	1198	1202	1206
12	1206	1211	1215	1219	1223	1227	1231	1236	1240	1244	1248	1252	1257	1261	1265	1269	1273	1277	1282	1286	1290
13	1290	1294	1298	1302	1307	1311	1315	1319	1323	1328	1332	1336	1340	1344	1349	1353	1357	1361	1365	1370	1374
14	1374	1378	1382	1386	1391	1395	1399	1403	1407	1412	1416	1420	1424	1428	1433	1437	1441	1445	1449	1454	1458
15	1458	1462	1466	1471	1475	1479	1483	1488	1492	1496	1500	1504	1509	1513	1517	1521	1526	1530	1534	1538	1543
16	1543	1547	1551	1555	1560	1564	1568	1572	1577	1581	1585	1589	1594	1598	1602	1607	1611	1615	1619	1624	1628
17	1628	1632	1636	1641	1645	1649	1653	1655	1662	1666	1671	1675	1679	1684	1688	1692	1697	1701	1705	1709	1714

Millivolts	°C per 0.005 mv
0	0.73
1	0.59
2	0.54
3	0.52
4	0.51
5	0.49
6	0.47
7	0.46
8	0.44
9	0.44
10	0.43
11	0.42
12	0.42
13	0.42
14	0.42
15	0.43
16	0.43
17	0.43

To interpolate between two printed values, add the increase in degrees per 0.005 mv (shown in the right-hand column) for each 0.005 mv above the lower printed value.

Example: 12.52 mv = 12.50 mv + 0.02 mv

$$0.02 \text{ mv} = \frac{0.020}{0.005} \times 0.42° = 1.68°C$$

$$12.50 \text{ mv} = 1248°C \qquad 12.52 \text{ mv} = 1248° + 1.68° = 1249.68°C \quad (1250°C)$$

* The Brown Instrument Co.

TABLE 53. PLATINUM–PLATINUM-10 PER CENT RHODIUM THERMOCOUPLE*
To convert millivolts to degrees Fahrenheit; cold junction at 32°F. Type HER.

Millivolts	0	0.05	0.10	0.15	0.20	0.25	0.30	0.35	0.40	0.45	0.50	0.55	0.60	0.65	0.70	0.75	0.80	0.85	0.90	0.95	1.00
											Degrees Fahrenheit										
0	32	48	64	79	94	108	122	136	150	163	176	189	201	214	226	238	250	262	273	285	296
1	296	308	319	330	341	352	363	374	384	395	406	416	427	437	448	458	468	479	489	499	509
2	509	519	529	539	549	559	569	579	589	599	608	618	628	638	647	657	666	676	686	695	705
3	705	714	724	733	743	752	761	771	780	790	799	808	818	827	836	846	855	864	873	883	892
4	892	901	910	920	929	938	947	956	965	974	983	992	1001	1010	1019	1028	1037	1046	1055	1064	1073
5	1073	1082	1090	1099	1108	1117	1126	1135	1143	1152	1161	1170	1178	1187	1196	1204	1213	1222	1230	1239	1248
6	1248	1256	1265	1273	1282	1290	1299	1307	1316	1324	1333	1341	1350	1358	1367	1375	1384	1392	1400	1409	1417
7	1417	1426	1434	1442	1450	1459	1467	1475	1484	1492	1500	1508	1517	1525	1533	1541	1549	1557	1566	1574	1582
8	1582	1590	1598	1606	1614	1622	1630	1639	1647	1655	1663	1671	1679	1687	1695	1703	1711	1719	1727	1735	1742
9	1742	1750	1758	1766	1774	1782	1790	1798	1806	1813	1821	1829	1837	1845	1852	1860	1868	1876	1884	1891	1899
10	1899	1907	1915	1922	1930	1938	1945	1953	1961	1968	1976	1984	1991	1999	2007	2014	2022	2029	2037	2045	2052
11	2052	2060	2067	2075	2083	2090	2098	2105	2113	2120	2128	2136	2143	2151	2158	2166	2173	2181	2188	2196	2204
12	2204	2211	2219	2226	2234	2241	2249	2256	2264	2271	2279	2286	2294	2301	2309	2316	2324	2331	2339	2347	2354
13	2354	2362	2369	2377	2384	2392	2399	2407	2414	2422	2429	2437	2444	2452	2460	2467	2475	2482	2490	2497	2505
14	2505	2512	2520	2528	2535	2543	2550	2558	2565	2573	2580	2588	2596	2603	2611	2618	2626	2634	2641	2649	2656
15	2656	2664	2672	2679	2687	2693	2702	2710	2717	2725	2732	2740	2748	2755	2763	2771	2778	2786	2793	2801	2809
16	2809	2816	2824	2832	2839	2847	2855	2862	2870	2878	2885	2893	2901	2908	2916	2924	2931	2939	2947	2954	2962
17	2962	2970	2977	2985	2993	3001	3008	3016	3024	3032	3039	3047	3055	3062	3070	3078	3086	3093	3101	3109	3117

Millivolts	°F. per 0.005 mv.
0	1.32
1	1.07
2	0.98
3	0.94
4	0.91
5	0.88
6	0.85
7	0.83
8	0.80
9	0.79
10	0.77
11	0.76
12	0.75
13	0.75
14	0.76
15	0.77
16	0.77
17	0.78

To interpolate between two printed values, add the increase in degrees per 0.005 mv (shown in the right-hand column) for each 0.005 mv above the lower printed value.

Example: 14.73 mv = 14.70 mv + 0.03 mv

$$0.03 \text{ mv} = \frac{0.03}{0.005} \times 0.76° = 4.56°F$$

14.70 mv = 2611°F

14.73 mv = 2611° + 4.56° = 2615.56°F

* The Brown Instrument Co.

TABLE 54. CONVERSION OF CENTIGRADE AND FAHRENHEIT SCALES*

Temp., °C.	For temperatures above 0°C.									
	0	1	2	3	4	5	6	7	8	9
0	32.0	33.8	35.6	37.4	39.2	41.0	42.8	44.6	46.4	48.2
10	50.0	51.8	53.6	55.4	57.2	59.0	60.8	62.6	64.4	66.2
20	68.0	69.8	71.6	73.4	75.2	77.0	78.8	80.6	82.4	84.2
30	86.0	87.8	89.6	91.4	93.2	95.0	96.8	98.6	100.4	102.2
40	104.0	105.8	107.6	109.4	111.2	113.0	114.8	116.6	118.4	120.2
50	122.0	123.8	125.6	127.4	129.2	131.0	132.8	134.6	136.4	138.2
60	140.0	141.8	143.6	145.4	147.2	149.0	150.8	152.6	154.4	156.2
70	158.0	159.8	161.6	163.4	165.2	167.0	168.8	170.6	172.4	174.2
80	176.0	177.8	179.6	181.4	183.2	185.0	186.8	188.6	190.4	192.2
90	194.0	195.8	197.6	199.4	201.2	203.0	204.8	206.6	208.4	210.2
100	212.0	213.8	215.6	217.4	219.2	221.0	222.8	224.6	226.4	228.2
110	230.0	231.8	233.6	235.4	237.2	239.0	240.8	242.6	244.4	246.2
120	248.0	249.8	251.6	253.4	255.2	257.0	258.8	260.6	262.4	264.2
130	266.0	267.8	269.6	271.4	273.2	275.0	276.8	278.6	280.4	282.2
140	284.0	285.8	287.6	289.4	291.2	293.0	294.8	296.6	298.4	300.2
150	302.0	303.8	305.6	307.4	309.2	311.0	312.8	314.6	316.4	318.2
160	320.0	321.8	323.6	325.4	327.2	329.0	330.8	332.6	334.4	336.2
170	338.0	339.8	341.6	343.4	345.2	347.0	348.8	350.6	352.4	354.2
180	356.0	357.8	359.6	361.4	363.2	365.0	366.8	368.6	370.4	372.2
190	374.0	375.8	377.6	379.4	381.2	383.0	384.8	386.6	388.4	390.2
200	392.0	393.8	395.6	397.4	399.2	401.0	402.8	404.6	406.4	408.2
210	410.0	411.8	413.6	415.4	417.2	419.0	420.8	422.6	424.4	426.2
220	428.0	429.8	431.6	433.4	435.2	437.0	438.8	440.6	442.4	444.2
230	446.0	447.8	449.6	451.4	453.2	455.0	456.8	458.6	460.4	462.2
240	464.0	465.8	467.6	469.4	471.2	473.0	474.8	476.6	478.4	480.2
250	482.0	483.8	485.6	487.4	489.2	491.0	492.8	494.6	496.4	498.2
260	500.0	501.8	503.6	505.4	507.2	509.0	510.8	512.6	514.4	516.2
270	518.0	519.8	521.6	523.4	525.2	527.0	528.8	530.6	532.4	534.2
280	536.0	537.8	539.6	541.4	543.2	545.0	546.8	548.6	550.4	552.2
290	554.0	555.8	557.6	559.4	561.2	563.0	564.8	566.6	568.4	570.2
300	572.0	573.8	575.6	577.4	579.2	581.0	582.8	584.6	586.4	588.2
310	590.0	591.8	593.6	595.4	597.2	599.0	600.8	602.6	604.4	606.2
320	608.0	609.8	611.6	613.4	615.2	617.0	618.8	620.6	622.4	624.2
330	626.0	627.8	629.6	631.4	633.2	635.0	636.8	638.6	640.4	642.2
340	644.0	645.8	647.6	649.4	651.2	653.0	654.8	656.6	658.4	660.2
350	662.0	663.8	665.6	667.4	669.2	671.0	672.8	674.6	676.4	678.2
360	680.0	681.8	683.6	685.4	687.2	689.0	690.8	692.6	694.4	696.2
370	698.0	699.8	701.6	703.4	705.2	707.0	708.8	710.6	712.4	714.2
380	716.0	717.8	719.6	721.4	723.2	725.0	726.8	728.6	730.4	732.2
390	734.0	735.8	737.6	739.4	741.2	743.0	744.8	746.6	748.4	750.2
400	752.0	753.8	755.6	757.4	759.2	761.0	762.8	764.6	766.4	768.2
410	770.0	771.8	773.6	775.4	777.2	779.0	780.8	782.6	784.4	786.2
420	788.0	789.8	791.6	793.4	795.2	797.0	798.8	800.6	802.4	804.2
430	806.0	807.8	809.6	811.4	813.2	815.0	816.8	818.6	820.4	822.2
440	824.0	825.8	827.6	829.4	831.2	833.0	834.8	836.6	838.4	840.2
450	842.0	843.8	845.6	847.4	849.2	851.0	852.8	854.6	856.4	858.2
460	860.0	861.8	863.6	865.4	867.2	869.0	870.8	872.6	874.4	876.2
470	878.0	879.8	881.6	883.4	885.2	887.0	888.8	890.6	892.4	894.2
480	896.0	897.8	899.6	901.4	903.2	905.0	906.8	908.6	910.4	912.2
490	914.0	915.8	917.6	919.4	921.2	923.0	924.8	926.6	928.4	930.2
500	932.0	933.8	935.6	937.4	939.2	941.0	942.8	944.6	946.4	948.2
510	950.0	951.8	953.6	955.4	957.2	959.0	960.8	962.6	964.4	966.2
520	968.0	969.8	971.6	973.4	975.2	977.0	978.8	980.6	982.4	984.2
530	986.0	987.8	989.6	991.4	993.2	995.0	996.8	998.6	1000.4	1002.2
540	1004.0	1005.8	1007.6	1009.4	1011.2	1013.0	1014.8	1016.6	1018.4	1020.2
550	1022.0	1023.8	1025.6	1027.4	1029.2	1031.0	1032.8	1034.6	1036.4	1038.2
560	1040.0	1041.8	1043.6	1045.4	1047.2	1049.0	1050.8	1052.6	1054.4	1056.2
570	1058.0	1059.8	1061.6	1063.4	1065.2	1067.0	1068.8	1070.6	1072.4	1074.2
580	1076.0	1077.8	1079.6	1081.4	1083.2	1085.0	1086.8	1088.6	1090.4	1092.2
590	1094.0	1095.8	1097.6	1099.4	1101.2	1103.0	1104.8	1106.6	1108.4	1110.2
600	1112.0	1113.8	1115.6	1117.4	1119.2	1121.0	1122.8	1124.6	1126.4	1128.2
610	1130.0	1131.8	1133.6	1135.4	1137.2	1139.0	1140.8	1142.6	1144.4	1146.2
620	1148.0	1149.8	1151.6	1153.4	1155.2	1157.0	1158.8	1160.6	1162.4	1164.2
630	1166.0	1167.8	1169.6	1171.4	1173.2	1175.0	1176.8	1178.6	1180.4	1182.2
640	1184.0	1185.8	1187.6	1189.4	1191.2	1193.0	1194.8	1196.6	1198.4	1200.2

* The Brown Instrument Co.

TABLE 54. CONVERSION OF CENTIGRADE AND FAHRENHEIT SCALES.*—(Continued)

Temp., °C.	For temperatures above 0°C.									
	0	1	2	3	4	5	6	7	8	9
650	1202.0	1203.8	1205.6	1207.4	1209.2	1211.0	1212.8	1214.6	1216.4	1218.2
660	1220.0	1221.8	1223.6	1225.4	1227.2	1229.0	1230.8	1232.6	1234.4	1236.2
670	1238.0	1239.8	1241.6	1243.4	1245.2	1247.0	1248.8	1250.6	1252.4	1254.2
680	1256.0	1257.8	1259.6	1261.4	1263.2	1265.0	1266.8	1268.6	1270.4	1272.2
690	1274.0	1275.8	1277.6	1279.4	1281.2	1283.0	1284.8	1286.6	1288.4	1290.2
700	1292.0	1293.8	1295.6	1297.4	1299.2	1301.0	1302.8	1304.6	1306.4	1308.2
710	1310.0	1311.8	1313.6	1315.4	1317.2	1319.0	1320.8	1322.6	1324.4	1326.2
720	1328.0	1329.8	1331.6	1333.4	1335.2	1337.0	1338.8	1340.6	1342.4	1344.2
730	1346.0	1347.8	1349.6	1351.4	1353.2	1355.0	1356.8	1358.6	1360.4	1362.2
740	1364.0	1365.8	1367.6	1369.4	1371.2	1373.0	1374.8	1376.6	1378.4	1380.2
750	1382.0	1383.8	1385.6	1387.4	1389.2	1391.0	1392.8	1394.6	1396.4	1398.2
760	1400.0	1401.8	1403.6	1405.4	1407.2	1409.0	1410.8	1412.6	1414.4	1416.2
770	1418.0	1419.8	1421.6	1423.4	1426.2	1427.0	1428.8	1430.6	1432.4	1434.2
780	1436.0	1437.8	1439.6	1441.4	1443.2	1445.0	1446.8	1448.6	1450.4	1452.2
790	1454.0	1455.8	1457.6	1459.4	1461.2	1463.0	1464.8	1466.6	1468.4	1470.2
800	1472.0	1473.8	1475.6	1477.4	1479.2	1481.0	1482.8	1484.6	1486.4	1488.2
810	1490.0	1491.8	1493.6	1495.4	1497.2	1499.0	1500.8	1502.6	1504.4	1506.2
820	1508.0	1509.9	1511.6	1513.4	1515.2	1517.0	1518.8	1520.6	1522.4	1524.2
830	1526.0	1527.8	1529.6	1531.4	1533.2	1535.0	1536.8	1538.6	1540.4	1542.2
840	1544.0	1545.8	1547.6	1549.4	1551.2	1553.0	1554.8	1556.6	1558.4	1560.2
850	1562.0	1563.8	1565.6	1567.4	1569.2	1571.0	1572.8	1574.6	1576.4	1578.2
860	1580.0	1581.8	1583.6	1585.4	1587.2	1589.0	1590.8	1592.6	1594.4	1596.2
870	1598.0	1599.8	1601.6	1603.4	1605.2	1607.0	1608.8	1610.6	1612.4	1614.2
880	1616.0	1617.8	1619.6	1621.4	1623.2	1625.0	1626.8	1628.6	1630.4	1632.2
890	1634.0	1635.8	1637.6	1639.4	1641.2	1643.0	1644.8	1646.6	1648.4	1650.2
900	1652.0	1653.8	1655.6	1657.4	1659.2	1661.0	1662.8	1664.6	1666.4	1668.2
910	1670.0	1671.8	1673.6	1675.4	1677.2	1679.0	1680.8	1682.6	1684.4	1686.2
920	1688.0	1689.8	1691.6	1693.4	1695.2	1697.0	1698.8	1700.6	1702.4	1704.2
930	1706.0	1707.8	1709.6	1711.4	1713.2	1715.0	1716.8	1718.6	1720.4	1722.2
940	1724.0	1725.8	1727.6	1729.4	1731.2	1733.0	1734.8	1736.6	1738.4	1740.2
950	1742.0	1743.8	1745.6	1747.4	1749.2	1751.0	1752.8	1754.6	1756.4	1758.2
960	1760.0	1761.8	1763.6	1765.4	1767.2	1769.0	1770.8	1772.6	1774.4	1776.2
970	1778.0	1779.8	1781.6	1783.4	1785.2	1787.0	1788.8	1790.6	1792.4	1794.2
980	1796.0	1797.8	1799.6	1801.4	1803.2	1805.0	1806.8	1808.6	1810.4	1812.2
990	1814.0	1815.8	1817.6	1819.4	1821.2	1823.0	1824.8	1826.6	1828.4	1830.2
1000	1832.0	1833.8	1835.6	1837.4	1839.2	1841.0	1842.8	1844.6	1846.4	1848.2
1010	1850.0	1851.8	1853.6	1855.4	1857.2	1859.0	1860.8	1862.6	1864.4	1866.2
1020	1868.0	1869.8	1871.6	1873.4	1875.2	1877.0	1878.8	1880.6	1882.4	1884.2
1030	1886.0	1887.8	1889.6	1891.4	1893.2	1895.0	1896.8	1898.6	1900.4	1902.2
1040	1904.0	1905.8	1907.6	1909.4	1911.2	1913.0	1914.8	1916.6	1918.4	1920.2
1050	1922.0	1923.8	1925.6	1927.4	1929.2	1931.0	1932.8	1934.6	1936.4	1938.2
1060	1940.0	1941.8	1943.6	1945.4	1947.2	1949.0	1950.8	1952.6	1954.4	1956.2
1070	1958.0	1959.8	1961.6	1963.4	1965.2	1967.0	1968.8	1970.6	1972.4	1974.2
1080	1976.0	1977.8	1979.6	1981.4	1983.2	1985.0	1986.8	1988.6	1990.4	1992.2
1090	1994.0	1995.8	1997.6	1999.4	2001.2	2003.0	2004.8	2006.6	2008.4	2010.2
1100	2012.0	2013.8	2015.6	2017.4	2019.2	2021.0	2022.8	2024.6	2026.4	2028.2
1110	2030.0	2031.8	2033.6	2035.4	1037.2	2039.0	2040.8	2042.6	2044.4	2046.2
1120	2048.0	2049.8	2051.6	2053.4	2055.2	1057.0	2058.8	2060.6	2062.4	2064.2
1130	2066.0	2067.8	2069.6	2071.4	2073.2	2075.0	2076.8	2078.6	2080.4	2082.2
1140	2084.0	2085.8	2087.6	2089.4	2091.2	2093.0	2094.8	2096.6	2098.4	2100.2
1150	2102.0	2103.8	2105.6	2107.4	2109.2	2111.0	2112.8	2114.6	2116.4	2118.2
1160	2120.0	2121.8	2123.6	2125.4	2127.2	2129.0	2130.8	2132.6	2134.4	2136.2
1170	2138.0	2139.8	2141.6	2143.4	2145.2	2147.0	2148.8	2150.6	2152.4	2154.2
1180	2156.0	2157.8	2159.6	2161.4	2163.2	2165.0	2166.8	2168.6	2170.4	2172.2
1190	2174.0	2175.8	2177.6	2179.4	2181.2	2183.0	2184.8	2186.6	2188.4	2190.2
1200	2192.0	2193.8	2195.6	2197.4	2199.2	2201.0	2202.8	2204.6	2206.4	2208.2
1210	2210.0	2211.8	2213.6	2215.4	2217.2	2219.0	2220.8	2222.6	2224.4	2226.2
1220	2228.0	2229.8	2231.6	2233.4	2235.2	2237.0	2238.8	2240.6	2242.4	2244.2
1230	2246.0	2247.8	2249.6	2251.4	2253.2	2255.0	2256.8	2258.6	2260.4	2262.2
1240	2264.0	2265.8	2267.6	2269.4	2271.2	2273.0	2274.8	2276.6	2278.4	2280.2

* The Brown Instrument Co.

TABLE 54. CONVERSION OF CENTIGRADE AND FAHRENHEIT SCALES.*—(Continued)

Temp., °C.	For temperatures above 0°C.									
	0	1	2	3	4	5	6	7	8	9
1250	2282.0	2283.8	2285.6	2287.4	2289.2	2291.0	2292.8	2294.6	2296.4	2298.2
1260	2300.0	2301.8	2303.6	2305.4	2307.2	2309.0	2310.8	2312.6	2314.4	2316.2
1270	2318.0	2319.8	2321.6	2323.4	2325.2	2327.0	2328.8	2330.6	2332.4	2334.2
1280	2336.0	2337.8	2339.6	2341.4	2343.2	2345.0	2346.8	2348.6	2350.4	2352.2
1290	2354.0	2355.8	1357.6	2359.4	2361.2	2363.0	2364.8	2366.6	2368.4	2370.2
1300	2372.0	2373.8	2375.6	2377.4	2379.2	2381.0	2382.8	2384.6	2386.4	2388.2
1310	2390.0	2391.8	2393.6	2395.4	2397.2	2399.0	2400.8	2402.6	2404.4	2406.2
1320	2408.0	2409.8	2411.6	2413.4	2415.2	2417.0	2418.8	2420.6	2422.4	2424.2
1330	2426.0	2427.8	2429.6	2431.4	2433.2	2435.0	2436.8	2438.6	2440.4	2442.2
1340	2444.0	2445.8	2447.6	2449.4	2451.2	2453.0	2454.8	2456.6	2458.4	2460.2
1350	2462.0	2463.8	2465.6	2467.4	2469.2	2471.0	2472.8	2474.6	2476.4	2478.2
1360	2480.0	2481.8	2483.6	2485.4	2487.2	2489.0	2490.8	2492.6	2494.4	2496.2
1370	2498.0	2499.8	2501.6	2503.4	2505.2	2507.0	2508.8	2510.6	2512.4	2514.2
1380	2516.0	2517.8	2519.6	2521.4	2523.2	2525.0	2526.8	2528.6	2530.4	2532.2
1390	2534.0	2535.8	2537.6	2539.4	2541.2	2543.0	2544.8	2546.6	2548.4	2550.2
1400	2552.0	2553.8	2555.6	2557.4	2559.2	2561.0	2562.8	2564.6	2566.4	2568.2
1410	2570.0	2571.8	2573.6	2575.4	2577.2	2579.0	2580.8	2582.6	2584.4	2586.2
1420	2588.0	2589.8	2591.6	2593.4	2595.2	2597.0	2598.8	2600.6	2602.4	2604.2
1430	2606.0	2607.8	2609.6	2611.4	2613.2	2615.0	2616.8	2618.6	2620.4	2622.2
1440	2624.0	2625.8	2627.6	2629.4	2631.2	2633.0	2634.8	2636.6	2638.4	2640.2
1450	2642.0	2643.8	2645.6	2647.4	2649.2	2651.0	2652.8	2654.6	2656.4	2658.2
1460	2660.0	2661.8	2663.6	2665.4	2667.2	2669.0	2670.8	2672.6	2674.4	2676.2
1470	2678.0	2679.8	2681.6	2683.4	2685.2	2687.0	2688.8	2690.6	2692.4	2694.2
1480	2696.0	2697.8	2699.6	2701.4	2703.2	2705.0	2706.8	2708.6	2710.4	2712.2
1490	2714.0	2715.8	2717.6	2719.4	2721.2	2723.0	2724.8	2726.6	2728.4	2730.2
1500	2732.0	2733.8	2735.6	2737.4	2739.2	2741.0	2742.8	2744.6	2746.4	2748.2
1510	2750.0	2751.8	2753.6	2755.4	2757.2	2759.0	2760.8	2762.6	2764.4	2766.2
1520	2768.0	2769.8	2771.6	2773.4	2775.2	2777.0	2778.8	2780.6	2782.4	2784.2
1530	2786.0	2787.8	2789.6	2791.4	2793.2	2795.0	2796.8	2798.6	2800.4	2802.2
1540	2804.0	2805.8	2807.6	2809.4	2811.2	2813.0	2814.8	2816.6	2818.4	2820.2
1550	2822.0	2823.8	2825.6	2827.4	2829.2	2831.0	2832.8	2834.6	2836.4	2838.2
1560	2840.0	2841.8	2843.6	2845.4	2847.2	2849.0	2850.8	2852.6	2854.4	2856.2
1570	2858.0	2859.8	2861.6	2863.4	2865.2	2867.0	2868.8	2876.6	2872.4	2874.2
1580	2876.0	2877.8	2879.6	2881.4	2883.2	2885.0	2886.8	2888.6	2890.4	2892.2
1590	2894.0	2895.8	2897.6	2899.4	2901.2	2903.0	2904.8	2906.6	2908.4	2910.2
1600	2912.0	2913.8	2915.6	2917.4	2919.2	1921.0	2922.8	2924.6	2926.4	2928.2
1610	2930.0	2931.8	2933.6	2935.4	2937.2	2939.0	2940.8	2942.6	2944.4	2946.2
1620	2948.0	2949.8	2951.6	2953.4	2955.2	2957.0	2958.8	2960.6	2962.4	2964.2
1630	2966.0	2967.8	2969.6	2971.4	2973.2	2975.0	2976.8	2978.6	2980.4	2982.2
1640	2984.0	2985.8	2987.6	2989.4	2991.2	2993.0	2994.8	2996.6	2998.4	3000.2
1650	3002.0	3003.8	3005.6	3007.4	3009.2	3011.0	3012.8	3014.6	3016.4	3018.2
1660	3020.0	3021.8	3023.6	3025.4	3027.2	3029.0	3030.8	3032.6	3034.4	3036.2
1670	3038.0	3039.8	3041.6	3043.4	3045.2	3047.0	3048.8	3050.6	3052.4	3054.2
1680	3056.0	3057.8	3059.6	3061.4	3063.2	3065.0	3066.8	3068.6	3070.4	3072.2
1690	3074.0	3075.8	3077.6	3079.4	3081.2	3083.0	3084.8	3086.6	3088.4	3090.2
1700	3092.0	3093.8	3095.6	3097.4	3099.2	3101.0	3102.8	3104.6	3106.4	3108.2
1710	3110.0	3111.8	3113.6	3115.4	3117.2	3119.0	3120.8	3122.6	3124.4	3126.2
1720	3128.0	3129.8	3131.6	3133.4	3135.2	3137.0	3138.8	3140.6	3142.4	3144.2
1730	3146.0	3147.8	3149.6	3151.4	3153.2	3155.0	3156.8	3158.6	3160.4	3162.2
1740	3164.0	3165.8	3167.6	3169.4	3171.2	3173.0	3174.8	3176.6	3178.4	3180.2
1750	3182.0	3183.8	3185.6	3187.4	3189.2	3191.0	3192.8	3194.6	3196.4	3198.2
1760	3200.0	3201.8	3203.6	3205.4	3207.2	3209.0	3210.8	3212.6	3214.4	3216.2
1770	3218.0	3219.8	3221.6	3223.4	3225.2	3227.0	3228.8	3230.6	3232.4	3234.2
1780	3236.0	3237.8	3239.6	3241.4	3243.2	3245.0	3246.8	3248.6	3250.4	3252.2
1790	3254.0	3255.8	3257.6	3259.4	3261.2	3263.0	3264.8	3266.6	3268.4	3270.2

For interpolation	°C	0.1	0.2	0.3	0.4	0.5	0.6	0.7	0.8	0.9	1.0
	°F	0.18	0.36	0.54	0.72	0.90	1.08	1.26	1.44	1.62	1.80

Degrees centigrade $\times 1.8 + 32$ = degree Fahrenheit

$$\frac{\text{Degrees Fahrenheit} - 32}{1.8} = \text{degrees centigrade}$$

* The Brown Instrument Co.

TABLE 55. SPECTRAL EMISSIVITY (E_λ) OF SOME MATERIALS, SURFACE UNOXIDIZED*
Wave length of radiation = 0.65 micron

Element	Solid	Liquid	Element	Solid	Liquid
Beryllium.........	0.61	0.61	Titanium..........	0.63	0.65
Carbon...........	0.80–0.93		Tungsten..........	0.43	
Chromium.........	0.34	0.39	Uranium..........	0.54	0.34
Cobalt............	0.36	0.37	Vanadium.........	0.35	0.32
Columbium.......	0.37	0.40	Yttrium...........	0.35	0.35
Copper...........	0.10	0.15	Zirconium........	0.32	0.30
Erbium...........	0.55	0.38	Steel.............	0.35	0.37
Gold.............	0.14	0.22	Cast iron.........	0.37	0.40
Iridium...........	0.30		Constantan.......	0.35	
Iron.............	0.35	0.37	Monel............	0.37	
Manganese.......	0.59	0.59	Chromel P		
Molybdenum......	0.37	0.40	(90% Ni, 10% Cr)	0.35	
Nickel...........	0.36	0.37	80% Ni, 20% Cr...	0.35	
Palladium........	0.33	0.37	60% Ni, 24% Fe,		
Platinum.........	0.30	0.38	16% Cr..........	0.36	
Rhodium..........	0.24	0.30	Alumel		
Silver............	0.07	0.07	(95% Ni; Bal., Al,		
Tantalum.........	0.49		Mn, Si)..........	0.37	
Thorium..........	0.36	0.40	90% Pt, 10 Rh.....	0.27	

* From "Temperature, Its Measurement and Control in Science and Industry," Reinhold Publishing Corporation, New York, 1941.

TABLE 56. SPECTRAL EMISSIVITY (E_λ) OF SOME OXIDES*
Wave length of radiation = 0.65 micron

Material	Range of observed values	Probable value for the oxide formed on smooth metal
Aluminum oxide	0.22–0.40	0.30
Beryllium oxide	0.07–0.37	0.35
Cerium oxide	0.58–0.80	
Chromium oxide	0.60–0.80	0.70
Cobalt oxide	....	0.75
Columbium oxide	0.55–0.71	0.70
Copper oxide	0.60–0.80	0.70
Iron oxide	0.63–0.98	0.70
Magnesium oxide	0.10–0.43	0.20
Nickel oxide	0.85–0.96	0.90
Thorium oxide	0.20–0.57	0.50
Tin oxide	0.32–0.60	
Titanium oxide	....	0.50
Uranium oxide	....	0.30
Vanadium oxide	....	0.70
Yttrium oxide	....	0.60
Zirconium oxide	0.18–0.43	0.40
Alumel (oxidized)	....	0.87
Cast iron (oxidized)	....	0.70
Chromel P (90% Ni, 10% Cr, oxidized)	....	0.87
80% Ni, 20% Cr (oxidized)	....	0.90
60% Ni, 24% Fe, 16% Cr (oxidized)	....	0.83
55% Fe, 37.5% Cr, 7.5% Al (oxidized)	....	0.78
70% Fe, 23% Cr, 5% Al, 2% Co (oxidized)	....	0.75
Constantan (55% Cu, 45% Ni, oxidized)	....	0.84
Carbon steel (oxidized)	....	0.80
Stainless steel (18–8, oxidized)	....	0.85
Porcelain	0.25–0.50	

Note: The emissivity of oxides and oxidized metals depend to a large extent upon the roughness of the surface. In general, higher values of emissivity are obtained on the rougher surfaces.

* From "Temperature, Its Measurement and Control in Science and Industry," Reinhold Publishing Corporation, New York, 1941.

TABLE 57. TOTAL EMISSIVITY (E_t) OF SOME MISCELLANEOUS MATERIALS*

Material	Temperature, °C	Emissivity
Aluminum (oxidized)....................	200	0.11
	600	0.19
Brass (oxidized).........................	200	0.61
	600	0.59
Calorized copper........................	100	0.26
	500	0.26
Calorized copper (oxidized)..............	200	0.18
	600	0.19
Calorized steel (oxidized)...............	200	0.52
	600	0.57
Cast iron (strongly oxidized).............	40	0.95
	250	0.95
Cast iron (oxidized).....................	200	0.64
	600	0.78
Copper (oxidized).......................	200	0.60
	1000	0.60
Fire brick..............................	1000	0.75
Gold enamel............................	100	0.37
Iron (oxidized).........................	100	0.74
	500	0.84
	1200	0.89
Iron (rusted)..........................	25	0.65
Lead (oxidized).........................	200	0.63
Monel (oxidized).......................	200	0.43
	600	0.43
Nickel (oxidized).......................	200	0.37
	1200	0.85
Silica brick............................	1000	0.80
	1100	0.85
Steel (oxidized)........................	25	0.80
	200	0.79
	600	0.79
Steel plate (rough).....................	40	0.94
	400	0.97
Wrought iron (dull oxidized).............	25	0.94
	350	0.94
20% Ni, 25% Cr, 55% Fe (oxidized).......	200	0.90
	500	0.97
60% Ni, 12% Cr, 28% Fe (oxidized).......	270	0.89
	560	0.82
80% Ni, 20% Cr (oxidized)..............	100	0.87
	600	0.87
	1300	0.89

Note: Most values in the above table are uncertain by 10 to 30 per cent.   In many cases, value depends upon particle size.

* From "Temperature, Its Measurement and Control in Science and Industry," Reinhold Publishing Corporation, New York, 1941. *

TABLE 58. TOTAL EMISSIVITY (E_t) OF SOME METALS, SURFACE UNOXIDIZED*

Material	25°C	100°C	500°C	1000°C	1500°C	2000°C
Aluminum....	0.022	0.028	0.060			
Bismuth......	0.048	0.061				
Carbon.......	0.081	0.081	0.079			
Chromium....	....	0.08				
Cobalt.......	....	....	0.13	0.23		
Columbium...	....	....	....	....	0.19	0.24
Copper.......	....	0.02	....	(Liquid, 0.15)		
Gold.........	....	0.02	0.03			
Iron..........	....	0.05				
Lead.........	....	0.05				
Mercury......	0.10	0.12				
Molybdenum..	....	....	....	0.13	0.19	0.24
Nickel........	0.045	0.06	0.12	0.19		
Platinum.....	0.037	0.047	0.096	0.152	0.191	
Silver........	....	0.02	0.035	....		
Tantalum.....	....	....	....	....	0.21	0.26
Tin..........	0.043	0.05				
Tungsten.....	0.024	0.032	0.071	0.15	0.23	0.28
Zinc..........	....	(0.05 at 300°C)				
Brass.........	0.035	0.035				
Cast iron.....	....	0.21	....	....	(Liquid, 0.29)	
Steel.........	....	0.08	....	....	(Liquid, 0.28)	

* From "Temperature, Its Measurement and Control in Science and Industry," Reinhold Publishing Corporation, New York, 1941.

TABLE 59.   TRUE TEMPERATURES vs. APPARENT TEMPERATURES AS OBSERVED WITH
AN OPTICAL PYROMETER*

Temperatures, in degrees centigrade, are for $\lambda = 0.65$ micron, $c_2 = 14,320$ micron degrees
Observed temperatures in headings, true temperatures in the body of the table

Spectral emissivity	700	800	900	1000	1100	1200	1300	1400	1600	1800	2000
0.05	848	983	1123	1266	1415	1569	1728	1849	2240	2614	3017
0.10	810	935	1064	1195	1330	1468	1609	1754	2056	2373	2708
0.15	789	909	1032	1157	1284	1414	1546	1682	1960	2250	2552
0.20	774	891	1010	1130	1253	1378	1504	1633	1897	2170	2453
0.25	763	878	993	1111	1230	1350	1473	1597	1850	2111	2379
0.30	755	867	980	1095	1211	1329	1448	1568	1814	2065	2322
0.35	747	858	969	1082	1196	1311	1427	1545	1783	2027	2276
0.40	741	850	960	1071	1183	1296	1410	1525	1758	1996	2237
0.45	736	843	952	1062	1172	1283	1395	1508	1736	1968	2204
0.50	731	837	945	1053	1162	1272	1382	1493	1717	1945	2175
0.55	726	832	939	1046	1153	1261	1370	1480	1700	1923	2149
0.60	722	827	933	1039	1145	1252	1360	1467	1685	1905	2126
0.65	719	823	927	1032	1138	1244	1350	1457	1671	1888	2106
0.70	716	819	923	1027	1131	1236	1341	1447	1659	1872	2087
0.75	712	815	917	1021	1125	1229	1333	1437	1647	1858	2069
0.80	710	812	914	1017	1119	1222	1325	1429	1636	1844	2054
0.85	707	809	910	1012	1114	1216	1318	1421	1626	1832	2039
0.90	704	805	907	1008	1109	1210	1312	1413	1617	1821	2025
0.95	702	803	903	1004	1104	1205	1306	1407	1608	1810	2012
1.00	700	800	900	1000	1100	1200	1300	1400	1600	1800	2000

* From "Temperature, Its Measurement and Control in Science and Industry," Reinhold Publishing Corporation, New York, 1941.

Table 60. True Temperatures vs. Apparent Temperatures as Observed with a Total Radiation Pyrometer*

Observed temperatures in headings, true temperatures in body of table

Total emis- sivity	100	200	300	400	600	800	1000	1200	1400	1600	1800
0.05	422	686	916	1137	1567	1993	2317	2841	3264	3687	4110
0.10	316	536	728	913	1275	1632	1989	2345	2701	3057	3413
0.15	264	460	633	799	1126	1449	1771	2093	2415	2736	3058
0.20	231	410	571	725	1029	1330	1629	1929	2228	2527	2827
0.25	207	375	526	672	958	1243	1526	1809	2093	2376	2658
0.30	189	347	491	630	904	1175	1446	1717	1987	2258	2528
0.35	175	325	463	596	860	1121	1381	1642	1902	2162	2422
0.40	164	307	439	568	823	1075	1327	1579	1830	2082	2333
0.45	154	291	419	544	791	1036	1281	1525	1769	2014	2258
0.50	146	278	402	523	763	1002	1240	1478	1716	1954	2192
0.55	138	266	387	505	739	972	1204	1437	1669	1902	2134
0.60	132	255	373	489	718	945	1173	1400	1628	1855	2082
0.65	126	246	361	474	698	921	1144	1367	1590	1813	2036
0.70	121	238	350	461	680	900	1119	1337	1556	1775	1993
0.75	117	230	340	448	664	880	1095	1310	1525	1740	1955
0.80	113	223	331	437	649	861	1073	1284	1496	1707	1919
0.85	109	217	322	427	636	844	1053	1261	1469	1678	1886
0.90	106	211	314	417	623	828	1034	1239	1445	1650	1855
0.95	103	205	307	408	611	814	1016	1219	1422	1624	1827
1.00	100	200	300	400	600	800	1000	1200	1400	1600	1800

* From "Temperature, Its Measurement and Control in Science and Industry," Reinhold Publishing Corporation, New York, 1941.

TABLE 61. INDICATING TEMPERATURE LEVELS OF AVAILABLE TEMPIL MATERIALS
Mean accuracy—1 per cent of stated temperature

Tempilstiks		Tempil pellets		Tempilaq	
°F	°C	°F	°C	°F	°C
125	52	125	52	125	52
138	59	150	65	150	65
150	65	175	80	175	80
163	73	200	93	200	93
175	80	225	107	225	107
188	87	250	121	250	121
200	93	275	135	275	135
213	100	300	149	300	149
225	107	325	163	325	163
238	114	350	176	350	176
250	121	400	204	400	204
263	128	450	232	450	232
275	135	500	260	500	260
288	142	550	288	550	288
300	149	600	316	600	316
313	156	650	343	650	343
325	163	700	371	700	371
338	170	750	399	750	399
350	176	800	427	800	427
363	184	850	454	850	454
375	190	900	482	900	482
388	198	950	510	950	510
400	204	1000	538	1000	538
450	232	1050	566	1050	566
500	260	1100	593	1100	593
550	288	1150	621	1150	621
600	316	1200	649	1200	649
650	343	1250	677	1250	677
700	371	1300	704	1300	704
750	399	1350	732	1350	732
800	427	1400	760	1400	760
850	454	1450	788	1450	788
900	482	1500	815	1500	815
950	510	1550	843	1550	843
1000	538	1600	871	1600	871
1050	566	1700	927		
1100	593				
1150	621				
1200	649				
1250	677				
1300	704				
1350	732				
1400	760				
1450	788				
1500	815				
1550	843				
1600	871				

INDEX

A

Aberration, lateral chromatic, 84
longitudinal chromatic, 83–84, 86
correction for, 85
spherical, 84–86
correction for, 86
Abnormality, in steels, 303, 304
Abrasive hardness, 213
Abrasive wheels, for spark testing, 313
Abrasives, metallographic, 17–21
alumina (see Alumina abrasive)
alundum, 18
characteristics of, 17
chromic oxide, 21
diamond dust, 17
iron oxide rouge, 21
for lead laps, 13
magnesium oxide, 18
miscellaneous, 21
Absorption, radiation, 370
Achromatic microscope objectives, 86–87
proper light filter for, 87
AISI "H" steels, 310
Alcohol, isopropyl, 23
Allison Company, The, 5
Allotropy, 385
Aloxite emery papers, 8
Alteration of structure, improper cutoff
procedure for, 5, 6
Alumel, electrolytically etched, 76
Alumina abrasive, 19–21
C-730, 20
characteristics of, 19
crystal structure of, 19
for galvanized sheets, 42
Gamal, 19
kinds of, 19
levigation of, 20
Linde grades of, 19
Precisionite, 19
preparation of, 20–21
for tin and tin alloys, 38

Aluminum and its alloys, etching reagents
for, macroscopic examination, table,
424
microscopic examination, table, 424–
426
metallographic preparation of, 33–34
Alundum abrasive, 18
American Optical Company, 13
ASTM austenitic grain-size standards,
charts, 283–290
data related to, table, 291
ASTM grain-size index, 282
ASTM nonferrous grain-size standards,
charts, 294–298
Amplifier eyepieces, 101–102
Ampliplans, 101
Homals, 101
in projection, 99, 101
Analysis, magnetic, 317, 401
thermal, 385–407
Angular aperture, 91
Annealing of thermocouple elements, 336
Anodic constituent during etching, 64–65
Antimony, supercooling of, 388
Aperture, angular, 91
numerical, 90–93
Aperture diaphragm, 117–118
effect on resolution, 117, 119
position of, 118
purpose of, 117, 118
setting of, 118
Apochromatic microscopic objectives, 87–
88
proper light filters for, 88
Apparent temperatures vs. true temperatures, optical pyrometry, 364, 494
radiation pyrometry, 375, 495
Aquanite A, 49
Armstrong-Vickers hardness number, equation for, 228
table of values, 454
Armstrong-Vickers hardness test, 224–229
condition of test surface for, 229